SPECTROCHEMICAL ANALYSIS BY X-RAY FLUORESCENCE

SPECTROCHEMICAL ANALYSIS BY X-RAY FLUORESCENCE

Rudolf O. Müller
CIBA-GEIGY AG
Basel, Switzerland

Translated from German by
Klaus Keil
Director, Institute of Meteoritics
Department of Geology
University of New Mexico
Albuquerque, New Mexico

℗ SPRINGER SCIENCE+BUSINESS MEDIA, LLC 1972

Rudolf O. Müller, after completing his studies at the Eidgenössische Technische Hochschule in Zurich and at the University of Bern, became a staff scientist in 1959 at the CIBA-GEIGY AG, Basel. He is in charge of the X-ray fluorescence laboratory for spectrochemical analysis and production control as well as of the laboratory for X-ray diffraction and X-ray structure analysis. Dr. Müller is particularly known for his work on basic theoretical aspects of quantitative X-ray fluorescence analysis and its application to spectrochemical determination.

The original German text, published by R. Oldenbourg, Munich and Vienna, in 1967, as *Spektrochemische Analysen mit Röntgenfluoreszenz,* has been corrected by the author for the present edition.

Library of Congress Catalog Card Number 70-107540

ISBN 978-1-4684-1799-9   ISBN 978-1-4684-1797-5 (eBook)
DOI 10.1007/978-1-4684-1797-5

© 1972 Springer Science+Business Media New York
Originally published by Plenum Press, New York in 1972
Softcover reprint of the hardcover 1st edition 1972

Preface

In recent years the x-ray fluorescence technique has become increasingly important in modern analysis and production control; it can be classified as a spectroscopical method for the determination of the elemental composition. Many articles treat this method; however, there exists no modern textbook suitable for the beginner as well as the practician and theoretician. In this monograph the author intends to fill this need to present the principles of x-ray fluorescence analysis and to develop a theoretical understanding of the technique. Both principles and theory will be treated extensively, for they are the basis for successful practical application of the method. X-ray fluorescence, on the other hand, is often carried out exclusively because of its practical usefulness. For this reason theoretical investigations are used exclusively as a basis for practical work and the multitude of applications, which constitute the value of the x-ray fluorescence method, will be explained on the basis of simple theory.

The idea to write this monograph originated and developed when efforts to train coworkers required a more complete treatise. I would like to thank the CIBA Aktiengesellschaft in Basel, where this work originated, for generous support and permission to publish the book. The head of the Physics Department, Dr. E. Ganz, and my colleagues have contributed to this book by providing a stimulating working atmosphere. I am grateful to my associates, in particular Messrs. E. Eng, S. Gasser, and H. R. Walter, for assistance in setting up the x-ray fluorescence laboratory over the past seven years. H. O. Meyer read part of the manuscript.

Basel, May 1966   RUDOLF MÜLLER

Contents

Part II

Quantitative Analysis

Part III
Examples of Applications and Abstracts

Introduction

In recent years x-ray fluorescence analysis, together with other analytical techniques, has been applied in industry to research as well as continuous production control. Hardly any other analytical technique is so universally suited for qualitative and quantitative determinations as is x-ray fluorescence analysis. This is largely due to the fact that analysis is independent of composition and chemical bond of the sample. Solid, powdered, and liquid samples can be analyzed. On the basis of x-ray spectra, Coster and Hevesy (1923) discovered element hafnium ($Z = 72$) and have detected its occurrence in natural zircons; and Noddack, Tacke, and Berg (1925) discovered the element rhenium ($Z = 75$) in columbite concentrates.

In 1895 W. C. Röntgen, while working with cathode-ray tubes, discovered an hitherto unknown and invisible radiation which occurs together with the cathode rays; because of its unknown nature he named it x-radiation. The physical properties of these rays were described, in a complete and qualitatively correct manner, by Röntgen in the first two *Sitzungsberichte der Würzburger Physikalischen-Medicinischen Gesellschaft*. After Röntgen's discovery, x-rays were the subject of numerous investigations by various researchers. In particular, C. G. Barkla proved indirectly that the radiation emitted from a sample consists of several spectra of different wavelengths, which he named K and L spectra. In the summer of 1912 M. Laue, together with M. Friedrich and P. Knipping, made the fundamental discovery of the interference of x-rays which occurs when the rays are scattered by a three-dimensional ordered crystal lattice. This experiment shows that x-rays are electromagnetic waves of a wavelength on the order of 10^{-8} cm. In England, W. H. and W. L. Bragg used monochromatic x-rays for diffraction by crystal plates. In 1913 they interpreted diffraction as a reflection on selected net planes in the crystal according to the equation $2d \sin \theta = n\lambda$. Shortly thereafter, Laue showed that diffraction and reflection are two different interpretations of one and the same phenomenon, and that

1

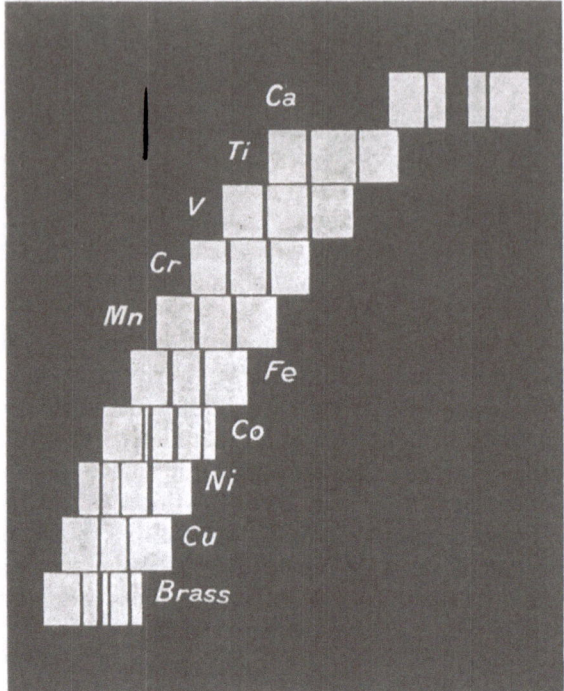

Fig. I.1. X-ray spectra of the elements calcium ($Z = 20$) to zinc ($Z = 30$) as a function of wavelength. Every spectrum consists of an intense α line (always to the right) and a weaker β line of shorter wavelength. The orderly decrease of wavelength with increasing atomic number is clearly visible. Recorded photographically. Original after Moseley (1913).

the Bragg equation can be derived directly from the Laue equation for interference on three-dimensional lattices. In 1913 N. Bohr published two papers on the constitution of atoms and molecules. Spectral lines were interpreted as electron transitions between the various energy levels of the electrical field around the nucleus and were documented numerically for the light elements. Stimulated by this work and by Bragg's experiments, H. G. J. Moseley, in the same year, systematically studied the x-ray spectra of the elements calcium through zinc. He found that the frequency of emission lines is proportional to the square of the charge of the nucleus. This regularity can be understood on the basis of Bohr's atom theory extended to the heavy nuclei. Moseley had already shown that the spectrum of brass contains the spectra of the elements copper and zinc and, hence, that the chemical composition of a substance can be determined accurately with the aid of x-ray spectra. This was the beginning of the application of x-ray spectroscopy to chemical analysis. For a long time, however, instrumental and preparative procedures remained complicated. Only progress in the indirect excitation of spectra by the "white" radiation of an x-ray tube and recording of spectra by counters made possible the broad analytical application of the method. Modern x-ray spectroscopy can be said to have begun after 1950 when work with new hafnium–zirconium and tantalum–niobium alloys as well as uranium–thorium–plutonium reactor fuels required new and more accurate analytical methods.

According to Bohr's model, the atom can be thought to consist of a positively charged nucleus and negatively charged electrons. The electrons

Fig. I.2. Commercial spectrometer for x-ray fluorescence analysis. (1) X-ray tube; (2) sample chamber; (3) sample port and knob of the sample changer; (4) crystal chamber; (5) lever for crystal changer; (6) scintillation counter; (7) large-angle goniometer; and (8) high-voltage generator for x-ray tube. (Courtesy of C. H. F. Müller-Philips Company.)

Fig. I.3. View of a programmed x-ray fluorescence unit (cover removed). Elements (wavelengths) to be determined and optimal measurement conditions can be preselected so that, after the insertion of the sample, the measurement is carried out automatically. Results are printed out or can be further reduced by an electronic computer. (1) X-ray tube; (2) sample chamber with sample port and sample changer; (3) crystal chamber; (4) scintillation counter; and (5) large-angle goniometer with automatic drive. (Photograph courtesy of Siemens & Halske Company.)

circle around the nucleus in discrete shells and are bound to the nucleus by Coulomb forces. These shells are designated K, L, M, N, etc. shells, where the K shell is the one closest to the nucleus. The number of positive charges in the nucleus is equal to the atomic number of the atom in the periodic system of the elements. The origin of characteristic x-ray spectra can briefly be described as follows: When sufficient energy is introduced into the atom an electron may be knocked out of one of the inner shells. The atom is then in an excited (ionized) state and, within 10^{-8} sec, returns to the ground state. The place of the missing electron is filled by an electron from a neighboring outer shell whose place, in turn, is filled by an electron from a still outer shell. The atom thus returns to the ground state in steps. In every step, i.e. in every electron jump, an electron from a higher energy level goes into a lower energy level emitting excess energy in form of an x-ray quantum. The energy of the emitted radiation is characteristic for the atomic number of the emitting element as well as for the particular electron transitions taking place within the electron shell of the atom. By measuring, respectively, the energy or the wavelength of the emitted radiation the particular element can be identified unambiguously. The energy which is necessary to knock out an electron from one of the inner shells may be introduced into the atom either by collision with an high energy electron or by the absorption of an energy-rich photon (x-ray quantum). Correspondingly, there are two ways of exciting characteristic x-ray spectra. First, the sample may be bombarded by electrons which are accelerated by high voltage or, second, the sample may be irradiated by x-ray or gamma rays. Nowadays, the second method is frequently applied. In analogy to the optical case, this technique is referred to as fluorescence, which is responsible for the name x-ray fluorescence analysis for the technique of spectrochemical analysis with x-rays.

The aforementioned method for the production of characteristic spectra, namely excitation through electron bombardment, is the one that is used in x-ray tubes to produce monochromatic x-rays for diffraction investigations, and in the electron microprobe. If this method is to be used for spectrochemical analysis, then the sample has to be prepared in form of a plate which is used as the anode in an evacuated x-ray tube. This arrangement has been used by Moseley, Siegbahn, and others for early systematic investigations of characteristic spectra. With this method Coster and Hevesy (1923) demonstrated the existence of the element hafnium. Glocker and Schreiber (1928) excited spectra by the so-called cold excitation method using the radiation of an x-ray tube and, thus were the first to introduce the technique which is now referred to as x-ray fluorescence analysis. The term x-ray fluorescence was coined by Schreiber (1929).

In modern x-ray fluorescence analysis the sample is irradiated by polychromatic radiation from an x-ray tube. In this process, elements in the sample are excited to emit their characteristic x-ray radiation. This secondary radiation consists of several lines which are diffracted by a crystal plate (analyzing crystal) and separated into the individual wavelengths. According to Bragg's diffraction condition, $2d \sin \theta = n\lambda$, where d is the lattice parameter and n is an integer, 1, 2, 3, etc., the radiation of the wavelength λ will only be diffracted by the lattice plane of a crystal when the glancing angle θ is such as to fulfill the diffraction condition. In the diffraction process, the radiation is deflected by the angle 2θ from the direction of incidence. If this angle is measured, then the wavelength of the diffracted radiation can be calculated. Neglecting for the moment the higher order reflections there exists, for every wavelength, only one value of the glancing angle 2θ. Every spectrum emitted by an element consists of only a few characteristic lines. The K spectrum, which is emitted when the original ionization of the atom took place in the K shell, consists largely of the strong K_α line and the weaker K_β line. The K_β line is reduced in intensity by about a factor of four in comparison to the K_α line. Both lines always occur together, for they represent two possible electron transitions; the weaker line corresponds to the less frequent electron transition. The L spectrum, which originates when the original ionization took place in the L shell, is somewhat richer in lines and consists of two strong L_α and $L_{\beta 1}$ lines and several weaker $L_{\beta 2}$, $L_{\beta 3}$, L_γ, etc. lines. Again, all lines of the L spectrum are visible simultaneously. It is a simple task to unambiguously identify elements on the basis of their emission lines and, thus, to determine the qualitative composition of a sample.

For quantitative analysis not only the wavelength but also the intensity of the emitted radiation has to be measured using a proportional or scintillation counter. The larger the relative intensity of the radiation the higher is the content of the respective element in the sample. For accurate quantitative analysis comparative samples of known composition (so-called calibration standards) are used; they serve, first of all, to determine the relationship between fluorescent intensity and concentration of the individual elements. The fluorescent intensity of an element is not strictly proportional to its concentration but deviates more or less from a linear relationship because of the interaction of the radiation with other elements in the sample. If, for example, the element is associated with a strongly absorbing substance, then the resulting fluorescent intensity is less than when the associated component has a small absorption coefficient and the radiation is only weakly absorbed. Other than by absorption, the fluorescent intensity of an element is also affected by secondary or interelemental excitation. An associated element, whose own fluorescent radiation is of higher energy than the absorption edge of another element, can excite the element in question to additional emission. As a result the fluorescent intensity of that element increases over the intensity that would be expected from its concentration. For mixtures consisting of only a few components, interactions between individual components are easily understood. In mixtures consisting of many (n) components, however, mutual interactions are best formulated as a linear system of equations that consists of as many members as there are components, and that contains $n + 1$ equations. Certain samples have to be chemically decomposed by either a solvent or a flux before they are suited for quantitative x-ray fluorescence analysis. The magnitude of the interactions between components is reduced due to element dilution in the decomposed sample. In order to further lower interelement effects a strongly absorbing substance, such as barite, may be added to the sample. Frequently, a standard element is added to the sample in known amounts and relative intensities are measured in relation to the internal standard; this method simplifies considerably the quantitative determination of concentration. Occasionally, reference standards of compositions similar to that of the sample are used as external standards. Measured intensities are recalculated to reduced normalized intensities, and the composition of the sample is determined by comparative calculation. Some researchers prefer, on the basis of calibration standards, to numerically determine once and for all the individual interaction factors; in the analysis of an unknown sample the respective system of equations is then solved mathematically and the desired contents are obtained directly. Today, x-ray fluorescence is a well-established tool in the analysis of reactor metals and reactor fuels, steel and special alloys, light-metals, cement and rocks, oil and oil products, and in the chemical industry. Special analyses and production control measurements are carried out concentrating in the range of 0.1 to 100% as well as of trace elements in the parts per million range. Programmed x-ray fluorescence units now allow largely automatic analysis in production control.

In the first part of this book the physical and instrumental principals of qualitative and quantitative x-ray fluorescence analysis are treated as far as is necessary for an understanding of the method and its successful application.* The second part of the book is devoted entirely to quantitative analysis. Prerequisites and practical procedures for quantitative analysis are discussed in detail and the term "regression coefficient" is introduced to explain quantitative analysis. The regression coefficient allows one to summarize into one number all interactions of two components and to describe that interaction sufficiently. In a multicomponent system it is therefore possible to illustrate the interactions of components by a linear system of equations and to treat the interactions mathematically. On the basis of this system of equations various possibilities for quantitative analysis are unambiguously and critically explained. The third part of the book contains approximately 200 summaries of papers dealing with x-ray fluorescence analysis; these abstracts are classified according to subject matter and illustrate the great variety of applications of x-ray fluorescence analysis.[†]

---

* Questions concerning the physical nature and the properties of x-rays are treated extensively in the book by Compton and Allison entitled *X-Rays in Theory and Experiment* and in the 30th volume of the *Handbuch der Physik*, entitled *X-Ray Radiation*. A book by Blochin, *Physics of X-Ray Radiation*, is also available.

† Papers selected are the ones which were easily accessible to the author. A more complete list of titles is contained in Liebhafsky and Winslow (1958); Liebhafsky, Winslow, and Pfeiffer (1960, 1962); Campbell and Brown (1964); as well as in the *Spectrochemical Abstracts* which appear in regular intervals and are edited presently by Van Someren, Lachman, and Birks. Talks presented at the Annual Denver (Colorado) Conference on the application of x-ray analysis are contained in *Advances in X-ray Analysis*; a number of these talks are later published in American scientific journals.

PART I
PRINCIPLES AND QUALITATIVE ANALYSIS

Chapter 1

Absorption and Scattering of X-Rays

X-rays are absorbed by materials in differing degrees; absorption coefficients are defined to quantitatively describe the magnitudes of this process. Let us assume that when an x-ray beam passes through a thin layer of material a fraction dN/N of the pulse rate N is absorbed. This fraction is proportional to the thickness dx of the layer

$$\frac{dN}{N} = - \mu \, dx$$

where μ is a proportionality factor. The negative sign indicates that x-rays are reduced in intensity when passing through matter. If the factor μ is independent of x, we then obtain by integration

$$N = N_0 \exp \left[-\mu x\right]$$

where N_0 is the pulse rate of the primary incident radiation and N the pulse rate after passage through the layer of the thickness x. The term μ is known as the linear absorption coefficient and, in the following, is therefore always written as μ_l. It is

$$\mu_l = - \frac{1}{N} \frac{dN}{dx} \qquad [\mu_l] = cm^{-1}$$

where μ_l describes the magnitude of absorption after passage through one centimeter of matter.

Let us now consider an x-ray beam of 1 cm^2 cross section. The fraction dN/N of an x-ray beam of the pulse rate N which is absorbed when passing through a thin layer, is also proportional to the mass dm of this layer

$$\frac{dN}{N} = -\mu_m \, dm \qquad [\mu_m] = cm^2/g$$

The proportionality factor μ_m describes the magnitude of absorption per gram per square centimeter of material. After integration the relationship between pulse rate N_0 of the incident radiation and the pulse rate N after passage through the mass m per unit area is as follows:

$$N = N_0 \exp\left[-\mu_m m\right]$$

A simple relation exists between the linear absorption coefficient and the absorption coefficient per unit mass. When passing through the unit length, the beam also passes through the mass m. If the density of the layer is ρ, then the mass per unit length is $m = \rho$, and this gives

$$\frac{\mu_l}{\varrho} = \mu_m$$

The designation μ/ϱ for the mass absorption coefficient is therefore frequently found in the literature. In order to avoid confusion with the linear absorption coefficient the mass absorption coefficient is always designated μ_m. For simplicity, however, the symbol μ is also used provided there is no danger of confusion with the linear absorption coefficient. The mass absorption coefficient is independent of the state of aggregation of the absorber. Water vapor, water, and ice have the same mass absorption coefficient (as defined per unit mass of matter), while the linear absorption coefficients (as defined per unit length) are different.

The mass absorption coefficient is, however, dependent upon the wavelength. Figure 1.1 illustrates graphically the mass absorption coefficient of molybdenum as a function of the wavelength. The orderly change of the absorption coefficient is interrupted in places by the so-called absorption

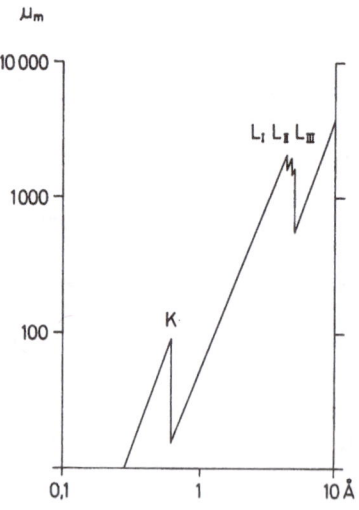

Fig. 1.1. Graphical presentation of the mass absorption coefficient of molybdenum as a function of wavelength.

Table 1.1. Calculation of Mass Absorption Coefficients of 18/8-Cr–Ni-Steel for the Wavelength $\lambda = 0.71$Å

Weight %		μ (0.71 Å) of the elements	Proportion
Cr	18	31.1 =	560
Ni	8	46.6 =	373
Fe	74	38.5 =	2849
	100		3782　μ (0.71 Å) = 37.8

edges. The absorption edges occur at certain wavelengths which are characteristic for every element; they correspond to the energy levels of the electrons in the atom and are named after the electron shells, i.e., K, L_{I}, L_{II}, L_{III} edges. The relationship between the mass absorption coefficient μ, the atomic number of the absorber Z, and the wavelength of the absorbed radiation λ, may be approximately expressed by the following equation:

$$\mu_m \cong cZ^m \lambda^n \approx cZ^3 \lambda^3$$

The factor c has different values depending upon whether λ has a shorter wavelength than the K absorption edge, whether it is between the K and L absorption edges, or whether it is between the L and M absorption edges. This equation applies strictly only to the true absorption and not to the total absorption coefficient. The proportion of scattering in comparison to true absorption is, however, usually negligible. If the absorbing substance consists of different elements, then the individual elements i contribute to the absorption in accordance with their concentration C_i, and the mass absorption coefficient μ is the weighted sum of the absorption coefficients of the individual elements

$$\mu (\lambda) = \frac{\Sigma\, C_i \mu_i (\lambda)}{\Sigma\, C_i}$$

An example for the calculation of the mass absorption coefficient of a compound from the coefficients of the individual elements is given in Table 1.1.

In the absorption process an electron is taken out of the electron configuration and assumes the total energy of the absorbed x-ray quantum. Electrons of all energy levels are eligible for such absorption. The absorption coefficient μ is the sum of the partial absorption efficiencies or the partial absorption coefficients of the individual energy levels

$$\mu = \mu_K + \mu_{LI} + \mu_{LII} + \mu_{LIII} + \cdots$$

Table 1.2. Energies of Absorption Edges. [Values after Sandström (1957) p. 226, and Tomboulian (1957) p. 303; additional values after Hill, Church, and Mihelich (1952).]

		K	L_I	L_{II}	L_{III}			K	L_I	L_{II}	L_{III}
		keV	keV	keV	keV			keV	keV	keV	keV
3	Li	0.055			0.003	46	Pd	24.349	3.602	3.329	3.172
4	Be	0.112			0.002	47	Ag	25.512	3.805	3.522	3.350
5	B	0.186			0.003	48	Cd	26.711	4.019	3.727	3.538
6	C	0.282			0.002	49	In	27.937	4.236	3.937	3.729
7	N	0.400			0.008	50	Sn	29.200	4.464	4.156	3.928
8	O	0.531			0.009	51	Sb	30.491	4.698	4.381	4.132
9	F	0.692				52	Te	31.813	4.939	4.612	4.342
10	Ne	0.874			0.019	53	I	33.170	5.190	4.853	4.559
11	Na	1.07	0.064	0.031	0.031	54	Xe	34.551	5.448	5.103	4.783
12	Mg	1.303	0.089	0.050	0.050	55	Cs	35.983	5.722	5.359	5.011
13	Al	1.559	0.115	0.073	0.073	56	Ba	37.443	5.981	5.624	5.247
14	Si	1.842	0.150	0.103	0.102	57	La	38.932	6.272	5.897	5.490
15	P	2.142	0.185	0.129	0.128	58	Ce	40.447	6.553	6.169	5.729
16	S	2.470	0.224	0.164	0.163	59	Pr	41.995	6.853	6.445	5.969
17	Cl	2.820	0.269	0.199	0.197	60	Nd	43.577	7.135	6.728	6.215
18	A	3.200	0.287	0.246	0.244	61	Pm	45.19	7.431	7.022	6.462
19	K	3.609	0.377	0.298	0.295	62	Sm	46.835	7.743	7.340	6.720
20	Ca	4.038	0.439	0.350	0.346	63	Eu	48.498	8.055	7.633	6.981
21	Sc	4.496	0.506	0.411	0.406	64	Gd	50.228	8.377	7.930	7.243
22	Ti	4.964	0.565	0.460	0.454	65	Tb	51.983	8.710	8.252	7.515
23	V	5.464	0.631	0.519	0.512	66	Dy	53.840	9.053	8.588	7.796
24	Cr	5.987	0.679	0.581	0.572	67	Ho	55.600	9.390	8.912	8.067
25	Mn	6.537	0.752	0.650	0.639	68	Er	57.457	9.749	9.263	8.357
26	Fe	7.111	0.841	0.720	0.708	69	Tm	59.376	10.120	9.616	8.650
27	Co	7.709	0.929	0.794	0.779	70	Yb	61.313	10.487	9.979	8.944
28	Ni	8.331	1.010	0.871	0.854	71	Lu	63.306	10.867	10.346	9.242
29	Cu	8.981	1.102	0.953	0.933	72	Hf	65.347	11.267	10.736	9.558
30	Zn	9.661	1.197	1.045	1.022	73	Ta	67.406	11.678	11.129	9.874
31	Ga	10.395	1.30	1.171	1.144	74	W	69.519	12.098	11.540	10.202
32	Ge	11.100	1.42	1.245	1.214	75	Re	71.678	12.528	11.960	10.537
33	As	11.867	1.53	1.359	1.324	76	Os	73.866	12.959	12.380	10.866
34	Se	12.656	1.66	1.475	1.434	77	Ir	76.107	13.419	12.819	11.210
35	Br	13.470	1.79	1.599	1.552	78	Pt	78.386	13.885	13.266	11.557
36	Kr	14.32	1.92	1.729	1.674	79	Au	80.723	14.354	13.734	11.919
37	Rb	15.201	2.067	1.866	1.806	80	Hg	83.113	14.848	14.213	12.286
38	Sr	16.106	2.217	2.008	1.941	81	Tl	85.529	15.345	14.696	12.658
39	Y	17.037	2.372	2.155	2.080	82	Pb	88.014	15.864	15.205	13.039
40	Zr	17.995	2.530	2.305	2.221	83	Bi	90.471	16.390	15.715	13.419
41	Nb	18.989	2.700	2.467	2.374	88	Ra	103.9	19.235	18.482	15.444
42	Mo	20.002	2.867	2.629	2.524	90	Th	109.630	20.470	19.696	16.299
44	Ru	22.117	3.226	2.966	2.838	92	U	115.610	21.758	20.939	17.164
45	Rh	23.218	3.410	3.144	3.002	94	Pu	121.2	23.097	22.262	18.066

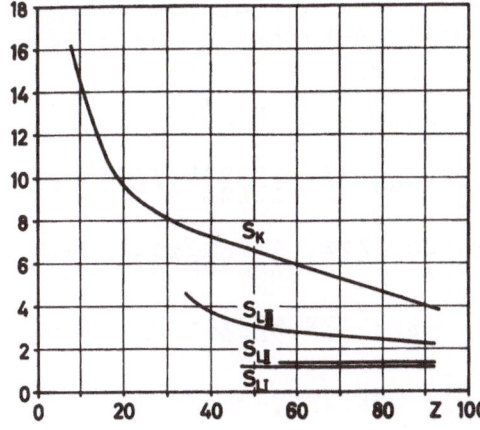

Fig. 1.2. Value of absorption jump ratio at K and L edges as a function of atomic number [graphically averaged for values given by various authors in Compton and Allison (1935), p. 528; and Landolt-Börnstein (1950), p 319].

The absorption jump edge S_K of the K absorption edge is defined as the ratio

$$S_K = \frac{\mu_K + \mu_{LI} + \mu_{LII} + \mu_{LIII} + \cdots}{\mu_{LI} + \mu_{LII} + \mu_{LIII} + \cdots}$$

Analogously, the absorption jump edge S_{LI} of the L_I absorption edge is defined as

$$S_{LI} = \frac{\mu_{LI} + \mu_{LII} + \mu_{LIII} + \cdots}{\mu_{LII} + \mu_{LIII} + \cdots}$$

The absorption jump edge is a measure of that part of the absorbed radiation that is absorbed exclusively by the respective energy level. This fraction changes with the atomic number of the element (Figure 1.2).

The absorption coefficient consists of the true photoelectric absorption coefficient τ and the scattered portion σ

$$\mu = \tau + \sigma$$

Photoelectric absorption constitutes the major portion of the absorption coefficient, while the scattered portion σ is usually small in comparison to the true absorption. The scattered portion σ corresponds to diffuse scattering of a beam when passing through matter and, hence, is not due to true absorption. In the scattering process, part of the radiation is scattered inelastically and shifted towards longer wavelengths. As a result the scattered radiation consists of a coherently scattered unmodified portion

Table 1.3. Mass Absorption Coefficients of the Elements for Various Wavelengths. Values for the wavelength 0.2 Å after Victoreen (1949); for H (Z = 1) to Bi (Z = 83) in the wavelength range 0.4 to 2.5 Å after Milledge (1962); for Th (Z = 90) to U (Z = 92), and for H (Z = 1) to Bi (Z = 83) in the wavelength range 3 to 10 Å from the author's graphical presentations, which represent averages of values determined by various authors.

λ [Å]	0.20	0.40	0.60	0.71	0.80	1.00	1.20	1.40	1.60	1.80	2.00	2.50	3.00	4.00	5.00	6.00	8.00	10.00
1 H	0.33	0.36	0.37	0.38	0.38	0.40	0.41	0.42	0.44	0.47	0.49	0.59	0.74	1.21	1.98	3.14	6.90	13.1
2 He	0.16	0.18	0.20	0.21	0.22	0.24	0.28	0.34	0.41	0.50	0.61	1.00	1.58	3.47	6.57	11.2	26.2	50.8
3 Li	0.14	0.17	0.19	0.22	0.24	0.32	0.43	0.58	0.78	1.04	1.36	2.49	4.18	9.64	18.6	31.9	74.9	145
4 Be	0.15	0.18	0.24	0.30	0.36	0.53	0.80	1.17	1.65	2.28	3.06	5.79	9.85	23.0	44.4	76.2	178	344
5 B	0.16	0.21	0.30	0.39	0.49	0.78	1.22	1.83	2.65	3.69	4.98	9.52	16.3	38.0	73.5	126	293	564
6 C	0.17	0.26	0.45	0.63	0.81	1.40	2.28	3.50	5.12	7.19	9.76	18.8	32.1	74.9	144	246	569	1082
7 N	0.18	0.31	0.63	0.92	1.23	2.21	3.67	5.69	8.38	11.8	16.1	31.0	52.9	123	236	402	919	1731
8 O	0.19	0.38	0.87	1.31	1.79	3.31	5.55	8.66	12.8	18.0	24.6	47.3	80.7	187	357	602	1359	2524
9 F	0.19	0.46	1.16	1.80	2.49	4.67	7.90	12.4	18.3	25.8	35.2	67.7	110	253	480	804	1784	3251
10 Ne	0.21	0.59	1.56	2.47	3.44	6.49	11.0	17.3	25.5	36.0	49.0	93.9	158	361	678	1125	2447	4350
11 Na	0.22	0.72	2.01	3.21	4.49	8.52	14.5	22.7	33.5	47.3	64.3	123	202	457	850	1393	2950	5075
12 Mg	0.25	0.89	2.56	4.11	5.75	11.0	18.6	29.1	42.9	60.4	82.0	156	268	598	1097	1775	3634	341
13 Al	0.27	1.08	3.20	5.16	7.25	13.8	23.5	36.8	54.1	76.0	103	195	327	721	1304	2072	280	520
14 Si	0.31	1.32	3.97	6.44	9.05	17.3	29.3	45.9	67.5	94.7	128	241	406	889	1591	2507	402	740
15 P	0.34	1.58	4.85	7.89	11.1	21.2	36.0	56.1	82.4	115	156	292	477	1037	1803	240	552	1050
16 S	0.39	1.89	5.87	9.55	13.4	25.6	43.4	67.6	99.0	138	186	347	583	1230	2107	318	730	1410
17 Cl	0.42	2.23	7.00	11.4	16.0	30.6	51.6	80.2	117	163	219	404	654	1343	248	412	920	1700
18 A	0.45	2.62	8.27	13.5	18.9	36.0	60.6	93.9	137	190	253	462	703	154	285	475	1060	1950
19 K	0.54	3.07	9.71	15.8	22.2	42.0	70.6	109	158	218	290	523	851	206	372	600	1300	2380
20 Ca	0.63	3.56	11.3	18.3	25.7	48.4	80.9	124	179	246	326	577	964	252	455	728	1560	2800
21 Sc	0.67	4.09	13.0	21.1	29.5	55.5	92.4	141	203	277	364	635	143	303	546	878	1820	3320
22 Ti	0.73	4.69	14.9	24.2	33.9	63.5	105	161	229	312	409	74.6	179	364	638	1000	2050	3560
23 V	0.79	5.34	17.0	27.5	38.5	71.9	119	181	257	348	454	88.1	194	400	690	1110	2260	3700
24 Cr	0.91	6.03	19.2	31.1	43.4	80.9	133	202	286	386	500	103	210	433	760	1200	2350	4100
25 Mn	0.96	6.76	21.5	34.7	48.4	89.9	147	222	313	419	62.7	120	240	480	820	1260	2540	4350
26 Fe	1.13	7.54	23.9	38.5	53.6	99.1	162	242	338	53.7	72.9	139	255	510	880	1400	2800	4700
27 Co	1.25	8.37	26.4	42.5	59.0	108	176	261	362	62.1	84.3	160	270	548	960	1500	3000	5230
28 Ni	1.42	9.25	29.1	46.6	64.6	118	190	279	50.9	71.6	97.1	184	296	600	1050	1580	3300	5620
29 Cu	1.50	10.2	31.9	50.9	70.4	128	204	40.0	58.9	82.9	112	213	319	690	1210	1880	3540	5590
30 Zn	1.65	11.2	34.8	55.4	76.3	137	217	45.6	67.2	94.4	128	242	350	700	1200	1890	3760	6400
31 Ga	1.73	12.2	37.8	60.1	82.5	147	32.8	51.3	75.5	106	144	272	385	760	1410	2150	4050	5520
32 Ge	1.90	13.3	41.0	64.8	88.7	157	36.5	57.1	84.1	118	160	301	410	820	1500	2310	4300	4000
33 As	2.00	14.5	44.3	69.7	95.0	166	40.4	63.1	92.7	130	176	330	440	879	1610	2510	4460	5010
34 Se	2.20	15.7	47.7	74.7	101	26.1	44.4	69.2	102	142	192	359	472	940	1710	2700	3900	2160
35 Br	2.43	16.9	51.2	79.8	108	31.2	48.5	75.6	111	155	208	388	508	1020	1880	2560	1280	2310
36 Kr	2.58	18.3	54.8	84.9	114	33.8	52.7	82.0	120	167	225	416	572	1190	2000	820	820	1330
37 Rb	2.80	19.6	58.4	90.0	120	36.5	57.1	88.7	129	180	241	444	590	1180	2150	2010	1440	2640
38 Sr	3.03	21.0	62.1	95.0	19.2	39.4	61.6	95.5	139	193	258	472	640	1260	2320	2600	1520	2810
39 Y	3.35	22.5	65.7	100	20.8	42.4	66.3	102	149	206	275	499	686	1490	1890	788	2640	3040
40 Zr	3.55	24.0	69.3	15.9	22.4		71.1	110	159	220	292	525	742			847	1210	3390

	10.00	8.00	6.00	5.00	4.00	3.00	2.50	2.00	1.80	1.60	1.40	1.20	1.00	0.80	0.71	0.60	0.40	0.20
41 Nb	3420	1900	912	520	1600	796	551	309	233	169	117	76.1	45.4	24.0	17.1	73.0	25.5	3.80
42 Mo	3660	2050	981	590	1710	858	576	326	247	180	125	81.3	48.6	25.8	18.4	76.5	27.1	4.05
43 Tc	3900	2200	1030	650	1850	920	601	343	261	190	133	86.6	52.0	27.6	19.7	12.1	28.6	4.30
44 Ru	4170	2360	1090	704	1540	960	629	362	276	202	141	92.4	55.5	29.6	21.1	13.0	30.2	4.55
45 Rh	4470	2500	1160	748	1060	1050	657	382	292	214	150	98.4	59.3	31.6	22.6	13.9	31.8	4.85
46 Pd	4790	2710	1230	800	480	1140	685	402	308	227	159	105	63.2	33.8	24.1	14.9	33.4	5.13
47 Ag	5140	2890	1340	848	390	1250	714	423	325	240	169	111	67.3	36.0	25.8	15.9	35.0	5.50
48 Cd	5460	3200	1450	900	507	1340	744	445	343	254	179	118	71.6	38.4	27.5	17.0	36.5	5.55
49 In	5740	3190	1520	940	542	1110	772	466	360	268	189	125	76.1	40.9	29.3	18.1	38.0	5.80
50 Sn	5900	3360	1580	982	561	790	796	487	378	282	199	132	80.7	43.4	31.1	19.2	39.5	6.32
51 Sb	6200	3500	1660	1040	688	284	823	508	396	296	210	140	85.5	46.1	33.1	20.5	6.85	6.40
52 Te	6460	3650	1730	1080	618	296	839	527	412	310	221	147	90.3	48.8	35.0	21.7	7.27	6.76
53 I	6700	3700	1800	1140	646	312	908	543	428	323	231	155	95.1	51.6	37.1	23.0	7.70	7.29
54 Xe	4120	2640	1460	1010	632	347	504	558	442	335	241	162	100	54.4	39.2	24.3	8.16	7.54
55 Cs	7250	4200	2050	1290	730	356	211	571	455	347	251	170	105	57.3	41.3	25.7	8.22	7.70
56 Ba	5400	4400	2160	1380	790	391	234	582	468	359	261	177	110	60.3	43.5	27.2	8.62	8.10
57 La	5600	4600	2250	1440	840	415	256	619	480	371	271	185	115	63.4	45.8	28.6	9.12	8.22
58 Ce	5800	4830	2360	1570	885	441	277	647	492	383	281	193	121	66.6	48.2	30.2	9.63	8.10
59 Pr	6000	5000	2480	1600	920	465	297	389	502	394	292	201	126	70.0	50.7	31.8	10.2	8.55
60 Nd	4500	3900	2590	1670	982	483	317	175	550	405	302	209	132	73.4	53.2	33.4	10.7	8.88
61 Pm	3300	3500	2690	1740	1040	515	337	187	332	416	312	217	138	76.9	55.9	35.1	11.3	9.10
62 Sm	3400	3600	2850	1830	1090	545	356	199	349	448	323	226	144	80.6	58.6	36.9	11.9	9.40
63 Eu	3500	3700	2920	1910	1130	558	374	210	158	464	333	235	150	84.3	61.5	38.8	12.5	9.90
64 Gd	2800	2700	3020	1990	1180	575	392	221	167	286	344	244	157	88.3	64.4	40.7	13.1	10.5
65 Tb	2170	2200	3210	2060	1230	610	410	232	176	299	356	253	163	92.3	67.5	42.7	13.8	10.6
66 Dy	2270	2260	3420	2200	1310	630	427	243	185	135	374	263	170	96.5	70.6	44.7	14.5	11.2
67 Ho	2380	2330	2500	2290	1350	665	444	254	193	141	234	272	177	101	73.9	46.9	15.2	11.6
68 Er	2490	2400	2570	2350	1420	690	460	265	202	148	245	282	185	105	77.3	49.1	16.0	12.3
69 Tm	2500	1950	2740	2510	1490	720	477	276	211	155	255	293	192	110	80.8	51.4	16.8	12.4
70 Yb	2620	1570	2060	2600	1530	745	493	287	220	162	113	290	200	115	84.5	53.8	17.6	12.8
71 Lu	2700	1670	2100	1980	1590	805	509	299	229	168	118	189	208	120	88.2	56.2	18.4	3.05
72 Hf	2820	1730	1660	1340	1660	825	525	310	238	175	123	197	215	124	91.7	58.6	19.2	3.25
73 Ta	2930	1840	1700	1380	1730	890	541	321	247	183	128	205	222	129	95.4	61.1	20.1	3.40
74 W	3010	1920	1740	1480	1800	925	557	333	256	190	134	213	229	134	99.1	63.6	21.0	3.50
75 Re	3140	1980	1420	1120	1910	970	573	344	266	197	139	92.1	214	139	103	66.1	21.9	3.60
76 Os	3250	2060	1170	890	1980	1010	589	356	275	205	145	95.9	219	143	106	68.5	22.8	3.70
77 Ir	3390	2170	1200	910	1450	1070	605	368	285	212	150	99.8	152	147	110	71.0	23.7	3.80
78 Pt	3500	2240	1240	1200	1500	1130	621	380	295	228	156	104	158	151	113	73.3	24.6	4.25
79 Au	3620	2290	1300	1250	1550	1190	637	393	306	237	162	108	165	153	115	75.2	25.5	4.40
80 Hg	3700	2380	1350	1300	1150	1310	653	406	316	245	168	112	171	155	117	76.9	26.3	4.60
81 Tl	3960	2430	1410		1250	1380	670	419	327	254	175	117	71.3	156	119	78.6	27.2	4.80
82 Pb	4020	2560	1700		920	720	687	432	338	263	181	121	74.2	148	120	80.1	28.1	4.90
83 Bi	4340		1790		730	750	705	446	350	350	188	126	77.2	109	100	81.3	29.1	5.10
90 Th	4470	3470	1890		760		930	660	328	355	250	160	95.0	52	101	110	42.0	5.50
91 Pa	4650	3570			790		725	683	540	380	250	162	100	53.0	116	102	41.5	5.60
92 U	4820	3690					755				260	170	101	55.0		79.0	45.0	6.50
λ [Å]	10.00	8.00	6.00	5.00	4.00	3.00	2.50	2.00	1.80	1.60	1.40	1.20	1.00	0.80	0.71	0.60	0.40	0.20

and of an incoherently scattered wavelength-modified portion. The scattering coefficient σ is given

$$\sigma = \sigma_{coh} + \sigma_{incoh}$$

Incoherent scattering is also referred to as Compton scattering. It originates when an x-ray photon hits a loosely bound electron and changes its direction during the collision. The loosely bound electron receives a backscatter pulse and, hence, part of the original energy of the x-ray photon; after the collision the x-ray photon is of lower energy and of longer wavelength. Change in wavelength is a function of the deflection angle θ and is given as

$$\Delta\lambda = 0.024\,(1-\cos\theta)\ \ [\text{Å}]$$

When the primary x-ray spectrum of an x-ray tube is scattered diffusely by a sample consisting of light elements the individual spectral lines are split into a coherently and an incoherently scattered portion.

X-rays are scattered by the electrons. The scattering efficiency per unit mass is proportional to the number of electrons per unit mass

$$\sigma_m = \text{prop}\,\frac{Z'}{2\,Z'm}\,I_e = \text{prop}\,\frac{1}{2\,m}\,I_e$$

In this equation, I_e is the scattering power of an electron, Z' is the number of electrons per unit volume, while the mass per unit volume is $\approx 2Z'm$, where m is, respectively, the mass of a photon or neutron (the atomic weight of an element is approximately double the atomic number). It can be shown that the scattering power per unit mass (σ_m) is approximately constant

$$\sigma_m \approx 0.20\ \text{cm}^2/\text{g}$$

Characteristic Emission Spectra

X-rays are emitted when materials are irradiated by electrons or x-rays of sufficient energy. Systematic investigation of the emitted spectra was first carried out by Moseley (1913). He found that the spectra of the elements are rather similar and consist only of a very few lines; he also distinguished a *K* series and an *L* series (Figures 2.1 and 2.2). Moseley found a simple relationship between the wavelength of analog lines and the atomic number of the emitting element:

$$\sqrt{\frac{1}{\lambda}} = \text{prop } Z$$

$$Z = \text{atomic number}$$

This remarkable regularity is interpreted by Bohr's atom model.

We treat Bohr's theory of the atom in somewhat more detail because it is important for an understanding of x-ray spectra, and also because this theory is usually not treated extensively in papers on x-ray spectroscopy. In the simplified model the electrons are thought to circle in shells around the positively charged nucleus, where electrons of the various shells and levels have different energies. Transition of an electron from one shell into another results in a change of its energy and emission of the energy difference ΔE as an x-ray quantum of the energy $h\nu$

$$h\nu = \Delta E = E_a - E_e$$

In this equation E_a and E_e are, respectively, the energies in the starting and end levels; such electron transitions are possible when an electron is missing from the interior of an atom. These vacancies can be produced artificially by bombarding the atom with electrons or x-ray photons of sufficient energy. To calculate the released energy difference the electrons have to be viewed as particles of the mass m for which the centripetal force is equal to the

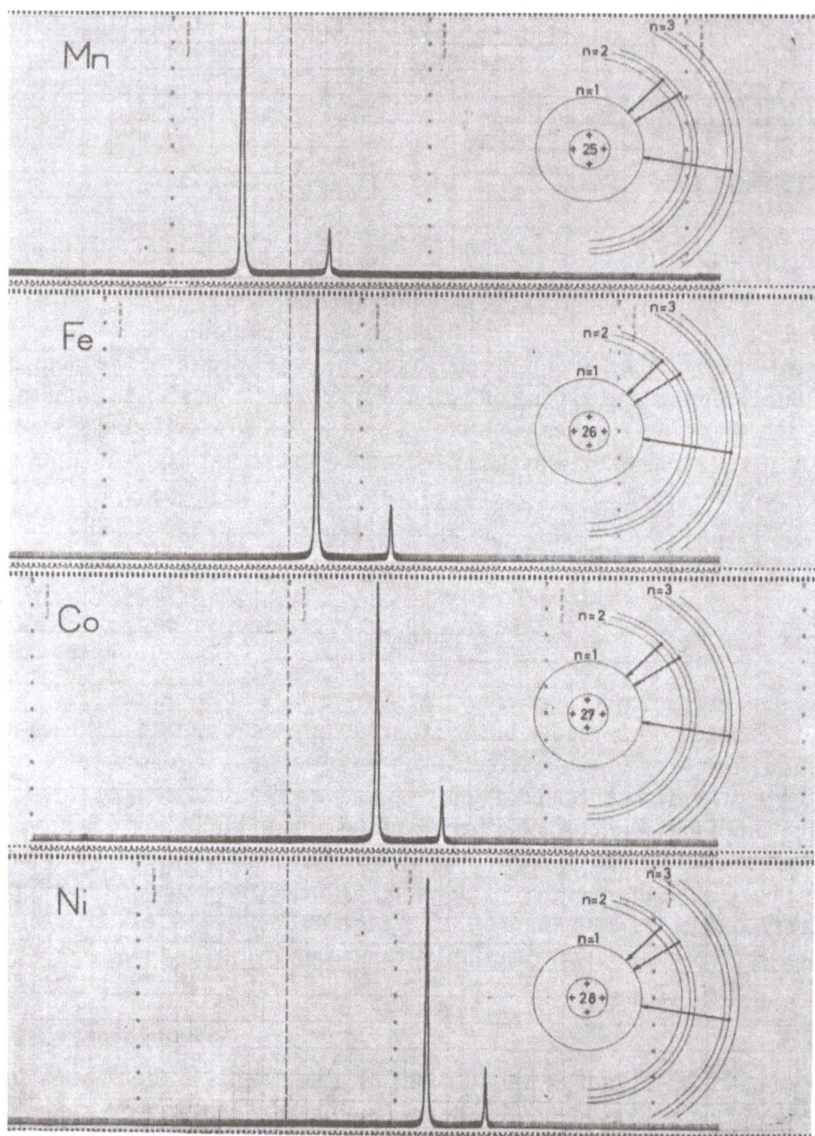

Fig. 2.1. K emission spectra of the elements manganese ($Z = 25$) to nickel ($Z = 28$), registered on a commercial spectrograph with a topaz crystal and scintillation counter. Schematic presentation of the respective electron transitions in the K series.

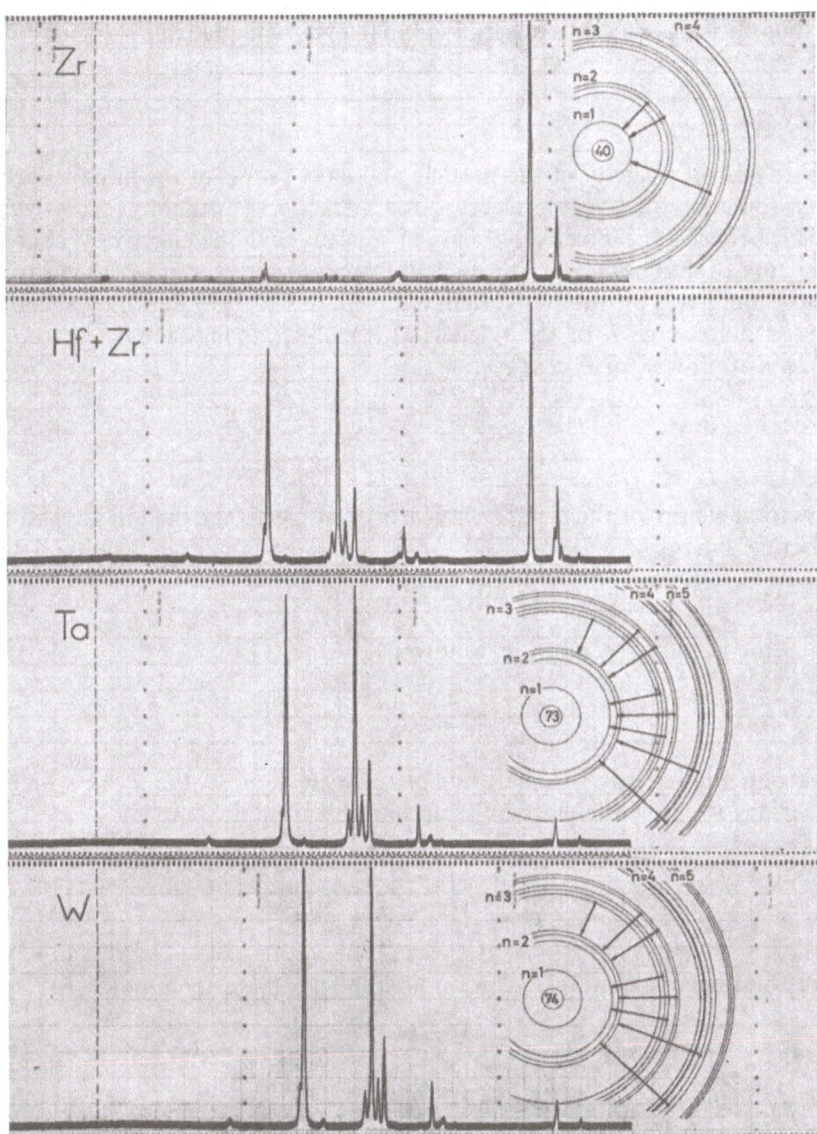

Fig. 2.2. *K* emission spectra of zirconium (*Z* = 40) and *L* emission spectra of the elements hafnium (*Z* = 72) to tungsten (*Z* = 74). The spectrum of hafnium is contaminated by additional lines of zirconium. Recorded on a commercial spectrograph, with a topaz crystal and scintillation counter. Schematic presentation of the respective electron transitions in the *K* and *L* series.

electrical attraction between nucleus and electron. The centripetal force acting on a mass m moving in a circular orbit is equal to

$$K = \frac{m v^2}{r}$$

where v is the velocity of the particle and r the radius of the circular orbit. This force is equal to the electrostatic attraction (Coulomb force) which exists between the positively charged nucleus and the negatively charged electrons of the charge e^-. The nucleus is surrounded by numerous electrons whose influence on the individual electron has to be taken into account: the partial charge F of the nucleus which affects an individual electron, is given as follows (for $F \cong Z$):

$$\frac{m v^2}{r} = \frac{F e^2}{r^2} \quad \rightarrow \quad m v^2 = \frac{F e^2}{r} \tag{2.1}$$

The total energy of the electron consists of, respectively, the kinetic and the potential energy

$$E_{tot} = E_{kin} + E_{pot} \tag{2.2}$$

The kinetic energy is given as follows:

$$E_{kin} = \frac{1}{2} m v^2 = \frac{1}{2} \frac{F e^2}{r}$$

The potential energy is equal, but of opposite sign, to the work which is necessary to move the electron from its orbit of radius r to ∞

$$E_{pot} = \int_{\infty}^{r} \frac{F e^2}{r^2} dr = -\frac{F e^2}{r}$$

The total energy of an electron on a circular orbit is then given as

$$E_{tot} = E_{kin} + E_{pot} = -\frac{1}{2} \frac{F e^2}{r} \tag{2.3}$$

Only discrete values are possible, however, for the energy of the electron where the length of the circular orbit is equal to a multiple of the "matter" wavelength Λ of the moving particle (condition after Bohr)

$$2 \pi r = n \Lambda$$

The matter wavelength is that wavelength which, on the basis of the dualism between wave and corpuscles, can be designated to a moving particle of the mass m and the velocity v. According to de Broglie the matter wavelength is given by the equation

$$\varLambda = \frac{h}{mv} \qquad \rightarrow 2\pi r = \frac{nh}{mv}$$

where h is Planck's constant.

When this equation is squared, the condition for a discrete circular orbit is given as follows:

$$4\pi^2 r^2 = \frac{n^2 h^2}{m^2 v^2}$$

After multiplying equation (2.1) by m we obtain

$$m^2 v^2 = \frac{m\,Fe^2}{r}$$

After substituting this value in the above equation we obtain for the radii of circular orbits

$$r = \frac{n^2 h^2}{4\pi^2 m\,Fe^2}$$

Using this value for r, equation (2.3) can be written to give the total energy

$$E_{\text{tot}} = -\frac{2\pi^2\,me^4 F^2}{n^2 h^2} \qquad (2.4)$$

In this equation $n = 1, 2, \ldots$ are the major quantum numbers, which for the K, L, M shells are $n = 1, 2, 3$, respectively. When an electron of the L shell, with $n = 2$, is transferred into the K shell, with $n = 1$, an energy difference is released as follows:

$$\varDelta E_{K\alpha} = \frac{2\pi^2\,me^4 F^2}{h^2}\left(\frac{1}{1^2} - \frac{1}{2^2}\right) \qquad (2.5)$$

The particular line in question is designated as the K_α line. For transfer from the M shell ($n = 3$) into the K shell ($n = 1$), it is

$$\varDelta E_{K\beta} = \frac{2\pi^2\,me^4 F^2}{h^2}\left(\frac{1}{1^2} - \frac{1}{3^2}\right) \qquad (2.6)$$

The corresponding line is designated as the K_β line; it is more energetic and shorter in wavelength than the K_α line. Since both electron transitions occur

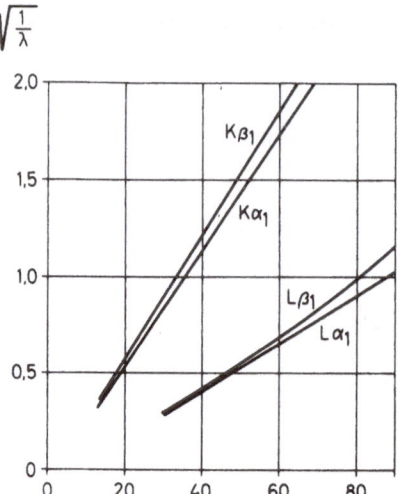

$\sqrt{\dfrac{1}{\lambda}}$

Fig. 2.3. Wavelengths λ of the most intense lines of the K and L series are plotted against the atomic number of the elements. According to Moseley's law a nearly linear relationship exists between $\sqrt{(1/\lambda)}$ and Z.

in the K shell, both lines belong to the spectrum of the K series. In case of the K series, electron transfer from the N shell ($n = 4$) is most improbable and, hence, the corresponding line is very weak. This line is designated as $K_{\beta 2}$ and is only observed for elements of atomic numbers larger than $Z = 31$ (gallium) or $Z = 33$ (arsenic): elements with lower atomic numbers have N shells with an insufficient number of electrons. The spectrum of the L series originates through electron transitions from the M or N shells into the L shell. The released energy difference for $n = 3$ or 4 is equal to

$$\Delta E = \frac{2\,\pi^2\,m\,e^4\,F^2}{h^2}\left(\frac{1}{2^2} - \frac{1}{n^2}\right) \tag{2.7}$$

A simple relationship exists between the energy ΔE of analogous lines and the atomic number of the emitting element

$$\Delta E = \text{prop } F^2 \cong \text{prop } Z^2 \tag{2.8}$$

With c as the velocity of light, the wavelength λ of the emitted line is calculated from the energy difference ΔE

$$\lambda = ch \mathbin{/} \Delta E \qquad \rightarrow \Delta E = \frac{ch}{\lambda}$$

Moseley's law is derived when the wavelength λ is introduced into equation (2.8) instead of ΔE

$$\sqrt{\frac{1}{\lambda}} \cong \text{prop } Z \tag{2.9}$$

The nearly linear relationship between $\sqrt{(1/\lambda)}$ and Z is shown in Figure 2.3. With σ as the shielding constant, the effective charge F of the nucleus is

$$F = (Z - \sigma)$$

The value for the shielding constant changes with the shell. For an individual electron in the K series σ is ≈ 0.5; for an individual electron in the L shell $\sigma = 5.5$.

This simple derivation does not take into account that the L and M shells have several energy levels. Several electron transitions are possible from, as well as into, these shells and, hence, the spectra of the L and M series are more complicated. In the L series, for example, the α_1 and α_2 lines, the $\beta_1, \beta_2, \beta_3, \beta_4, \ldots, \beta_{10}$ lines, and several γ lines, are distinguished.

In order to illustrate all electron transitions in an atom the energy-level scheme is frequently used. In the energy-level diagram all energy levels occupied by electrons, as well as all possible electron transitions of the K, L, M series, etc. are illustrated (Figure 2.4).

Intensities of the various lines within a series are different. Lines which correspond to electron transitions from neighboring shells are more intense because these electron transitions occur more frequently than those from more distant shells. Intensity of a line is expressed in terms of a relative line intensity, where the strongest line is arbitrarily set equal to 100. Line intensities can also be expressed in terms of the number of emitted photons or as the amount of the emitted energy. We choose the first possibility because the relative number of photons is proportional to the transition probabilities. Transition probabilities change only little with the atomic number of the elements. The relative intensity within the K series, for example, is $K_{\alpha 1} : K_{\alpha 2} : K_{\beta 1} \approx 4 : 2 : 1$; within the L series it is $L\alpha_1 : L_{\alpha 2} : L_{\beta 1} : L_{\beta 2} : L_{\beta 3} \approx 10 : 1 : 6 : 2 : 1$.

Because of the dead time of the counter, absorption and scattering by the analyzing crystal, and self-absorption of radiation in the sample it is difficult to accurately determine the relative line intensities. For these reasons the values given in the literature scatter widely.

X-ray lines originate through electron transitions in the inner-most electron shells and, hence, the wavelengths of the emission lines are largely independent of the chemical state of the emitting element (chemical state only affects the outermost shell). The chemical bonding energy is much too low to affect the energy levels of the inner electrons. Effects of chemical bonding on wavelength can, however, be observed in case of the light elements where electrons from the M shell participate in the chemical bonding. Wavelengths shifts between the pure element and its oxide for the elements K ($Z = 19$) to Cu ($Z = 29$) are approximately 10^{-4} Å, while for

Table 2.1. Wavelengths of the Most Important Emission Lines and Absorption Edges of the K Series. [Values after Sandström (1957) p. 182, 214; Tomboulian (1957) p. 271, 303.] The kX values were recalculated to Å values using the factor 1.0020 and supplemented with values from the *Handbook of Chemistry and Physics*.

	α_1 KL_{III}	α_2 KL_{II}	β_1 KM_{III}	β_3 KM_{II}	β_2 $KN_{II,III}$	β_4 $KN_{IV,V}$	β_5 $KM_{IV,V}$	Abs. Edge K
3 Li		240						226.5
4 Be		113						111
5 B		68						66.64
6 C		44.4						43.68
7 N		31.6						30.99
8 O		24						23.32
9 F		18.3						
10 Ne		14.65						14.301
11 Na		11.908	11.573	11.725				11.58
12 Mg		9.892	9.558	9.666				9.512
13 Al	8.339	8.340	7.960	8.058				7.951
14 Si	7.125	7.128	6.777	6.777				6.729
15 P	6.154	6.157	5.804					5.787
16 S	5.372	5.375	5.032					5.018
17 Cl	4.727	4.730	4.403					4.397
18 Ar	4.191	4.194	3.886					3.871
19 K	3.741	3.744	3.454				3.441	3.435
20 Ca	3.358	3.361		3.089			3.074	3.070
21 Sc	3.031	3.034		2.779			2.763	2.757
22 Ti	2.748	2.752		2.514			2.498	2.497
23 V	2.503	2.507		2.284			2.269	2.269
24 Cr	2.289	2.293		2.085			2.071	2.070
25 Mn	2.102	2.106		1.910			1.897	1.896
26 Fe	1.936	1.940		1.756			1.744	1.743
27 Co	1.789	1.793		1.621			1.609	1.608
28 Ni	1.658	1.662		1.500	1.489		1.489	1.488
29 Cu	1.540	1.544	1.392	1.392	1.381		1.381	1.380
30 Zn	1.435	1.439	1.295	1.295	1.284		1.285	1.283
31 Ga	1.340	1.344	1.208	1.208	1.196		1.198	1.195
32 Ge	1.254	1.258	1.129	1.129	1.117		1.119	1.117
33 As	1.176	1.180	1.057	1.058	1.045		1.049	1.045
34 Se	1.105	1.109	0.992	0.993	0.980		0.984	0.980
35 Br	1.040	1.044	0.933	0.933	0.920		0.925	0.920
36 Kr	0.980	0.984	0.878	0.879	0.866	0.865	0.871	0.866
37 Rb	0.925	0.930	0.829	0.829	0.816	0.815	0.822	0.816
38 Sr	0.875	0.879	0.783	0.783	0.771	0.770	0.776	0.770
39 Y	0.829	0.833	0.741	0.741	0.728	0.728	0.734	0.728
40 Zr	0.786	0.790	0.702	0.702	0.690	0.689	0.696	0.689
41 Nb	0.746	0.750	0.666	0.666	0.654	0.653		0.653
42 Mo	0.709	0.713	0.632	0.633	0.621	0.620	0.627	0.620
43 Tc	0.675	0.679	0.601	0.601	0.590	0.590		0.589
44 Ru	0.643	0.647	0.572	0.573	0.562	0.561	0.568	0.561
	α_1	α_2	β_1	β_3	β_2	β_4	β_5	K

Table 2.1 (*continued*)

		α_1	α_2	β_1	β_3	β_2	β_4	β_5	Abs. Edge
45	Rh	0.613	0.618	0.546	0.546	0.535	0.534	0.541	0.534
46	Pd	0.585	0.590	0.520	0.521	0.510		0.517	0.509
47	Ag	0.559	0.564	0.497	0.498	0.487	0.486	0.493	0.486
48	Cd	0.535	0.539	0.475	0.476	0.465			0.464
49	In	0.512	0.516	0.454	0.455	0.445	0.444	0.451	0.444
50	Sn	0.490	0.495	0.435	0.436	0.426	0.425	0.432	0.425
51	Sb	0.470	0.475	0.417	0.418	0.408	0.407	0.414	0.407
52	Te	0.451	0.456	0.400	0.401	0.391			0.390
53	I	0.433	0.438	0.384	0.384	0.375			0.374
54	Xe	0.416	0.420	0.368		0.360			0.359
55	Cs	0.400	0.405	0.354	0.355	0.346			0.345
56	Ba	0.385	0.390	0.341	0.341	0.333		0.338	0.331
57	La	0.371	0.375	0.328	0.329	0.320	0.319	0.325	0.318
58	Ce	0.357	0.362	0.316	0.316	0.308	0.307	0.313	0.307
59	Pr	0.344	0.349	0.304	0.305	0.297			0.300
60	Nd	0.332	0.336	0.293	0.294	0.286			0.285
61	Pm	0.320	0.324	0.283	0.283	0.276			0.274
62	Sm	0.309	0.314	0.273	0.274	0.266			0.265
63	Eu	0.298	0.303	0.264	0.264	0.257			0.256
64	Gd	0.288	0.293	0.254	0.255	0.248			0.247
65	Tb	0.279	0.283	0.246	0.247	0.240			0.238
66	Dy	0.269	0.274	0.238	0.238	0.232			0.231
67	Ho	0.261	0.265						0.223
68	Er	0.252	0.257	0.223	0.223	0.217			0.216
69	Tm	0.244	0.249	0.215	0.216				0.209
70	Yb	0.237	0.241	0.209	0.210	0.204			0.202
71	Lu	0.229	0.234	0.202	0.203	0.197			0.196
72	Hf	0.222	0.227	0.196	0.196	0.191			0.190
73	Ta	0.215	0.220	0.190	0.191	0.185	0.184	0.189	0.184
74	W	0.209	0.214	0.184	0.185	0.179	0.179	0.183	0.178
75	Re	0.203	0.208	0.179	0.180	0.174	0.174	0.178	0.173
76	Os	0.197	0.202	0.174	0.174	0.169	0.168	0.172	0.168
77	Ir	0.191	0.196	0.168	0.169	0.164	0.163	0.167	0.163
78	Pt	0.185	0.190	0.164	0.164	0.159	0.159	0.163	0.158
79	Au	0.180	0.185	0.159	0.160	0.155	0.154	0.158	0.153
80	Hg	0.175	0.180	0.154	0.155	0.150	0.150	0.153	0.149
81	Tl	0.170	0.175	0.150	0.151	0.146	0.145	0.149	0.145
82	Pb	0.165	0.170	0.146	0.147	0.142	0.141	0.145	0.141
83	Bi	0.161	0.166	0.142	0.143	0.138	0.138	0.141	0.137
84	Po								0.133
85	At								0.129
86	Rn								0.126
87	Fr								0.123
88	Ra								0.119
89	Ac								0.116
90	Th	0.133	0.138	0.117	0.118	0.114	0.114	0.117	0.113
91	Pa								0.110
92	U	0.126	0.131	0.111	0.112	0.108	0.108	0.111	0.107
		α_1	α_2	β_1	β_3	β_2	β_4	β_5	K

Table 2.2. Wavelengths of the Most Important Emission Lines and Absorption Edges of the L series. [Values were taken from Sandström (1957) p, 186, 214.] The kX values were recalculated to Å values using the factor 1.0020. Missing values for absorption edges were calculated from the energies given in Table 1.2 according to the formula $\lambda = 12.395/\text{keV}$.

	α_1 L_{III} M_V	α_2 L_{III} M_{IV}	β_1 L_{II} M_{IV}	β_2 L_{III} N_V	β_3 L_I M_{III}	β_4 L_I M_{II}	β_5 L_{III} $O_{IV,V}$	β_6 L_{III} N_I	β_7 L_{III} O_I	β_9 L_I M_V	β_{10} L_I M_{IV}
19 K											
20 Ca	36.323	35.952									
21 Sc	31.333	31.012									
22 Ti	27.395	27.014									
23 V	24.258	23.848									
24 Cr	21.673	21.282			19.429						
25 Mn	19.449	19.118			17.575						
26 Fe	17.567	17.254			15.741						
27 Co	15.968	15.657			14.268						
28 Ni	14.566	14.279			13.146						
29 Cu	13.330	13.053			12.094						
30 Zn	12.256	11.985			11.185						
31 Ga	11.290	11.023									
32 Ge	10.436	10.173									
33 As	9.671	9.414									
34 Se	8.990	8.735									
35 Br	8.375	8.125									
37 Rb	7.318	7.325	7.076		6.787	6.820		6.984			
38 Sr	6.862	6.869	6.624		6.367	6.402		6.519			
39 Y	6.448	6.455	6.212		5.983	6.018		6.094			
40 Zr	6.070	6.077	5.836	5.586	5.633	5.668		5.710			
41 Nb	5.724	5.732	5.492	5.238	5.310	5.345		5.361			
42 Mo	5.406	5.414	5.177	4.923	5.013	5.048		<5.048			
44 Ru	4.845	4.853	4.620	4.371	4.486	4.523		4.486			
45 Rh	4.597	4.605	4.374	4.130	4.252	4.289		4.241			
46 Pd	4.368	4.376	4.146	3.909	4.034	4.071		4.016		3.792	3.799
47 Ag	4.154	4.163	3.934	3.703	3.833	3.870		3.808		3.605	3.611
48 Cd	3.956	3.965	3.738	3.514	3.645	3.682		3.614		3.430	3.436
49 In	3.772	3.781	3.555	3.338	3.470	3.507		3.436		3.267	3.274
50 Sn	3.600	3.609	3.385	3.175	3.306	3.343		3.269	3.156	3.115	3.122
51 Sb	3.439	3.448	3.226	3.023	3.152	3.190		3.115	3.004	2.972	2.979
52 Te	3.289	3.298	3.077	2.882	3.009	3.046		2.971	2.863	2.839	2.847
53 I	3.148	3.158	2.937	2.750	2.874	2.912		2.837	2.730	2.713	2.721
55 Cs	2.892	2.902	2.683	2.511	2.628	2.666		2.593	2.485	2.478	2.492
56 Ba	2.776	2.785	2.568	2.404	2.516	2.555		2.482	2.380	2.376	2.387
57 La	2.666	2.675	2.459	2.303	2.410	2.449		2.379	2.275	2.282	2.290
58 Ce	2.561	2.570	2.356	2.209	2.311	2.349		2.281	2.181	2.188	2.196
59 Pr	2.463	2.473	2.258	2.119	2.217	2.255		2.190	2.092	2.100	2.107
	α_1	α_2	β_1	β_2	β_3	β_4	β_5	β_6	β_7	β_9	β_{10}

Table 2.2 (*continued*)

	γ_1 L_{II} N_{IV}	γ_2 L_I N_{II}	γ_3 L_I N_{III}	γ_4 L_I $O_{II,III}$	γ_5 L_{II} N_I	γ_6 L_{II} O_{IV}	l L_{III} M_I	η L_{II} M_I	Abs. Edges L_I	L_{II}	L_{III}
19 K							47.735	47.234	32.878	41.594	42.017
20 Ca							40.962	40.461	28.235	35.414	35.824
21 Sc							35.601	35.130	24.496	30.158	30.530
22 Ti							31.363	30.882	21.938	26.946	27.302
23 V							27.775	27.325	19.643	23.882	24.209
24 Cr							24.789	24.288	18.255	21.334	21.670
25 Mn							22.274	21.824	16.483	19.069	19.397
26 Fe							20.150	19.800	14.738	17.215	17.507
27 Co							18.297	17.896	13.342	15.611	15.911
28 Ni							16.708	16.313	12.272	14.231	14.514
29 Cu							15.296	14.900	11.248	13.006	13.285
30 Zn							14.053	13.692	10.355	11.858	12.131
31 Ga							12.950	12.594	9.535	10.585	10.835
32 Ge							11.946	11.610	8.729	9.956	10.210
33 As							11.106	10.732	8.107	9.124	9.367
34 Se							10.293	9.959	7.506	8.407	8.646
35 Br							9.583	9.253	6.925	7.752	7.986
37 Rb		6.045			6.755		8.363	8.041	5.998	6.643	6.863
38 Sr		5.644			6.296		7.836	7.517	5.583	6.172	6.387
39 Y		5.283			5.875		7.356	7.040	5.232	5.755	5.962
40 Zr	5.497	4.953			5.384		6.918	6.606	4.867	5.378	5.583
41 Nb	5.151	4.654			5.036		6.517	6.210	4.581	5.024	5.223
42 Mo	4.837	4.380			4.726		6.150	5.847	4.298	4.719	4.913
44 Ru	4.287	3.897			4.182		5.503	5.205	3.842	4.180	4.369
45 Rh	4.045	3.685			3.943		5.217	4.921	3.628·	3.942	4.130
46 Pd	3.724	3.489			3.822		4.952	4.660	3.435	3.724	3.908
47 Ag	3.522	3.312	3.306		3.616		4.707	4.418	3.251	3.513	3.701
48 Cd	3.335	3.137			3.425		4.480	4.193	3.085	3.326	3.504
49 In	3.162	2.980		2.926	3.249		4.268	3.983	2.926	3.147	3.324
50 Sn	3.001	2.832		2.777	3.085		4.071	3.789	2.777	2.982	3.156
51 Sb	2.851	2.695		2.640	2.932		3.888	3.607	2.639	2.830	3.000
52 Te	2.712	2.567		2.511	2.790		3.717	3.438	2.511	2.687	2.855
53 I	2.582	2.447		2.391	2.657		3.557	3.280	2.389	2.553	2.719
55 Cs	2.348	2.237	2.232	2.174	2.417		3.267	2.993	2.167	2.314	2.474
56 Ba	2.241	2.138	2.134	2.075	2.308		3.135	2.862	2.068	2.205	2.363
57 La	2.141	2.046	2.041	1.983	2.205		3.006	2.739	1.973	2.103	2.258
58 Ce	2.048	1.960	1.955	1.899	2.110		2.891	2.620	1.889	2.011	2.164
59 Pr	1.961	1.879	1.874	1.819	2.020		2.784	2.512	1.811	1.924	2.077
	γ_1	γ_2	γ_3	γ_4	γ_5	γ_6	ι	η	L_I	L_{II}	L_{III}

Table 2.2 (*continued*)

	α_1 L_{III} M_V	α_2 L_{III} M_{IV}	β_1 L_{II} M_{IV}	β_2 L_{III} N_V	β_3 L_I M_{III}	β_4 L_I M_{II}	β_5 L_{III} $O_{IV,V}$	β_6 L_{III} N_I	β_7 L_{III} O_I	β_9 L_I M_V	β_{10} L_I M_{IV}
60 Nd	2.370	2.380	2.167	2.035	2.126	2.167		2.103	2.009	2.016	2.023
61 Pm	2.282	2.292	2.080	1.956	2.042						
62 Sm	2.199	2.210	1.998	1.882	1.962	2.000		1.946	1.856	1.862	1.869
63 Eu	2.121	2.131	1.920	1.812	1.887	1.925	1.777	1.874	1.785	1.791	1.799
64 Gd	2.047	2.058	1.847	1.745	1.815	1.854	1.713	1.805	1.720	1.724	1.731
65 Tb	1.976	1.987	1.777	1.683	1.747	1.786	1.651	1.742	1.658		1.667
66 Dy	1.909	1.920	1.711	1.624	1.682	1.721	1.592	1.682	1.604	1.600	1.607
67 Ho	1.845	1.856	1.647	1.567	1.620	1.659	1.538	1.624			1.548
68 Er	1.784	1.796	1.587	1.514	1.561	1.601		1.567	1.494	1.485	1.494
69 Tm	1.727	1.738	1.530	1.464	1.506	1.545	1.435	1.516		1.434	1.441
70 Yb	1.672	1.683	1.476	1.415	1.452	1.491	1.387	1.466	1.395	1.384	1.391
71 Lu	1.619	1.630	1.424	1.370	1.401	1.440	1.342	1.419	1.349	1.335	1.343
72 Hf	1.569	1.580	1.374	1.326	1.353	1.392	1.297	1.374	1.306	1.290	1.298
73 Ta	1.522	1.533	1.327	1.284	1.307	1.346	1.256	1.331	1.264	1.246	1.253
74 W	1.476	1.487	1.282	1.245	1.263	1.302	1.215	1.290	1.224	1.205	1.212
75 Re	1.433	1.444	1.239	1.207	1.220	1.259	1.177	1.251	1.186	1.165	1.172
76 Os	1.391	1.402	1.197	1.170	1.179	1.218	1.140	1.213	1.149	1.126	1.133
77 Ir	1.351	1.363	1.158	1.135	1.141	1.180	1.106	1.178	1.115	1.090	1.097
78 Pt	1.313	1.324	1.120	1.102	1.104	1.142	1.073	1.143	1.082	1.055	1.062
79 Au	1.276	1.288	1.083	1.070	1.068	1.106	1.040	1.111	1.050	1.021	1.028
80 Hg	1.241	1.253	1.049	1.040	1.034	1.072	1.010	1.080	1.019	0.986	0.996
81 Tl	1.207	1.219	1.015	1.010	1.001	1.039	0.981	1.050	0.990	0.957	0.964
82 Pb	1.175	1.186	0.983	0.982	0.969	1.008	0.953	1.021	0.962	0.927	0.934
83 Bi	1.144	1.155	0.952	0.955	0.939	0.977	0.926	0.993	0.935	0.898	0.905
84 Po	1.114	1.126	0.922	0.929	0.909	0.947	0.899	0.967			
87 Fr	1.030		0.840	0.858							
88 Ra	1.005	1.016	0.814	0.835	0.803	0.841	0.806	0.871	0.816	0.769	0.775
90 Th	0.956	0.968	0.765	0.793	0.755	0.793	0.765	0.828	0.774	0.723	0.730
91 Pa	0.933	0.945	0.742	0.774	0.732	0.770	0.745	0.808	0.755	0.702	0.709
92 U	0.911	0.922	0.720	0.755	0.710	0.748	0.726	0.788	0.736	0.681	0.688
	α_1	α_2	β_1	β_2	β_3	β_4	β_5	β_6	β_7	β_9	β_{10}

Table 2.2 (*continued*)

	γ_1 L_{II} N_{IV}	γ_2 L_I N_{II}	γ_3 L_I N_{III}	γ_4 L_I $O_{II,III}$	γ_5 L_{II} N_I	γ_6 L_{II} O_{IV}	l L_{III} M_I	η L_{II} M_I	Abs. Edges		
									L_I	L_{II}	L_{III}
60 Nd	1.878	1.801	1.796	1.744	1.935	1.855	2.676	2.409	1.735	1.843	1.995
61 Pm	1.799								1.668	1.765	1.918
62 Sm	1.727	1.659	1.655	1.607	1.779	1.705	2.482	2.218	1.599	1.703	1.845
63 Eu	1.657	1.596	1.590	1.544	1.708	1.628	2.395	2.131	1.538	1.627	1.776
64 Gd	1.592	1.533	1.530	1.484	1.641	1.564	2.312	2.049	1.478	1.563	1.712
65 Tb	1.530	1.476	1.472	1.427	1.579	1.503	2.235	1.973	1.422	1.502	1.650
66 Dy	1.473	1.423	1.416	1.375	1.518	1.446	2.158	1.897	1.368	1.444	1.590
67 Ho	1.417	1.370	1.364	1.322	1.462	1.392	2.086	1.826	1.320	1.390	1.537
68 Er	1.364	1.321	1.315	1.276	1.406		2.019	1.756	1.271	1.339	1.483
69 Tm	1.315	1.274	1.268	1.229	1.356	1.290	1.955	1.696	1.225	1.289	1.433
70 Yb	1.268	1.229	1.222	1.185	1.306	1.243	1.896	1.635	1.182	1.243	1.386
71 Lu	1.222	1.185	1.179	1.143	1.260	1.199	1.836	1.578	1.140	1.198	1.341
72 Hf	1.179	1.144	1.138	1.103	1.215	1.155	1.781	1.523	1.100	1.155	1.297
73 Ta	1.138	1.105	1.099	1.065	1.173	1.114	1.728	1.471	1.062	1.114	1.256
74 W	1.098	1.068	1.062	1.028	1.132	1.074	1.678	1.421	1.025	1.074	1.215
75 Re	1.061	1.032	1.026	0.993	1.094	1.037	1.630	1.373	0.989	1.037	1.177
76 Os	1.025	0.998	0.992	0.960	1.057	1.001	1.585	1.328	0.957	1.001	1.141
77 Ir	0.991	0.965	0.959	0.928	1.022	0.967	1.541	1.284	0.924	0.967	1.106
78 Pt	0.958	0.934	0.928	0.897	0.988	0.934	1.499	1.243	0.893	0.934	1.073
79 Au	0.926	0.904	0.898	0.867	0.956	0.903	1.460	1.203	0.864	0.903	1.040
80 Hg	0.896	0.875	0.869	0.839	0.924	0.873	1.422	1.164	0.835	0.872	1.009
81 Tl	0.867	0.848	0.841	0.812	0.895	0.844	1.385	1.128	0.808	0.844	0.980
82 Pb	0.840	0.821	0.814	0.786	0.866	0.817	1.350	1.092	0.782	0.816	0.951
83 Bi	0.813	0.796	0.789	0.761	0.839	0.790	1.316	1.059	0.757	0.789	0.924
84 Po	0.787		0.764			0.764					
87 Fr	0.716										
88 Ra	0.695	0.682	0.675	0.650	0.718	0.673	1.167	0.907	0.645	0.671	0.803
90 Th	0.653	0.642	0.635	0.612	0.675	0.632	1.115	0.854	0.606	0.630	0.761
91 Pa	0.634	0.624	0.617	0.594	0.655	0.613	1.091	0.829			
92 U	0.615	0.604	0.598	0.576	0.635	0.595	1.067	0.805	0.569	0.592	0.722
	γ_1	γ_2	γ_3	γ_4	γ_5	γ_6	l	η	L_I	L_{II}	L_{III}

Table 2.3. Relative Intensity of Lines in the Emission Spectrum of the L Series. Listed are the relative transition probabilities of individual electron transitions, which were calculated from the relative line intensities according to the values by Sandström (1957), p. 237; and Sagel (1959), p. 48.

		α_1	α_2	β_1	β_2	β_3	β_4	β_6	β_7	β_9	β_{10}	γ_1	γ_2	γ_{3+6}	γ_4	γ_5	l	η
		L_{III}	L_{III}	L_{II}	L_{III}	L_I	L_I	L_{III}	L_{III}	L_I	L_I	L_{II}	L_I	L_I	L_I, L_{II}	L_{II}	L_{III}	L_{II}
		M_V	M_{IV}	M_{IV}	N_V	M_{III}	M_{II}	N_I	O_I	M_V	M_{IV}	N_{IV}	N_{II}	N_{III},O_{III}	O_{IV}	N_I	M_I	M_I
42	Mo	100	13	59	7.3	13	9.2					5.9						
45	Rh	100	13	58	11	11	7.4					6.6						
46	Pd	100	12	56	12	9.2	6.0					7.2					3.9	2.1
47	Ag	100	8.6	46	11	6.6	3.6	0.51		0.064	0.037	3.7	0.53	0.83		0.24	5.0	2.1
73	Ta	100	11	50	17	6.3	5.7					8.2	1.5		0.56	0.46	4.1	1.2
74	W	100	11	42	21	6.0	3.8	1.0		0.49	0.49	7.3	1.0	1.6	0.40	0.33	2.7	1.1
78	Pt	100	12	33	19	6.9	4.5					6.6					3.4	1.0
90	Th	100	12	42	22	2.6		1.2				9.6	1.0				4.2	1.6
92	U	100	11	39	23	3.3	3.4	1.4				8.1	1.0				2.8	0.88

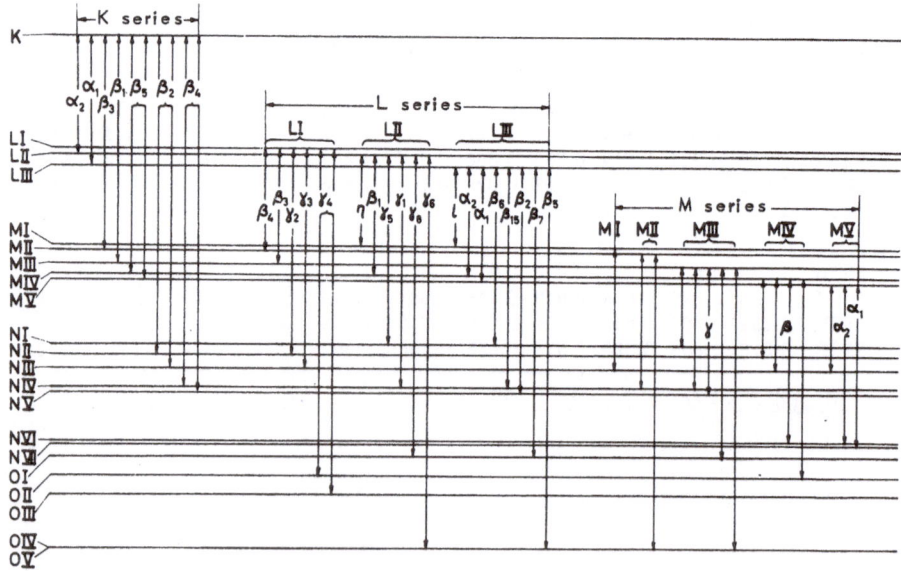

Fig. 2.4. Energy-level diagram. Electron transitions corresponding to frequently occurring lines are marked by arrows. $K, L,$ and M series of uranium (after Richtmyer and Kennard, 1942).

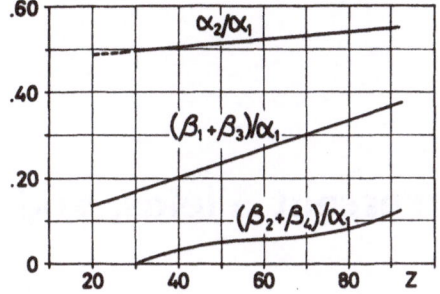

Fig. 2.5. Relative photon numbers of lines or relative transition probabilities in the spectrum of the K series as a function of the atomic number [calculated from relative line intensities and averaged over values given by Sandström (1957), p. 236, and Sagel (1959), p. 43].

Al ($Z = 11$) the AlK_α line shift ($\Delta\lambda$) amounts to 10^{-3} Å and the AlK_β $\Delta\lambda$ is 2×10^{-2} Å. The AlK_β line is more affected than the AlK_α line because the former corresponds to an electron transition from the M shell, which participates in the chemical bonding. Shifts in the wavelength are also connected with changes in the relative line intensities (Fischer and Baun, 1965). The wavelength of aluminum is also affected by the aluminum coordination number to oxygen.

Chapter 3

Photoelectrons, Fluorescent Yields, and Auger Electrons

When materials are irradiated by x-ray quanta a portion of the incident radiation is absorbed. In this process an electron of the atom is removed from its normal energy level and, as a photoelectron, absorbs the total energy of the x-ray quantum. The absorbed energy is, in part, consumed as the work required to remove the photoelectron from the atom (ionization work A). The remaining energy appears as kinetic energy of the photoelectron. The kinetic energy of the photoelectron is given as

$$\frac{1}{2} m v^2 = h \nu - A$$

where ν is the frequency of the absorbed radiation and h is Planck's constant. The production of photoelectrons may be seen in a Wilson cloud chamber. If the energy required to remove a photoelectron from the K shell of an atom is A_K, then removal is only possible when the energy of the absorbed quantum is $h\nu > A_K$. When this condition is not fulfilled, electrons cannot be removed from the K shell and, hence, fluorescence cannot occur.

To excite the characteristic radiation of copper and molybdenum, more than 9 kV and 20 kV, respectively, are required. A 50-kV instrument, for example, is capable of exciting the K series of elements to europium ($Z = 63$), whereas with a 100-kV instrument the K series of the elements up to lead and bismuth can be excited (Table 1.2).

A hole remains when a photoelectron is removed from an atom. An electron from a neighboring shell, however, transfers into the hole and, in this process, emits radiation of an energy corresponding to the energy difference between the two shells. When studying, in more detail, the energy emission of an electron during transfer from a higher energy level into a lower energy level, it is found that electron transfer is not always associated with emission of an x-ray quantum. The number of electron transitions per unit time which terminate in a particular energy level is termed n. Of all n

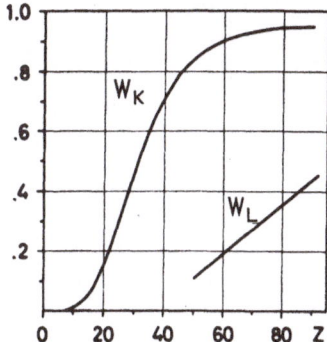

Fig. 3.1. Fluorescent yields W_K and W_L for the K and L shells of various elements [W_K after Roos (1957); and W_L averaged over values from Landolt-Börnstein I/1, p. 321].

only n_f transitions are associated with the emission of x-ray quanta while in $(n - n_f)$ cases no x-ray quanta are emitted. The fluorescent yield W is defined as the fraction n_f/n of all electron transitions which are associated with the emission of an x-ray quantum

$$W = \frac{n_f}{n}$$

Individual shells of an atom have different fluorescent yields, which are designated W_K, W_L, etc. The fluorescent yield W_K for ionizations in the K shell of aluminum is $W_K = 0.025$, for iron it is $W_K = 0.30$, and for silver it is $W_K = 0.82$. In case of aluminum only about 2% of all electron transitions which terminate in the K shell are associated with the emission of characteristic AlK_α or AlK_β lines.

The remaining $(n - n_f)$ transitions result in the emission of "Auger electrons." The Auger electrons originate in the outermost shells of the atom and are similar to the previously mentioned photoelectrons. An Auger electron absorbs the total energy which is released during an internal electron transition and escapes from the atom. Its kinetic energy is given as follows

$$\frac{1}{2} m v^2 = \Delta E - A$$

In this equation ΔE corresponds to the energy difference which is released during internal electron transition, and A to the ionization work required for the production of the Auger electron. Occurrence of Auger electrons can also be made visible in the Wilson cloud chamber.

Chapter 4

Qualitative Analysis

4.1. General Remarks

For qualitative analysis it is most advantageous to automatically scan the complete wavelength range while recording the fluorescent spectrum emitted from the sample. Commercial spectrographs are capable of registering x-ray spectra in the wavelength range from about 0.3 to 12 Å. This range covers the emission spectra of the elements from $Z = 11$ (sodium) to $Z = 92$ (uranium). Depending upon the wavelength range of interest, different analyzing crystals and detectors are employed. A LiF crystal, for example, is suited for measurement of the spectra of the elements from $Z = 20$ (calcium) to $Z = 92$ (uranium), while with an EDDT or PE crystal the spectra of the elements from $Z = 13$ (aluminum) to $Z = 26$ (iron) can be registered. Gypsum, ADP, or KHP are used as analyzing crystals to record the spectra of the elements $Z = 11$ (sodium) and $Z = 12$ (magnesium). Radiation of short wavelength is measured with scintillation counters and flow proportional counters are employed for the detection of radiation of long wavelength.

Two major emission spectra are distinguished, namely, the K spectrum (or K series) and the L spectrum (or L series). These spectra are named, respectively, after the K and L shells. An element emits the K spectrum when its K shell is excited, and emission of the L spectrum occurs when the L shell is excited. The K spectrum has two major lines, namely, the K_α and the K_β lines. The α line has the longer wavelength and the higher intensity (approximately four times the intensity of the β line; Fig. 4.1). K_α and K_β lines are actually doublets consisting of two individual lines that are very close in wavelength, namely, $K_{\alpha 1}$, $K_{\alpha 2}$ and $K_{\beta 1}$, $K_{\beta 3}$ (spin doublets). The wavelength difference between $K_{\alpha 1}$ and $K_{\alpha 2}$ lines is approximately 0.004 Å, and the difference between $K_{\beta 1}$ and $K_{\beta 3}$ lines is even less (approximately 0.0007 Å). These doublets often cannot be resolved in commercial spectrometers but appear as single lines. Only in case of high spectral resolution (second- and third-order reflections) can the $K_{\alpha 1}$ and $K_{\alpha 2}$ lines be

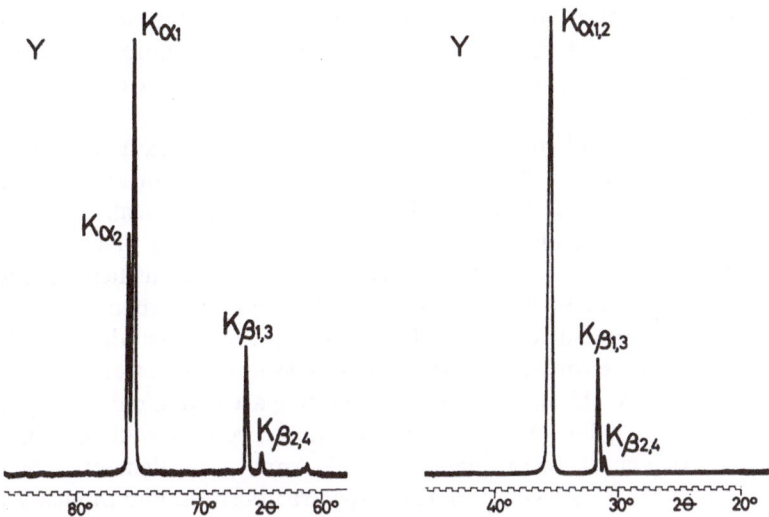

Fig. 4.1. K spectrum of yttrium ($Z = 39$). Right: first-order reflection with three lines, $K_{\alpha 1,2}$; $K_{\beta 1,3}$; $K_{\beta 2,4}$, which correspond to the electron transitions L_{II}, $L_{IV} \to K$; M_{II}, $M_{III} \to K$, and N_{II}, $N_{IV} \to K$. Left: spectrum recorded with high spectral resolution (second-order reflection); the doublet $K_{\alpha 1,2}$ is resolved into $K_{\alpha 1}$ and $K_{\alpha 2}$ lines (topaz diffraction crystal, scintillation counter, molybdenum x-ray tube).

resolved; even then, however, the $K_{\beta 1}$, $K_{\beta 3}$ doublet cannot be resolved. Occasionally, a third line is observed which, however, is very weak. It corresponds to the electron transition from the N into the K shell and, for this reason, has an even shorter wavelength than the $K_{\beta 1}$, $K_{\beta 3}$ line. This line is designated $K_{\beta 2,4}$. It is also a doublet (shielded doublet), the wavelength difference between $K_{\beta 2}$ and $K_{\beta 4}$ being approximately 0.0006 Å. The L spectrum consists of two intense lines, namely, the $L_{\alpha 1}$ and the $L_{\beta 1}$ lines; again, the β_1 line is of shorter wavelength than the α_1 line. Several lines of medium intensity ($L_{\beta 2}$, $L_{\beta 3}$, $L_{\beta 4}$, $L_{\gamma 1}$) and several weak lines such as $L_{\alpha 2}$, $L_{\beta 5}$, $L_{\beta 6}$, $L_{\gamma 2}$, $L_{\gamma 3}$, $L_{\gamma 6}$, L_{η}, L_{l} also occur (Figure 4.2). The greater number of lines in the L series in comparison to the K series is due to the fact that the K shell has only one energy level while the L shell is split up into three levels, L_{I}, L_{II}, and L_{III}; hence, a larger variety of electron transitions is possible in the L shell. If the energy of the primary x-ray radiation is sufficient to ionize the K shell, then the L and M shells are ionized as well. The corresponding L and M spectra, however, are of too long a wavelength to ordinarily be recorded simultaneously with the K spectrum. In some instances, however, the L spectrum may be observed together with the K spectrum (Figure 4.3). Recordings obtained from samples containing several elements show overlap of the various spectra of the elements involved (Figure 4.4). Every K spectrum by itself consists of a more intense K_α line and a somewhat weaker K_β line, and every L spectrum consists of two intense $L_{\alpha 1}$ and $L_{\beta 1}$ lines and weaker $L_{\beta 2}$, $L_{\beta 3}$, etc. lines.

Identification of individual spectra should best begin with the most intense line, i.e., with a K_α, L_α, or $L_{\beta 1}$ line. Wavelengths or reflection angles of all $K_{\alpha 1}$, $L_{\alpha 1}$, and $L_{\beta 1}$ lines should be scanned systematically. Once a line has tentatively been identified, then the remaining lines of the particular spectrum should be checked for their appropriate intensities. A line can only be considered to have been properly identified when the complete spectrum is identified. Once this has been achieved, then all lines of first, second, and higher order have to be identified in the spectrum in question. Correspondingly, the strongest lines remaining are compared to all possible $K_{\alpha 1}$, $L_{\alpha 1}$, and $L_{\beta 1}$ lines until all lines in the recording are identified. Comparative strip-chart recordings obtained from pure elements are often very helpful for identifying unknown spectra; position and intensity of all lines in the unknown spectrum can be determined by comparison with the strip charts of the pure elements.

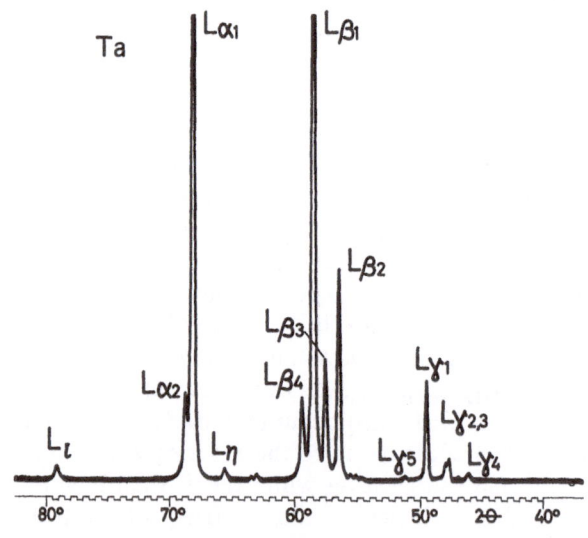

Fig. 4.2. L spectrum of tantalum ($Z = 73$). The reason for the large number of lines in the L spectrum is that the L shell consists of three energy levels and, hence, a larger variety of electron transitions from the M and N shells is possible (topaz diffraction crystal, scintillation counter, molybdenum x-ray tube).

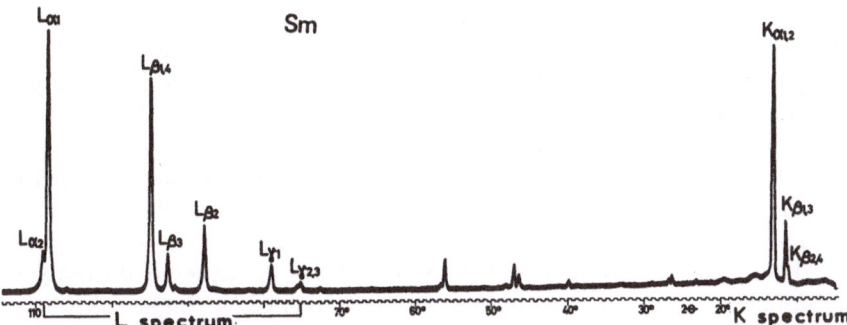

Fig. 4.3. Spectra of samarium ($Z = 62$). K and L spectra are recorded side by side (topaz diffraction crystal, scintillation counter, 54 kV, gold x-ray tube; the gold lines in the spectrum are weakly visible in the range of $2\theta = 40°$ to $60°$).

4.2. Anomalous Intensities of Lines in a Spectrum

Occasionally, anomalies are observed in the line intensities of a spectrum. These anomalies are the result of the particular composition of the sample and the properties of the spectrometer. Such anomalies originate, for example, when two lines of different spectra overlap; the β line of yttrium ($Z = 39$) and the α line of niobium ($Z = 41$) have nearly the same wavelength. In a strip-chart recording obtained from a mixture of yttrium and niobium, both lines occur at the same position. In the yttrium spectrum, the YK_β line is therefore too intense while in the niobium spectrum the intensity of the NbK_α line is too high. Anomalies also occur when lines of K and L spectra or of two different L spectra overlap. Examples for overlapping of two lines from different spectra are shown in Figures 4.5 and 4.6.

In this type of overlap both lines have the same energy and wavelength and, hence, cannot be resolved even by pulse height discrimination. Two other nonoverlapping lines of the spectrum have to be chosen for quantitative analysis. The following procedure may be used when intensity measurement of two overlapping lines of similar wavelength and energy cannot be avoided. The total intensity of the overlapping lines is measured first and then the intensity of a line whose value has been determined indirectly is subtracted from the total intensity. In the indirect determination of the intensity of a line, recourse is made to the fact that all emission lines of an element appear simultaneously and always with the same relative intensities. The intensity of a β line, for example, may be used to estimate the intensity of the corresponding α line, provided that the relative line intensities are known. The overlap of the PbL_α and AsK_α lines may serve to illustrate this procedure. In order to determine the intensity of the AsK_α line in lead-bearing samples, the intensity of thè Pb$L_{\beta 1}$ line must also be measured. The intensity ratio of the lines PbL_α/Pb$L_{\beta 1}$ is known to be 0.75, and, thus, the intensity of the PbL_α line can easily be calculated from the intensity of the corresponding Pb$L_{\beta 1}$ line. After correcting the intensity of the doublet (PbL_α + AsK_α) by this amount, the desired intensity of the AsK_α line is obtained. A similar method was described by Zemany (1960) for the correction of V–Cr–Mn lines in the analysis of steels.

The x-ray spectrum used for the excitation is scattered diffusely by the sample and appears to some extent in the recording. Light substances, such as organic compounds or water, scatter this spectrum particularly well. Part of the radiation is modified in the scattering process and shifted towards longer wavelengths (Compton scattering). It is for this reason that broader lines, with somewhat longer wavelengths, are observed in addition to the characteristic lines of the anode material; these broader lines correspond to the modified lines. The tube spectrum or its Compton-scattered portion may occasionally overlap a fluorescent line and cause anomalous intensities (Figure 4.6). The following are examples:

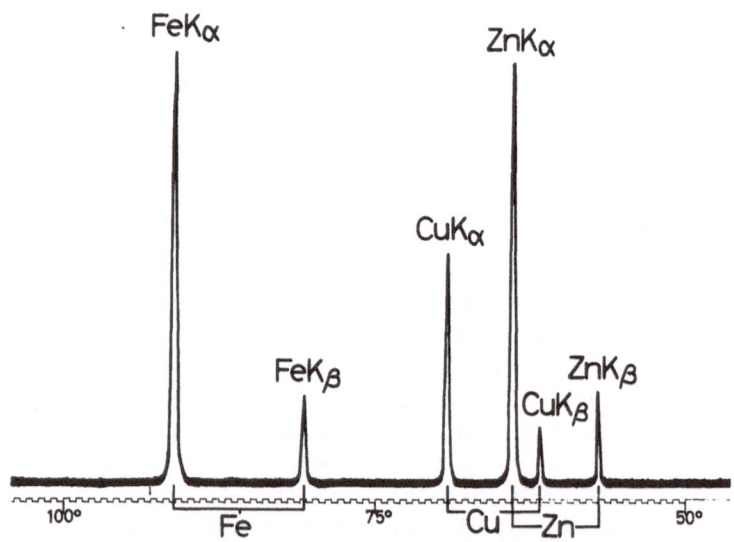

Fig. 4.4. Overlap of several K spectra obtained from a mixture. Every spectrum consists of a strong K_α line and a weaker K_β line of shorter wavelength. Intensity of the K_β line is approximately four times less than that of the K_α line (topaz diffraction crystal, scintillation counter, molybdenum x-ray tube).

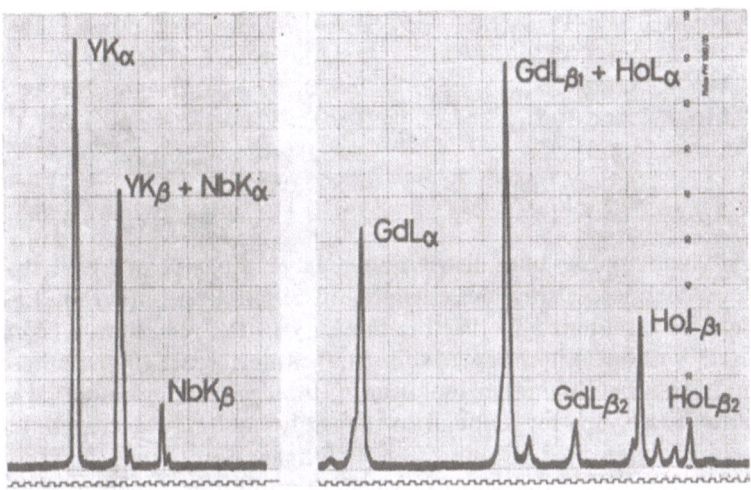

Fig. 4.5. Intensity anomalies. Lines of different spectra overlap resulting in anomalous line intensities. Overlapping is shown for the spectra of niobium and yttrium, and of holmium and gadolinium (topaz diffraction crystal, scintillation counter).

Overlapping lines	*Wavelength* Å
Compton scattering of WL_β and $PtL_{\alpha 1}$	1.29–1.34 and 1.313
Compton scattering of WL_α and $TaL_{\alpha 1}$	1.49–1.54 and 1.522
Compton scattering of MoK_α and $NbK_{\alpha 1}$	0.73–0.76 and 0.746
Compton scattering of AuL_β and $SeK_{\alpha 1}$	1.08–1.17 and 1.105

In addition to the characteristic lines of the anode material, other weak lines are occasionally observed as "contaminants" in the background radiation, such as lines of Fe, Ni, Cr, Cu, and Zn (from steel and brass), of Pb (from shielding sheets), and of W (from the filament). These lines are sometimes troublesome because they can be mistaken to indicate presence of small amounts of these elements in the sample. The apparatus, however, can be tested for this radiation using ultrapure samples and corrections can be made. Interference may also occur by lines of different order of reflection; a second-order line, for example, may overlap a first-order line of longer wavelength. In this case, the line of longer wavelength has approximately double the wavelength of the line of shorter wavelength, and the line of shorter wavelength is reflected in the second order. Examples of the overlapping of first- and second-order lines are given below.

Overlapping lines	*Wavelength* Å
$Zr_2K_{\alpha 2}$ and $HfL_{\alpha 1}$	$2 \times 0.786 = 1.572$ and 1.569
$Zr_2K_{\beta 1}$ and $HfL_{\beta 1}$	$2 \times 0.702 = 1.404$ and 1.374

$Nb_2K_{\alpha1}$ and $TaL_{\alpha1}$	$2 \times 0.746 = 1.492$ and 1.522
$Nb_2K_{\beta1}$ and $TaL_{\beta1}$	$2 \times 0.666 = 1.332$ and 1.327
$Mo_2K_{\alpha1}$ and $ReL_{\alpha1}$	$2 \times 0.709 = 1.418$ and 1.433
$Mo_2K_{\beta1}$ and $ReL_{\beta1}$	$2 \times 0.632 = 1.264$ and 1.239
$Cu_4K_{\alpha1}$ and $PK_{\alpha1}$	$4 \times 1.540 = 6.160$ and 6.154
$Cr_4K_{\beta1}$ and $AlK_{\alpha1}$	$4 \times 2.085 = 8.340$ and 8.339
Ag_2L_{α} and $AlK_{\alpha2}$	$2 \times 4.154 = 8.308$ and 8.339

The overlapping lines listed above have different wavelengths and quantum energies. Using pulse-height discrimination, either of the two lines may be eliminated so that the intensity of the remaining line can be measured without interference (Fig. 4.7). Interference by second-order lines may also be eliminated by lowering the accelerating potential of the x-ray tube to a value low enough so that only the long-wavelength part of the spectrum is excited to fluorescence. Furthermore, filters may be used to selectively absorb the short-wavelength radiation. Analyzing crystals made of germanium or silicon may be used to suppress second-order reflections of lighter elements. These crystals have diamond structure and may be cut so that the second-order reflection is eliminated.

Occasionally, intensity anomalies are caused by the presence of other components in the sample, which selectively absorb individual lines of the spectrum. This is the case when the absorption edge of the associated element in the matrix is located between the individual lines of a spectrum. A good

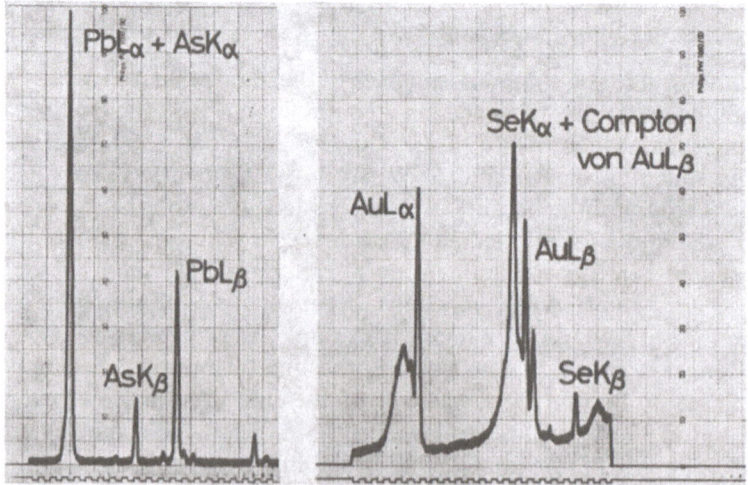

Fig. 4.6. Intensity anomalies. Overlapping of the spectra of lead and arsenic (left) and of the spectrum of selenium with an incoherently scattered line of the gold tube (right) (topaz diffraction crystal, scintillation counter).

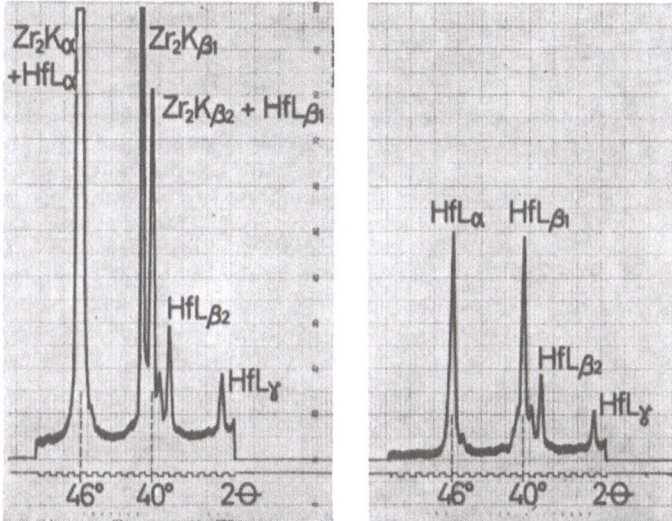

Fig. 4.7. Intensity anomalies. Interference of lines by higher order reflections of other strong lines (left). Pulse height analysis and discrimination may be employed to eliminate interfering lines (right). Emission spectrum of hafnium in a mixture with large quantities of zirconium (LiF diffraction crystal, scintillation counter, gold x-ray tube).

example is the spectrum of iron when it is associated with manganese. The wavelength of the K_α line of iron ($Z = 26$) (1.94 Å) is longer and the $\text{Fe}K_\beta$ line is shorter than the absorption edge of manganese ($Z = 25$) (1.90 Å). The $\text{Fe}K_\beta$ line is more strongly absorbed by manganese than is the $\text{Fe}K_\alpha$ line, and hence, the $\text{Fe}K_\beta$ line appears too weak in comparison to the $\text{Fe}K_\alpha$ line (Figure 4.8). Further examples for selective absorption by associated elements are the following: selective absorption of the $\text{Nb}K_\beta$ line of niobium ($Z = 41$) by yttrium ($Z = 39$) and zirconium ($Z = 40$), which causes the $\text{Nb}K_\beta$ line to appear weak in comparison to the $\text{Nb}K_\alpha$ line; and selective absorption of the K_β line of zinc ($Z = 30$) by ever-present copper ($Z = 29$). The fluorescent radiation may also be weakened by the emitting element itself. When comparing the L spectra of the elements thorium and uranium it is found that the intensity ratio of the two major lines $L_{\beta 1}/L_{\alpha 1}$ in the uranium spectrum is different from that of thorium (Figure 4.9). In the case of thorium ($Z = 90$), both major lines are longer in wavelength than the L_{III} absorption edge and, hence, both lines are absorbed about equally as strongly: in the thorium spectrum both $L_{\alpha 1}$ and $L_{\beta 1}$ lines appear with approximately the same intensity. The L_α line of element uranium ($Z = 92$)

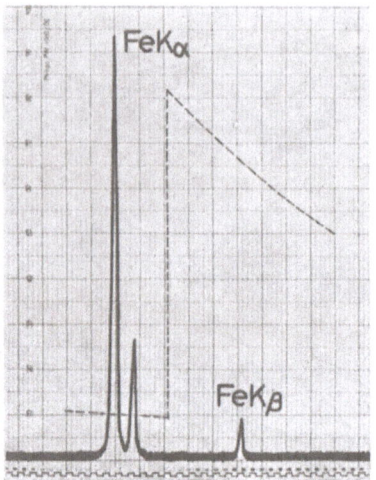

Fig. 4.8. Intensity anomalies. The short-wavelength part of a spectrum may be more strongly absorbed than the long-wavelength part due to the presence of certain elements in the mixture. As a result relative line intensities appear anomalous. Emission spectrum of iron in a mixture with manganese (the MnK_β line is visible to the right of the FeK_α line). The dashed line represents the absorption coefficient of manganese as a function of wavelength (topaz diffraction crystal, scintillation counter, molybdenum x-ray tube).

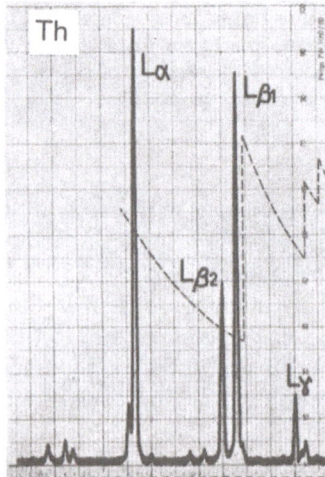 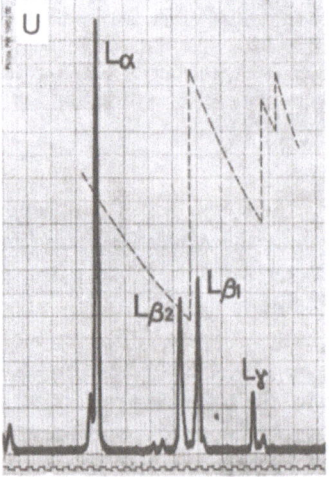

Fig. 4.9. Intensity anomalies. All emission lines of an element are absorbed by the element itself. Left: emission spectrum of thorium exhibiting proper relative line intensities. Right: emission spectrum of uranium with anomalously weak $L_{\beta 1}$ line. The L_{III} absorption edge of uranium is of longer wavelength than the $L_{\beta 1}$ line and, hence, this line is strongly absorbed by uranium itself and appears weaker. Dashed lines are the absorption coefficients of thorium and uranium, respectively (topaz diffraction crystal, scintillation counter, gold x-ray tube).

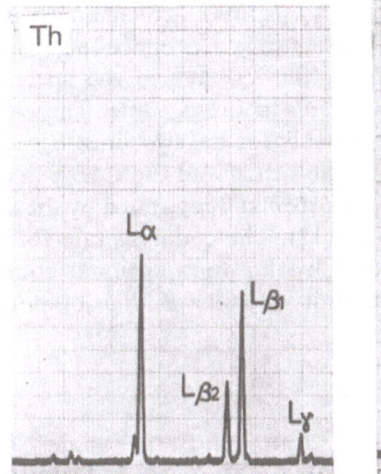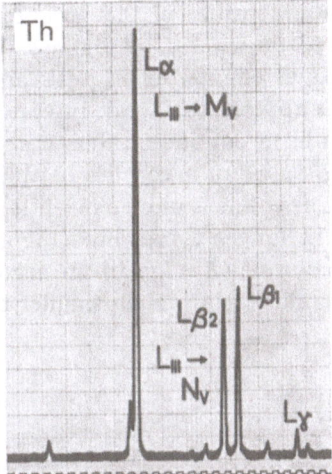

Fig. 4.10. Intensity anomalies. The characteristic radiation of the x-ray tube may cause selective excitation of certain energy levels of an atom. The particular lines of the resulting spectrum are therefore unusually intense. Left: thorium excited by a gold x-ray tube. Right: thorium excited by a molybdenum x-ray tube. The MoK_α line may cause additional excitation of the L_{III} energy level of the thorium and, hence, more frequent occurrence of the transitions $L_{III} \rightarrow M_V$ and $L_{III} \rightarrow N_V$ (topaz diffraction crystal, scintillation counter).

is longer and the $L_{\beta 1}$ line is shorter in wavelength than the L_{III} absorption edge. As a result, the $L_{\beta 1}$ line is more strongly absorbed than the $L_{\alpha 1}$ line, and the former appears considerably weaker than would be expected (Menix, Halteman, and Garcia, 1963).

Anomalies in the relative line intensities of a spectrum may also be caused by the particular x-ray tube and counter employed. The characteristic radiation of a molybdenum tube, for example, has a quantum energy sufficient to ionize the L_{III} energy level of the elements thorium and uranium. It is, however, too weak to excite the neighboring energy levels L_{II} and L_I. Therefore, lines representing electron transitions in the L_{III} level appear stronger while lines representing electron transitions in the L_{II} and L_I levels have normal intensities (Figure 4.10). The characteristic $AuL_{\beta 1}$ line of a gold tube, for example, causes additional excitation of the L_{III} energy level of element tungsten and of the L_{III} and L_{II} levels of tantalum. Therefore, the intensities of the $WL_{\alpha 1}$ and $WL_{\beta 2}$ lines are enhanced in comparison to that of the $WL_{\beta 1}$ line. In the case of tantalum, on the other hand, the intensities of the $L_{\alpha 1}$, $L_{\beta 1}$, $L_{\beta 2}$, L_{γ} lines are enhanced in comparison to the $L_{\beta 3}$ and $L_{\beta 4}$ lines. When using a proportional or a flow proportional counter filled with argon, the $L_{\alpha 1}$ line in the spectrum of cadmium appears too weak in comparison to the $L_{\beta 1}$ line. The efficiency of argon-filled counters differs considerably for the two lines because the absorption edge of argon lies between the $CdL_{\alpha 1}$ and the $CdL_{\beta 1}$ lines. The shorter $CdL_{\beta 1}$ line is more efficiently absorbed and registered by the counter than is the longer $CdL_{\alpha 1}$ line (Figure 4.11). The counting rate for the K_{α} line of chlorine ($Z = 17$) in an argon-filled flow proportional counter is, for the same reason, much smaller than it is for the K_{α} line of potassium ($Z = 19$).

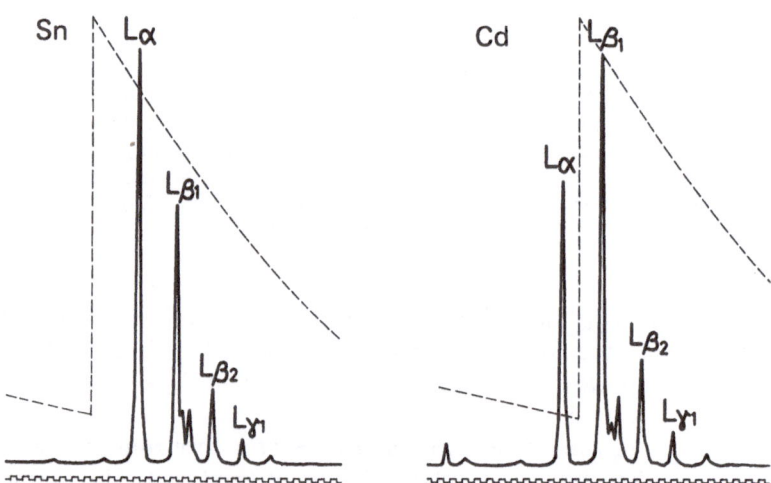

Fig. 4.11. Intensity anomalies. Depending upon the gas used in a gas-filled counter, different wavelengths are recorded with different efficiencies. The long-wavelength part of the spectrum is recorded less efficiently when the absorption edge of the counter gas lies between two lines of the spectrum and, hence, relative line intensities appear anomalous. Emission spectra of tin (*left*) and cadmium (*right*) recorded with an argon-filled flow proportional counter. The dashed line is the absorption coefficient of argon as a function of wavelength. In the spectrum of cadmium, the absorption edge of the counter gas is located between the L_{α} and the $L_{\beta 1}$ lines and, hence, the L_{α} line is registered less efficiently than the $L_{\beta 1}$ line (PE diffraction crystal, chromium x-ray tube).

Chapter 5

Fluorescent Intensity of a Pure Element

5.1. Derivation of the Intensity Formula

Fluorescence of an element in an x-ray spectrometer is caused by the primary radiation of an x-ray tube. An x-ray spectrometer can be said to consist of three major parts, namely, the one that produces the fluorescence, the one that analyzes it according to wavelength, and the one that registers the line intensity and the deflection angle (Figure 10.1). The intensity registered in the readout system depends on several factors, such as, the anode material of the particular x-ray tube and the tube efficiency. Fluorescent intensity also depends upon the incident and emergence angles of the radiation. The slit width of the collimator determines that fraction of the fluorescent radiation which eventually reaches the analyzing crystal. Depending upon the scattering properties of the crystal, a larger or smaller fraction of the radiation is scattered and registered in the counter. The intensity of the radiation registered in the electronic readout system is further influenced by the efficiency of the counter. Other than on instrumental factors, the fluorescent intensity also depends upon the element itself.

In order to calculate the fluorescent intensity, we consider first the portion of the x-ray path from the x-ray tube to the exit slit of the collimator. This calculation can be divided into three parts: the decrease in intensity of incident primary radiation in reaching a layer dx at a depth x in the sample; the excitation of fluorescent radiation in dx; the decrease in intensity of the excited fluorescent radiation on its way out of the sample and until it passes through the collimator. The x-ray tube used to excite fluorescent radiation emits a polychromatic spectrum; however, for the present we shall limit our calculations to the fluorescent radiation in the wavelength range between λ and $\lambda + d\lambda$, where the spectral intensity reaching the sample is $N_0(\lambda)$ photons per cm^2 per sec, and the average angle of incidence of the radiation is φ (Figure 5.1).

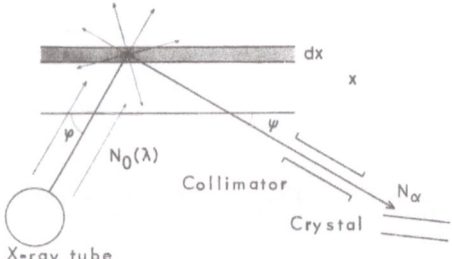

Fig. 5.1. Schematic of beam path for
the calculation of fluorescent intensity.

Let us consider the wavelength interval $d\lambda$. When penetrating the
sample to the layer dx at a depth x the primary number of photons $N_0(\lambda)d\lambda$
decreases exponentially; the intensity $N(\lambda, x)\,d\lambda$ at the depth x is equal to

$$N\,(\lambda, x)\,d\lambda = N_0(\lambda)d\lambda \exp\;[-\mu_l(\lambda)\,x\,/\sin\varphi] \qquad (5.1)$$

where $\mu_l(\lambda)$ is the linear absorption coefficient for the wavelength λ.
When passing through the layer dx the absorbed number of photons
$dN(\lambda, x)$ is given as follows:

$$dN\,(\lambda, x) = N\,(\lambda, x)\,d\lambda\,\mu_l(\lambda)\,dx\,/\sin\varphi \qquad (5.2)$$

All energy levels of the atom contribute to this absorption and the absorption coefficient μ is the sum of the partial absorption coefficients of the
individual energy levels

$$\mu = \mu_K + \mu_{LI} + \mu_{LII} + \mu_{LIII} + \cdots$$

The absorption jump ratio S_K is

$$S_K = \frac{\mu_K + \mu_{LI} + \mu_{LII} + \mu_{LIII} + \cdots}{\mu_{LI} + \mu_{LII} + \mu_{LIII} + \cdots}$$

For the moment, we are only interested in the intensity of the K spectrum
and, therefore, only in the number of photons which are absorbed by the
K energy level. This fraction is equal to

$$\frac{\mu_K}{\mu_K + \mu_{LI} + \mu_{LII} + \mu_{LIII} + \cdots}\,dN\,(\lambda, x) = \frac{S_K - 1}{S_K}\,dN\,(\lambda, x) \quad (5.3)$$

Every absorbed photon excites the K energy level. The excited state results
either in the emission of an Auger electron or in the emission of a fluorescence photon. The fluorescent yield W_K is the probability of the excited
state to result in the emission of fluorescent radiation. Fluorescent radiation
includes all lines of the K series, where the α line is equal to the fraction
p_α. The number of K_α fluorescent photons $dN_{K\alpha}(\lambda, x)$ which originates by
absorption of $dN(\lambda, x)$ primary photons is given as follows:

$$dN_{K\alpha}(\lambda, x) = \frac{S_K - 1}{S_K} W_K p_\alpha \, dN \, (\lambda, x)$$

$$= \frac{S_K - 1}{S_K} W_K p_\alpha \, \mu_l(\lambda) \frac{dx}{\sin \varphi} N_0(\lambda) \, d\lambda \, \exp\left[-\mu_l(\lambda) \, x / \sin \varphi\right] \quad (5.4)$$

Parameters which only depend upon the atomic number Z and, thus, are a constant for every element are summarized in factor E as follows:

$$E = \frac{S_K - 1}{S_K} W_K p_\alpha$$

X-ray fluorescent radiation which originates in layer dx is emitted centrosymmetrically in all directions. Only the fraction q, however, passes through the collimator and towards the crystal. The intensity of the fluorescent radiation of wavelength α decreases before emerging from the sample by a factor

$$\exp\left[-\mu_l(\alpha) \, x / \sin \psi\right]$$

In this equation $\mu_l(\alpha)$ is the linear absorption coefficient and ψ the emergence angle of the fluorescent radiation. Thus, we obtain the intensity of the fluorescent radiation that is produced by primary x-ray radiation of the wavelength λ in a layer dx at a depth x and reaches the analyzing crystal through the collimator. Fluorescent radiation is produced at all depths of the sample and, thus, we have to integrate over all x. The primary radiation is polychromatic and the fluorescent radiation is produced by all wavelengths between the continuum limit λ_0 and the absorption edge λ_K, of the element that is to be excited. The continuous x-ray spectrum of the tube is limited in wavelength towards the short-wavelength end, where the short-wavelength limit λ_0 depends upon the accelerating potential of the tube. The fluorescent intensity $N_{K\alpha}$ is then given by

$$N_{K\alpha} = \frac{q}{\sin \varphi} E \int_{\lambda_0}^{\lambda_K} \mu_l(\lambda) \, N_0(\lambda) \, d\lambda \cdot$$

$$\cdot \int_0^h dx \exp\left[-x \left(\mu_l(\lambda) / \sin \varphi + \mu_l(\alpha) / \sin \psi\right)\right] \quad (5.5)$$

This expression has a simple solution when the sample thickness h is large in comparison to the penetration depth. According to Koh and Caugherty (1952), sample thicknesses of about 0.03 mm are sufficient in the cases of chromium, iron, and nickel fluorescence. For thick samples, $N_{K\alpha}$ is given as follows:

$$N_{K\alpha} = \frac{q}{\sin \varphi} E \int_{\lambda_0}^{\lambda_K} \frac{\mu_l(\lambda) N_0(\lambda) \, d\lambda}{\mu_l(\lambda)/\sin \varphi + \mu_l(\alpha)/\sin \psi} \qquad (5.6)$$

The fluorescent intensity for thin layers is given as follows:

$$N_{K\alpha} = \frac{q}{\sin \varphi} E \int_{\lambda_0}^{\lambda_K} \frac{\mu_l(\lambda) N_0 \lambda \, d\lambda}{\mu_l(\lambda)/\sin \varphi + \mu_l(\alpha)/\sin \psi} \cdot$$

$$\cdot \{1 - \exp\left[-h\left(\mu_l(\lambda)/\sin \varphi + \mu_l(\alpha)/\sin \psi\right)\right]\} \qquad (5.7)$$

The intensity of thin layers of up to about 1000 Å can be calculated sufficiently by developing the exponential function of a row to the second term (Weyl, 1961).

$$\exp\left[-h\left(\mu_l(\lambda)/\sin \varphi + \mu_l(\lambda)/\sin \psi\right)\right] =$$

$$= 1 - h\left(\mu_l(\lambda)/\sin \varphi + \mu_l(\alpha)/\sin \psi\right) + \cdots$$

The fluorescent intensity in thin layers is given as

$$N_{K\alpha} = \frac{q}{\sin \varphi} E \int_{\lambda_0}^{\lambda_K} \mu_l(\lambda) N_0(\lambda) h \, d\lambda = \text{prop } h \qquad (5.8)$$

and is proportional to the thickness. The fluorescent intensity $N_{K\alpha}$ that passes through the collimator onto the analyzing crystal is reflected by the crystal and registered by the counter. The scattering efficiency of the crystal in regard to the incident line is designated S, and the counting efficiency of the counter is designated A. After passage through the linear amplifier, the discriminator, and other devices of the electronic readout system, the registered pulse rate N^* is given with a probability D as follows:

$$N^* = D A S N_{K\alpha} \qquad (5.9)$$

If the intensity of a line, other than that of the K_α line, is to be calculated, then the corresponding values for the particular line have to be introduced into the formula. The α line of the L series, for example, originates by electron transitions into the L_{III} energy level of the atom. Only part of the primary radiation absorbed in the layer dx is absorbed by this energy level.

The L_{III} energy level absorbs the following fractions (S_{LIII}, S_{LII}, and S_{LI} are the absorption jump ratios):

$$\frac{S_{LIII} - 1}{S_{LIII}} \qquad \text{when the wavelength } \lambda \text{ of the absorbed radiation is } \quad \lambda_{LII} < \lambda < \lambda_{LIII}$$

$$\frac{S_{LIII} - 1}{S_{LIII}S_{LII}} \qquad \text{when } \quad \lambda_{LI} < \lambda < \lambda_{LII}$$

$$\frac{S_{LIII} - 1}{S_{LIII}S_{LII}S_{LI}} \qquad \text{when } \quad \lambda < \lambda_{LI}$$

The fraction W_L, in accordance with the fluorescent yield, causes emission of fluorescent radiation. Only the fraction ψ_α of the radiation emitted by the L_{III} subseries consists of the fluorescent line L_α. The intensity, $dN_{L\alpha}$, which originates from the absorbed primary radiation $dN(\lambda, x)$ in the layer dx is, for the wavelength range $\lambda < \lambda_{LI}$, given by the equation

$$dN_{L\alpha}\,(\lambda, x) = dN\,(\lambda, x)\,\frac{S_{LIII} - 1}{S_{LIII}S_{LII}S_{LI}}\,W_L p_\alpha$$

The remaining calculation of the intensity for L lines is the same as that of K lines.

In general, fluorescence is produced by a broad range of the polychromatic continuous spectrum. All incident photons which interact with the sample are absorbed and, with the same probability, cause fluorescence. The average penetration depth, however, is different for the different wavelengths, and so is the decrease in intensity of the emitted radiation which originates at various depths. Normally, the sample is very thick in comparison to the penetration depth of the x-ray radiation (0.1 to 0.3 mm.) It is for this reason that all wavelengths of the spectrum contribute to the production of fluorescent radiation provided they are more energetic than the absorption edge (Figure 5.2). The results are different, however, if a very thin foil is irradiated. In this case the wavelengths close to the absorption edge would be more strongly absorbed than those which are further away from it and, hence, the fluorescent radiation would largely be produced by the primary radiation of a wavelength close to that of the absorption edge.

The formulas for fluorescent intensities of thick samples contain the linear absorption coefficients. It is often desirable to replace the linear absorption coefficients by the mass absorption coefficients. In order to transform the equation, the density ρ is introduced because the mass absorption coefficient μ_m is defined as $\mu_l/\rho = \mu_m$ (see Chapter 1)

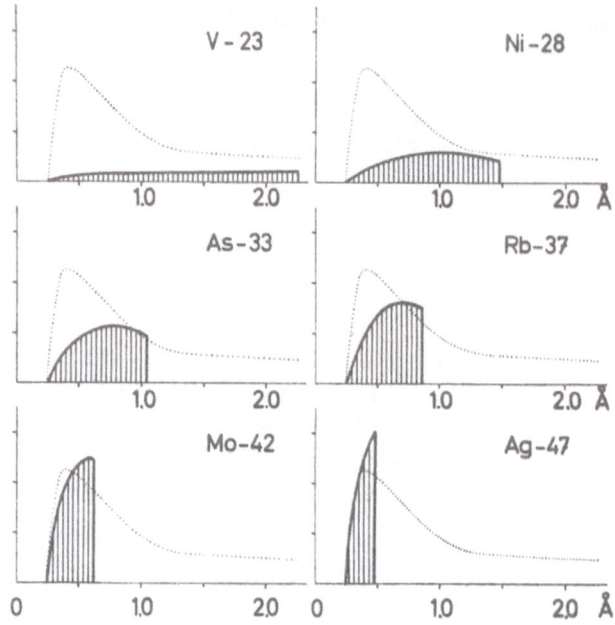

Fig. 5.2. Calculated fluorescent intensity produced by a certain wavelength range. It is apparent that in case of thick samples the fluorescent radiation is excited by all wavelength ranges of the primary continuous x-ray spectrum. Dotted line: continuous x-ray spectrum of a 50-kV molybdenum x-ray tube, plotted on an eleven-fold reduced scale (Müller, 1962b).

$$N_{K\alpha} = \frac{q}{\sin \varphi} E \int_{\lambda_0}^{\lambda_K} \frac{\mu_m(\lambda) N_0(\lambda) d\lambda}{\mu_m(\lambda)/\sin \varphi + \mu_m(\alpha)/\sin \psi}$$

5.2. Dependence of Fluorescent Intensity on the Anode Material of the Tube

To supplement the previous discussion we investigate the dependence of the fluorescent intensity on the anode material of a particular x-ray tube. An x-ray tube emits a polychromatic continuous x-ray spectrum whose short-wavelength limit λ_0 is determined by the accelerating potential of the tube. The characteristic lines of the anode material are superimposed onto the continuous x-ray spectrum (Figure 10.5). Fluorescence is produced only by that portion of the radiation which is shorter in wavelength than the absorption edge of the material which is to be excited. In order to investigate the dependence of the fluorescent intensity on the anode material of the tube, it is desirable to measure and compare the intensities of two tubes with different anode materials. In Figure 5.3. the resulting intensities of titanium ($Z = 22$) to uranium ($Z = 92$) are presented as a function of their atomic numbers. The fluorescent yields and, therefore, the fluorescent intensities of the light elements increase with increasing atomic number. The K series fluorescent intensity of medium–heavy elements, however, decreases with increasing atomic number because the K absorption edges of these elements quickly approach the short-wavelength limit λ_0 of the continuum, and the number of primary photons available for excitation in the interval between λ_0 and λ_K decreases quickly. Shift of the absorption edges towards shorter wavelengths further results in a tendency for the characteristic lines of anode materials of increasing atomic numbers to be located on the long-wavelength side of the absorption edges; hence, they cease to contribute to fluorescent excitation and discontinuity positions originate (for light elements the characteristic lines of the anode materials are on the short-wavelength side of the absorption edge, thus contributing heavily to fluorescent excitation). For example, the use of a molybdenum target x-ray tube results in a sharp discontinuity in the fluorescent intensity at $Z = 39$ (yttrium) and 40 (zirconium). A second "jump" occurs at $Z = 41$ (niobium) and 42 (molybdenum). The first jump is due to the inability of the strong $\text{Mo}K_\alpha$ line to excite fluorescence in yttrium or zirconium. The second jump is due to the inability of the $\text{Mo}K_\beta$ line to cause fluorescence in niobium and molybdenum. The intensity difference (before vs after the discontinuity) is proportional to the fraction of the tube

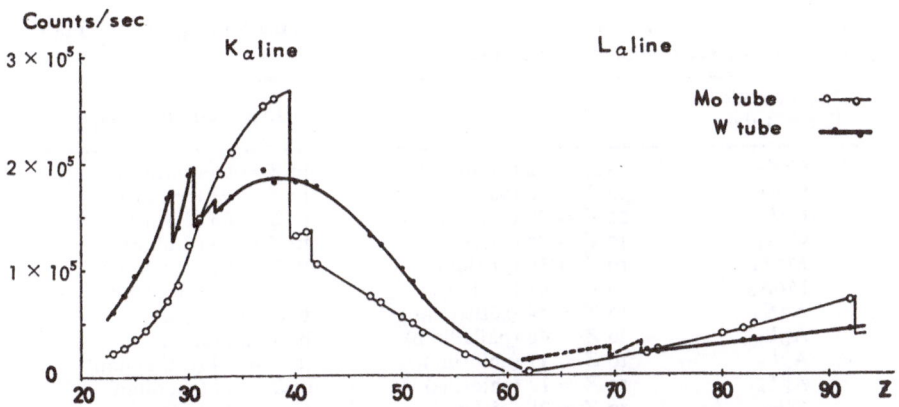

Fig. 5.3. Experimentally determined fluorescent intensities of the elements as a function of their atomic numbers using molybdenum and tungsten tubes. K_α and L_α lines [Philips x-ray spectrograph: 50-kV, 20-mA tube, topaz diffraction crystal, scintillation counter; (Müller, 1962b)].

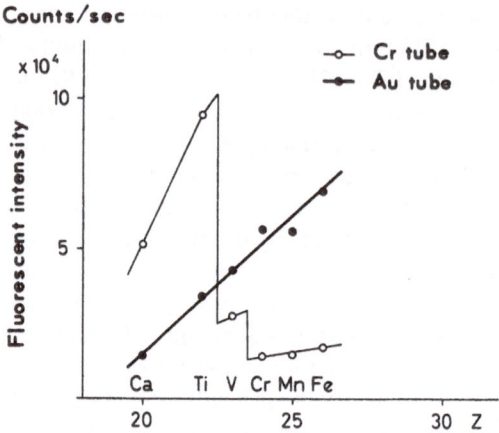

Fig. 5.4. Fluorescent intensities of the elements calcium ($Z = 20$) to iron ($Z = 26$) as a function of their atomic numbers. Excitation was by a chromium and a gold tube, $K\alpha$ lines (50 kV, PE diffraction crystal, argon flow proportional counter, vacuum; author's measurements).

intensity due to these two lines. In the most favorable case, 60% of the fluorescent radiation of molybdenum and copper tubes is produced by characteristic lines. In the case of light elements the characteristic radiation of a chromium tube may account for up to 85% of the total fluorescent radiation (Figure 5.4). Carter (1962) and Louis (1964) compared fluorescent

Table 5.1. Elements Which Can be Excited to Fluorescence in the K or L_{III} Energy Levels by the Characteristic Radiation of the Tube

Characteristic line of tube	In K energy level	In L_{III} energy level
CrK_α	to $Z = 22$ (titanium)	to $Z = 56$ (barium)
CrK_β	to $Z = 23$ (vanadium)	to $Z = 58$ (cerium)
CuK_α	to $Z = 27$ (cobalt)	to $Z = 66$ (dysprosium)
CuK_β	to $Z = 28$ (nickel)	to $Z = 69$ (thulium)
MoK_α	to $Z = 39$ (yttrium)	to $Z = 92$ (uranium)
MoK_β	to $Z = 41$ (niobium)	to $Z = 97$ (berkelium)
AgK_α	to $Z = 44$ (ruthenium)	beyond uranium
AgK_β	to $Z = 46$ (palladium)	beyond uranium
AgL_α	to $Z = 17$ (chlorine)	to $Z = 44$ (ruthenium)
$AgL_{\beta1}$	to $Z = 17$ (chlorine)	to $Z = 45$ (rhodium)
WL_α	to $Z = 28$ (nickel)	to $Z = 68$ (erbium)
$WL_{\beta1}$	to $Z = 30$ (zinc)	to $Z = 72$ (hafnium)
AuL_α	to $Z = 30$ (zinc)	to $Z = 72$ (hafnium)
$AuL_{\beta1}$	to $Z = 32$ (germanium)	to $Z = 77$ (iridium)

Table 5.2. Fluorescent Intensities (in counts/sec) of Some Elements Which Were Excited Exclusively by the Continuous Radiation of Tungsten or Molybdenum Tubes[a]

Element	With tungsten tube	With molybdenum tube	Ratio
Molybdenum ($Z = 42$)	35,570	21,020	1.69
Silver ($Z = 47$)	26,350	14,890	1.77
Cadmium ($Z = 48$)	24,590	13,740	1.79
Tellurium ($Z = 52$)	14,720	8,000	1.84
Barium ($Z = 56$)	7,460	4,000	1.86
	$Z = 74$	$Z = 42$	1.76

[a] (Müller, 1962b).

intensities of various light elements when excited by chromium and tungsten tubes. For light elements, the intensities obtained with a chromium tube are 3 to 4 times larger than those obtained with a tungsten tube.

Elements which can be excited by the characteristic tube radiation to fluoresce in the K and L energy levels are listed in Table 5.1. In the case of an x-ray fluorescence unit with open tube, any suitable metal may be chosen as an anode; thus, it is possible to select an anode material whose characteristic lines appear in a favorable position in regard to the absorption edge of the element that is to be analyzed. Rabillon (1961) selected the

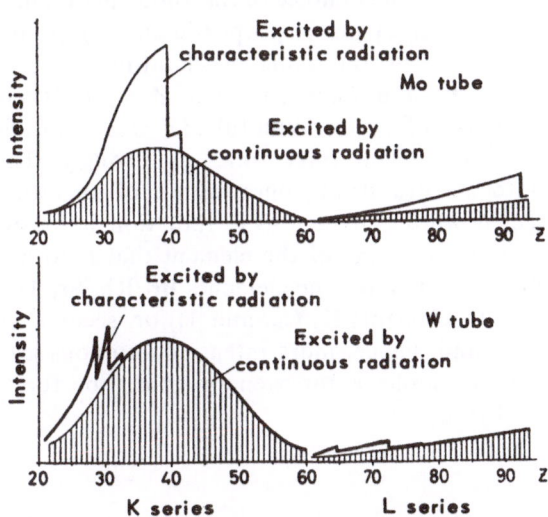

Fig. 5.5. Fluorescent intensities of the elements, separated into the two portions which originate, respectively, from the continuous x-ray spectrum and the characteristic radiation of the tube (Müller, 1962b).

Table 5.3. Relative Fluorescent Intensities of Some Elements Excited by X-Ray Tubes with Different Anode Materials (Rabillon, 1961). The indices K or L are to indicate that the characteristic K or L radiation of the tube contributes to the excitation. Values are arbitrarily normalized for the tungsten tube to be equal to 100.

Anode	Irradiated element								
	16 S	17 Cl	22 Ti	24 Cr	25 Mn	26 Fe	29 Cu	30 Zn	42 Mo
24-Cr	184^K	165^K	160^K						
26-Fe				171^K					
27-Co					150^K				
28-Ni						109^K			
29-Cu	107^K	90^K	102^K	108^K	111^K	106^K		31	41
42-Mo	62^K	52^K		43^K			48^K	57^K	
45-Rh	$84?^L$					.			$91?^K$
46-Pd	164^L	194^L							107^K
74-W	100^L	100^L	100^L	100^L	100^L	100^L	100^L	100^L	100
78-Pt							132^L		
79-Au								128^L	

proper anode materials for excitation of a number of elements which resulted in a two-fold gain in intensity in comparison to a tube with a heavy anode such as tungsten or gold (Table 5.3). An appreciable portion of the fluorescence is often produced by the continuous x-ray spectrum of the tube (Figure 5.5). The intensity of the continuous x-ray spectrum may therefore be increased by proper choice of the anode material. The intensity of the continuous x-ray spectrum is proportional to the atomic number of the anode material. For the same accelerating potential and tube current, the continuum of a tungsten tube ($Z = 74$) is approximately 3 times as intense as that of a chromium tube ($Z = 24$) and, hence, excites the fluorescence of an element three times as strongly. Therefore, two possibilities exist to achieve strong fluorescence of an element. First, a tube may be selected whose characteristic radiation is located favorably in regard to the absorption edge of the element that is to be excited (for example, a molybdenum tube for the elements Br, Rb, Sr, Th, and U, or a chromium tube for the elements K, Ca, and Ti) or, second, an element of high atomic number and, hence, more intense continuous x-ray radiation may be chosen (for example, a tungsten or gold tube for the elements V, Mn, Ag, Sn, and Ba).

Chapter 6

Fluorescent Intensity of an Element in Two- and Multicomponent Mixtures

6.1. Derivation of the Intensity Formula

The fluorescent intensity of an element in a mixture does not only depend upon its concentration but also upon the elements associated with it in the sample. The intensity of the fluorescent radiation of the element nickel in steel, for example, is six times lower than the intensity emitted by the same element when in aluminum and approximately thirty times lower when in an organic substance. Part of the primary tube radiation is absorbed by the associated elements and does not contribute to the excitation of secondary fluorescent radiation. Furthermore, absorption of secondary radiation in the sample depends upon the respective absorption coefficients, resulting in fluorescence of different intensities.

Let us assume the sample to be a mixture consisting of the elements A, B, C, ... whose concentrations by weight are C_A, C_B, C_C, ... It is further assumed that the sum of all concentrations is

$$C_A + C_B + C_C + \cdots = 1$$

The fluorescent intensity of element A, which is produced by the primary tube radiation and which is not additionally excited by the associated elements in the sample, is calculated (the fluorescent radiations of the associated elements are assumed to be at longer wavelengths than the absorption edge of element A). The intensity of the primary tube radiation in the wavelength range between λ and $\lambda + dx$ is $N_0(\lambda)\, d\lambda$ photons per cm² per second. The radiation is assumed to enter the sample surface under the angle φ. The primary radiation, which causes the fluorescence, exponentially decreases in intensity with increasing penetration depth. Every element contributes to the decrease in intensity according to its concentration and mass absorption coefficient. The mass absorption coefficient of a mixture is the weighted average of the absorption coefficients of the pure elements

$$\mu\left(\lambda\right) = C_A\,\mu_A\left(\lambda\right) + C_B\,\mu_B\left(\lambda\right) + C_C\,\mu_C\left(\lambda\right) + \cdots$$

where $\mu_A(\lambda)$, $\mu_B(\lambda)$, $\mu_C(\lambda)$, ... are the mass absorption coefficients of the elements for the wavelength λ. For the calculation in question an incident beam of 1-cm^2 cross section is considered. Before reaching the layer dx at a depth x the incident primary radiation passes through the mass of $1\,\text{cm}^2 \cdot x\rho/\sin\varphi$, where ρ is the average density (Figure 6.1). The initial intensity $N_0(\lambda)\,d\lambda$ of the primary radiation is reduced to the amount $N(\lambda, x)\,d\lambda$

$$N\left(\lambda, x\right) d\lambda = N_0\left(\lambda\right) d\lambda \exp\left[-\mu\left(\lambda\right) x\,\varrho/\sin\varphi\right] \qquad (6.1)$$

In the layer dx the amount $dN(\lambda, x)$ is absorbed by the mass $dx\rho/\sin\varphi$

$$dN\left(\lambda, x\right) = N\left(\lambda, x\right) d\lambda\,\mu\left(\lambda\right) dx\,\varrho/\sin\varphi \qquad (6.2)$$

As we are calculating the fluorescent intensity of the element A, only that fraction which is absorbed by the element A is considered. This fraction is given as follows:

$$\frac{C_A\,\mu_A\left(\lambda\right)}{\mu\left(\lambda\right)}$$

All energy levels of the atom contribute to the absorption.

We are interested for the moment only in the fluorescent intensity of the K spectrum and, hence, only in the number of photons which are absorbed by the K energy level of the element A; this fraction is $(S_K - 1)/S_K$. The excited state of the K energy level yields emission of x-ray fluorescent radiation with a certain probability W_K, which is the fluorescent yield. The fraction of the fluorescent radiation which is due to the α line is given by p_α. All constant parameters which, for the element A, depend only upon the atomic number are summarized in the factor E_A

$$E_A = \frac{S_K - 1}{S_K}\,W_K\,p_\alpha$$

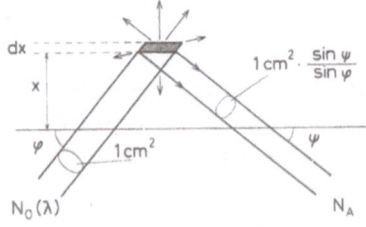

Fig. 6.1. Calculation of fluorescent intensity.

Part of the primary radiation, $dN(\lambda, x)$, which is absorbed in the layer dx, contributes to the fluorescent radiation and appears as the K_α line of the element A. The intensity dN_A of this line is as follows:

$$dN_A(\lambda, x) = dN(\lambda, x) \frac{C_A \mu_A(\lambda)}{\mu(\lambda)} E_A \tag{6.3}$$

$$= N(\lambda, x) \, d\lambda \, E_A \, C_A \, \mu_A(\lambda) \, \varrho \, dx / \sin \varphi \tag{6.4}$$

The fluorescent radiation which originates in the layer dx is emitted uniformally in all directions, and the fraction q passes through the collimator towards the crystal. Before emergence, the radiation passes through the mass $x\rho/\sin \varphi$ whereby the emerging beam has a cross section of 1 cm$^2 \cdot \sin$ $\psi \sin \varphi$, where ψ is the angle between the emerging radiation and the sample surface. The unit of the mass absorption coefficient is given in g^{-1} cm^2. The value of the mass absorption coefficient for a beam of a cross section of 1 cm$^2 \cdot \sin \psi / \sin \varphi$, is given as

$$\mu(\alpha) \frac{\sin \varphi}{\sin \psi}.$$

Before emerging from the sample the radiation of the wavelength α is absorbed exponentially by the factor

$$\exp[-\mu(\alpha) x \varrho / \sin \psi]$$

where the mass absorption coefficient $\mu(\alpha)$ is the weighted average of the mass absorption coefficients of the pure elements for the wavelength α:

$$\mu(\alpha) = C_A \mu_A(\alpha) + C_B \mu_B(\alpha) + C_C \mu_C(\alpha) + \cdots$$

We obtain the fluorescent intensity of element A which is produced in the layer dx by the primary tube radiation of the wavelength λ and which passes through the collimator onto the analyzing crystal. This radiation is produced at all depths x and by all wavelengths between the short wavelength limit of the continuum λ_0 of the tube spectrum and the absorption edge λ_A of element A. The fluorescent intensity, N_A, of the element A is then given by

$$N_A = \frac{q}{\sin \varphi} E_A C_A \int_{\lambda_0}^{\lambda_A} \mu_A(\lambda) \, \varrho \, N_0(\lambda) \, d\lambda \cdot$$

$$\cdot \int_0^h dx \exp[-x \varrho (\mu(\lambda) / \sin \varphi + \mu(\alpha) / \sin \psi)] \tag{6.5}$$

NiKα

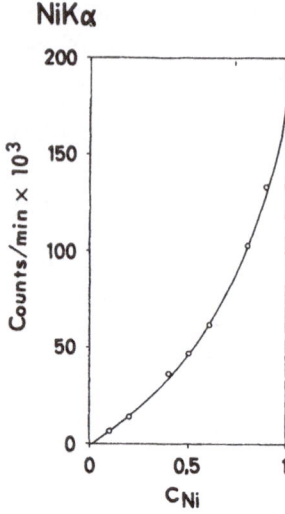

Fig. 6.2. Comparison of measured and calculated intensity of the NiKα line as a function of Ni concentration in iron–nickel alloys (Seeman *et al.*, 1961).

This expression has a simple solution when the sample thickness is very large in comparison to the penetration depth of the x-rays. For thick samples ($h = \infty$) it is

$$N_A = \frac{q}{\sin \varphi} E_A C_A \int_{\lambda_0}^{\lambda_A} \frac{\mu_A (\lambda)\, N_0 (\lambda)\, d\lambda}{\mu (\lambda) / \sin \varphi + \mu (\alpha) / \sin \psi} \qquad (6.6)$$

The mass absorption coefficients $\mu(\lambda)$ and $\mu(\alpha)$ consist of the absorption coefficients of the pure elements.

When introducing the absorption coefficients of the elements we obtain

$$N_A = \frac{q}{\sin \varphi} E_A C_A \cdot$$

$$\cdot \int_{\lambda_0}^{\lambda_A} \frac{\mu_A(\lambda)\, N_0(\lambda)\, d\lambda}{C_A \{\mu(\lambda) / \sin \varphi + \mu(\alpha) / \sin \psi\}_A + C_B \{\mu(\lambda) / \sin \varphi + \mu(\alpha) / \sin \psi\}_B + \cdots}$$
$$(6.7)$$

The contribution of the individual elements to the absorption can be abbreviated as follows:

$$\{\mu (\lambda) / \sin \varphi + \mu (\alpha) / \sin \psi\} \equiv \bar{\mu} (\alpha)$$

In thin samples of thickness h, where h is smaller than the penetration depth of the x-ray radiation, the fluorescent intensity of element A is given as follows:

$$N_A = \frac{q}{\sin \varphi} E_A C_A \int_{\lambda_0}^{\lambda_A} \frac{\mu_A(\lambda) N_0(\lambda) d\lambda}{C_A \bar\mu_A(\alpha) + C_B \bar\mu_B(\alpha) + \cdots} \cdot$$

$$\cdot \{1 - \exp [- h\varrho (C_A \bar\mu_A(\alpha) + C_B \bar\mu_B(\alpha) + \cdots)]\} \qquad (6.8)$$

The exponential function is developed to the second term and an approximate expression for the intensity emitted by thin layers is obtained

$$N_A \cong \frac{q}{\sin \varphi} E_A C_A h\varrho \int_{\lambda_0}^{\lambda_A} \mu_A(\lambda) N_0(\lambda) d\lambda \qquad (6.9)$$

In contrast to fluorescence of pure elements, where sample thicknesses of 0.05 to 0.10 mm are sufficient for maximum intensity, up to 5-mm thick samples are required, for example, for the analysis of a heavy metal in an organic substance. The intensity formula was first derived by Hamos (1945) and later discussed by Gillam and Heal (1953), Sherman (1954, 1955), Beattie and Brissey (1954), Seeman, Schmidt, and Stavenow (1961), and Müller (1962a). Seeman et al. and Müller calculated the fluorescent intensity with this formula for several two-component mixtures (Figures 6.2 and 6.3), while Criss and Birks (1968) calculated the relative intensities of alloyed components in steel and compared them to experimentally determined

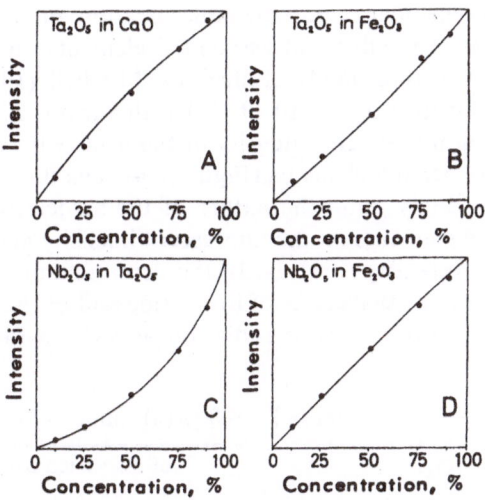

Fig. 6.3. Comparison of measured (···) to calculated (—) intensities as a function of concentration (Müller, 1962a).

values. These calculations, however, are too complicated for most routine laboratory work. For the most part, the derived formula serves as a starting point for further considerations. As the maximal and integral intensities are proportional for a line, it is common practice in analysis to measure only the maximal line intensity.

6.2. Intensity Formula for Low Concentrations

The formula for fluorescent intensity is simplified when the particular element A of the mixture occurs in very low concentrations (concentration range $C_A = 10^{-3}$ to 10^{-6}). The contribution of element A to the absorption of the x-ray radiation is proportional to its concentration and can therefore be neglected. The denominator of the equation for fluorescent intensity now contains only the contributions by the associated elements B, C, ..., instead of the sum of the contributions by the elements A, B, C, ...

$$N_A \cong -\frac{q}{\sin \varphi} \, E_A C_A \int_{\lambda_0}^{\lambda_A} \frac{\mu_A(\lambda) \, N_0(\lambda) \, d\lambda}{C_B \bar{\mu}_B(\alpha) + C_C \bar{\mu}_C(\alpha) + \cdots}$$

6.3. Effects of Associated Elements on the Fluorescent Intensity

The fluorescent intensity of an element in a mixture with other elements is not only dependent upon its concentration, properties, and the exciting radiation but also dependent upon the concentration and properties of the associated elements. The effects of associated elements on the fluorescent intensity of iron was thoroughly studied by Mitchell (1961). He mixed identical amounts of iron oxide (10 wt. %) with various metal oxides and plotted the resulting fluorescent intensity of the iron *vs* the atomic number of the respective associated elements (Figure 6.4). The fluorescent intensity of iron depends strongly upon the nature of the associated element. The effects of associated elements on the fluorescent intensity are best estimated with the aid of the intensity formula. In the denominator of this equation are the mass absorption coefficients of iron oxide and of the admixed metal oxides, both for the primary tube radiation as well as for the emerging iron radiation:

$$\underbrace{0.1\{\mu(\lambda)/\sin\varphi + \mu(\mathrm{Fe}_{K\alpha})/\sin\psi\}}_{\text{of Fe}_2\mathrm{O}_3} + \underbrace{0.9\{\mu(\lambda)/\sin\varphi + \mu(\mathrm{Fe}_{K\alpha})/\sin\psi\}}_{\text{of associated oxide}}$$

The larger the mass absorption coefficient

$$\{\mu(\lambda)/\sin \varphi + \mu(\mathrm{Fe}_{K\alpha})/\sin \psi\}$$

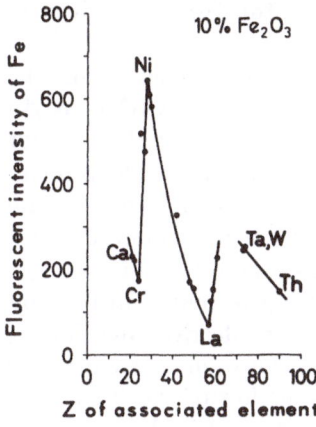

Fig. 6.4. Fluorescent intensity for 10% Fe_2O_3 in mixtures with various metal oxides. The intensity of the $FeK\alpha$ line is plotted *vs* the atomic number of the respective associated component (Mitchell, 1961).

of the associated oxide, the lower is the resulting fluorescent intensity of iron. Let us investigate the dependence of the values $\mu(FeK_\alpha)$ and $\mu(\lambda)$ from the atomic number of the associated elements. Starting with calcium, both values increase with increasing atomic number until they reach a first maximum at atomic number $Z = 24$ (chromium); correspondingly, the fluorescent intensity of iron in a mixture with chromium oxide is at a minimum. In the case of elements with atomic numbers larger than $Z = 24$ (chromium) the value for $\mu(FeK_\alpha)$ is much smaller than in the case of the immediately preceding elements. This is because the iron radiation is located on the long-wavelength side of their K absorption edges and, hence, is only absorbed. Furthermore, the K absorption edges of the associated elements starting with $Z = 27$ (cobalt) are located below the absorption edge of iron. As a result, the long-wavelength part of the primary tube radiation is absorbed less than the short-wavelength part and, hence, the average value for $\mu(\lambda)$ decreases with increasing atomic number. Furthermore, the fluorescent radiation of associated elements of atomic numbers larger than 27 cause additional fluorescence of the iron. The intensity of iron radiation in a mixture with nickel oxide is therefore at a maximum. For associated elements of still higher atomic numbers, the fluorescent intensity of iron decreases gradually, although occasional maximum and minimum values may occur at light atomic numbers. This is due to a shift in the L absorption edge of the associated elements from the long-wavelength side of the FeK_α line to the short-wavelength side which results in abrupt changes of the values for $\mu(FeK_\alpha)$ and $\mu(\lambda)$. The rule, that the fluorescent intensity of an element is smaller the heavier the associated component, has to be modified in certain instances.

6.4. Example of the Numerical Calculation and the Term Weighted—Average Wavelength

The numerical calculation of the relative fluorescent intensity from the intensity formula is illustrated here. We calculate the intensity of the NbK_α fluorescent line for a mixture of 25 wt. % Nb_2O_5 and 75 wt. % Ta_2O_5. For this calculation the intensity and spectral distribution $N_0(\lambda)$ of the primary tube radiation used for the excitation has to be known. The spectral distribution can be taken from published tube spectra in the literature or it can be determined experimentally by scattering of the tube radiation by a light matrix. However, only the relative intensity distribution in the tube spectrum is known; the factor which correlates the effective intensities to the experimentally determined values is not known even to an order of magnitude. In order to make possible the calculation of the fluorescent intensity, the relative intensity of the 100 % pure element is also calculated and the fluorescent intensity in the mixture is expressed as a fraction of the intensity obtained from the 100 % pure element. In our example, the spectrum of a molybdenum tube at 50 kV serves as a basis (Figure 10.5). As only relative intensities are to be calculated, the effect of the collimator, the factor $q/\sin \varphi$, the scattering efficiency S of the analyzing crystal, the efficiency A of the counter, and the electron registration efficiency D for the NbK_α line may be neglected. The value E, which includes the height of the absorption jump ratio, the fluorescent yield, and the relative line intensity, is a constant for the NbK_α line and will not have to be calculated for the relative intensities in question. The values for the mass absorption coefficients of the elements niobium and tantalum were taken from Table 1.3.

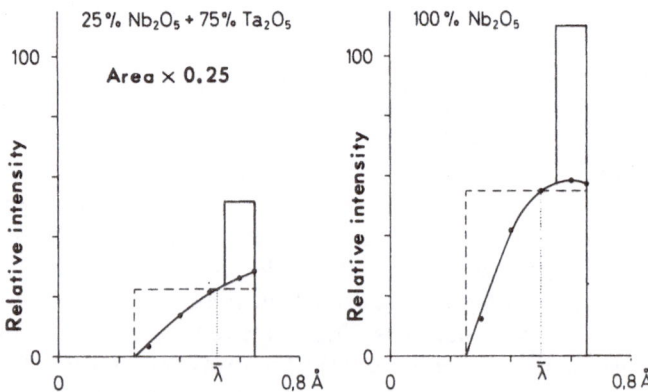

Fig. 6.5. Graphical presentation of the integral for the relative fluorescent intensity. In addition, the rectangle of equal area and the position of the weighted-average wavelength λ are shown.

Table 6.1. Numerical Calculation of the Function for Several Wavelengths λ

I	II	III	IV	V	VI	VII	VIII
λ	$N_0(\lambda)$	$\mu(\lambda)$ Nb_2O_5	II$\times$III	$\dfrac{\mu(\lambda)}{\sin\varphi}$	$\dfrac{\mu(0.75)}{\sin\psi}$	V+VI	0.25$\times$VII
0.65	88	62.2	5470	67.6	27.4	95.0	23.8
0.6	95	51.0	4840	55.4	27.4	82.8	20.7
0.5	108	31.5	3400	34.2	27.4	61.6	15.4
0.4	111	17.5	1940	19.0	27.4	46.4	11.6
0.3	58	7.4	430	8.0	27.4	35.4	8.8
0.25	0	4.9	0	5.3	27.4	32.7	8.2
MoK_β	80	61.8	4940	67.1	27.4	94.5	23.6

I	IX	X	XI	XII	XIII	XIV	XV
λ	$\mu(\lambda)$ Ta_2O_5	$\dfrac{\mu(\lambda)}{\sin\varphi}$	$\dfrac{\mu(0.75)}{\sin\psi}$	X+XI	0.75$\times$XII	VIII+XIII	IV: XIV
0.65	64.8	70.0	157	227.0	170.3	194.1	28.2
0.6	51.8	55.9	157	212.9	166.0	186.7	25.9
0.5	31.2	33.6	157	190.6	143.0	158.4	21.5
0.4	16.8	18.1	157	175.1	131.4	143.0	13.6
0.3	7.8	8.4	157	165.4	124.0	132.8	3.2
0.25	4.9	5.3	157	162.3	121.8	130.0	0
MoK_β	60.0	64.7	157	221.7	166.4	190.0	26.0

Because of a niobium content of 0.70, the mass absorption coefficient of Nb_2O_5 is smaller by that factor; for Ta_2O_5 this factor is 0.82. An average incidence angle of $\varphi = 67°$ and an average emergence angle of $\psi = 35°$ is assumed (Philips spectrometer). The relative fluorescent intensity of the NbK_α line ($\lambda = 0.75$ Å) in a mixture of 25 wt.% Nb_2O_5 with 75 wt.% Ta_2O_5 is given as follows:

$$N_{NbK_\alpha} = \text{prop } 0.25 \cdot$$

$$\cdot \int_{0.25}^{0.65} \frac{\mu(\lambda)_{Nb_2O_5}\, N_0(\lambda)\, d\lambda}{\underbrace{0.25\{\mu(\lambda)/\sin\varphi + \mu(Nb_{K\alpha})/\sin\psi\}}_{\text{From } Nb_2O_5} + \underbrace{0.75\{\mu(\lambda)/\sin\varphi + \mu(Nb_{K\alpha})/\sin\psi\}}_{\text{From } Ta_2O_5}}$$

The integration extends from the short-wavelength limit of the continuum (x-ray tube accelerating voltage = 50 kV; $\lambda = 0.25$ Å) to the K absorption edge of niobium ($\lambda = 0.65$ Å). Integration is carried out graphically by calculating the value of the function for several wavelengths λ and by measuring the area which is determined by these points (Table 6.1 and Figure 6.5). The calculated ratio of the intensity of 25% Nb_2O_5, in a mixture with 75% Ta_2O_5, to the intensity of pure Nb_2O_5 is 0.10, while the experimentally determined value is 0.11.

Kalman and Heller (1962) and Pluchery (1963) pointed out that according to the average-value theorem of the integral calculation, a wavelength λ must exist, for which the following equation is fulfilled:

$$\int_{\lambda_0}^{\lambda_K} \frac{\mu(\lambda)\,N_0(\lambda)\,d\lambda}{\mu(\lambda)/\sin\varphi + \mu(\alpha)/\sin\psi} = \frac{\mu(\bar{\lambda})\,N_0(\bar{\lambda})}{\mu(\bar{\lambda})/\sin\varphi + \mu(\alpha)/\sin\psi}\,(\lambda_K - \lambda_0)$$

The wavelength λ is referred to as the weighted-average wavelength. The integral can be thought of as being replaced by a rectangle of equal area, which has one side equal to the distance between lower and upper limit of the integral. The height of the rectangle corresponds to the value of the function at λ. It should be pointed out, however, that the weighted average is not located at one and the same wavelength for both mixture and pure elements, although deviations are usually small. In the present example, the weighted-average wavelength for the mixture is located at $\lambda = 0.52$ Å and for the pure component it is at $\lambda = 0.50$ Å.

Leroux et al. (1967b) experimentally determined the weighted-average wavelength for several elements. They found that the weighted-average wavelength is appreciably shorter than the absorption edge of the fluorescent element, although fluorescence is most effectively produced by wavelengths near the absorption edge.

Chapter 7

Interelemental or Secondary Excitation

In certain cases fluorescent excitation of an element not only results from the primary radiation of an x-ray tube but also from the fluorescent radiation of the associated elements.

Primary radiation ⎫   Fluorescent
of the x-ray tube ⎭ → radiation

Primary radiation ⎫   Fluorescent radiation   ⎫   Fluorescent
of the x-ray tube ⎭ → of the associated element ⎭ → radiation

For secondary excitation the fluorescent radiation of the associated elements must, however, be of shorter wavelength and of higher energy than the absorption edge of the element that is to be excited. Secondary excitation by associated elements adds to the excitation by the primary radiation of the tube. The intensity of the fluorescent radiation of elements which, in addition, are excited by associated elements is therefore larger than the intensity of elements which are only excited by the radiation from the x-ray tube. In contrast to direct excitation, where the primary radiation source (x-ray tube) is located outside the sample, interelemental excitation occurs from excitation centers which are distributed throughout the sample and act as point sources of secondary radiation (Figures 7.1 and 7.2). Let us assume a point source of radiation in a layer dx at a depth x in the sample and calculate the energy irradiated by the point source onto plane K located at a distance $|x-k|$ from the source. At distance r from the source the intensity Q (in photons/sec) decreases to the following amount:

$$\frac{Q}{4\,\pi r^2}\,\exp\,[-r\varrho\,\mu\,(\beta)] \qquad (7.1)$$

This equation accounts for the exponential absorption of the radiation of wavelength β by the mass between the source and plane K. Let us consider a cone of rays emanating from the source that intersects a plane K in an infinitesimal circle of points at distance r from the source. The number of photons which intersect plane K under angle θ is given as follows (Figure 7.2):

$$\frac{Q}{4\,\pi r^2}\exp\left[-r\varrho\mu(\beta)\right]2\,\pi\,u\,du\,\cos\theta \qquad (7.2)$$

When passing through layer dk, the following fraction of photons is absorbed by element A and results, with the probability E_A, in fluorescent radiation of wavelength α:

$$C_A\varrho\mu_A(\beta)dk\ /\ \cos\theta$$

We replace the terms r and u by the corresponding functions of $|x-k|$ and θ. The source Q emits radiation onto the plane through all angles θ and in layer dk, which is at a distance $|x-k|$ from the plane K, fluorescent radiation of intensity dN_α is produced:

$$dN_\alpha = \frac{Q}{2}\,C_A\varrho\mu_A(\beta)dk\int_{\theta=0}^{\theta=90^\circ}\tan\theta\exp\left[-\frac{|x-k|}{\cos\theta}\varrho\mu(\beta)\right]d\theta \qquad (7.3)$$

By substituting $t = 1/\cos\theta$, this function can be transformed into a well-known integral given in the literature (Integral Exponential Functions, Jahnke, Emde, and Lösch, 1960; Handbook of Chemistry and Physics):

$$dN_\alpha = \frac{Q}{2}\,C_A\varrho\mu_A(\beta)dk\int_{1}^{\infty}\frac{1}{t}\exp\left[-|x-k|\varrho\mu(\beta)t\right]dt \qquad (7.4)$$

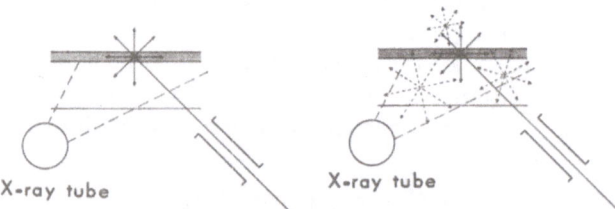

Fig. 7.1. Excitation of fluorescence by primary radiation of the x-ray tube (*left*) and by fluorescent radiation of the associated elements (*right*). In the first case the center of excitation is located outside the sample, while in case of interelemental excitation the secondary excitation centers are located within the sample.

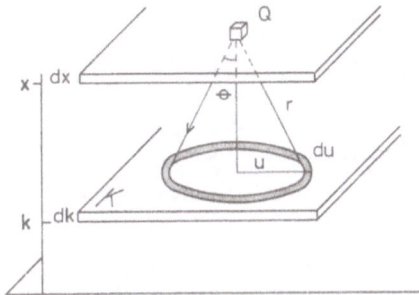

Fig. 7.2. Calculation of interelemental excitation.

Fluorescent radiation is emitted in all directions. However, only the fraction q passes through the collimator and towards the analyzing crystal as a parallel beam. When passing from a depth k to the emergence point from the sample, the fluorescent radiation is exponentially absorbed by the atoms of the sample (ψ = emergence angle)

$$\exp\left[-k\varrho\mu(\alpha)/\sin\psi\right] *$$

For polychromatic excitation, the intensity of the source Q at a depth x is given as follows (Chapter 6):

$$Q = \frac{E_B C_B \varrho}{\sin\varphi} \int_{\lambda_0}^{\lambda_B} N_0(\lambda)\,\mu_B(\lambda)\,\exp\left[-x\varrho\mu(\lambda)/\sin\varphi\right]dx\,d\lambda$$

The source Q produces fluorescence of the element A at all distances $|x - k|$; for the source itself, all values of x apply. The total fluorescent intensity of the element A which is produced by the associated element B is given as follows (Gillam and Heal, 1952; Sherman, 1955; Renaud, 1963):

$$N_A = \frac{q}{2\sin\varphi}\,E_A C_A \mu_A(\beta) \int_{\lambda_0}^{\lambda_B} \frac{E_B C_B \mu_B(\lambda)\,N_0(\lambda)\,d\lambda}{\mu(\lambda)/\sin\varphi + \mu(\alpha)/\sin\psi}\,L$$

$$\text{with } L = \left[\frac{\ln\left(1+\mu(\alpha)/\mu(\beta)\sin\varphi\right)}{\mu(\alpha)/\sin\varphi} + \frac{\ln\left(1+\mu(a)/\mu(\beta)\sin\psi\right)}{\mu(\lambda)/\sin\psi}\right] \quad (7.5)$$

---

* Dual meaning of α: in Chapters 2, 4, 5, and 6, α is used as the symbol for the strongest line in the emission spectrum of a series, for example K_α and K_β lines. In the present and the remaining chapters, however, α is the fluorescent wavelength of the element A just as β or γ designate the fluorescent wavelengths of the elements B or C.

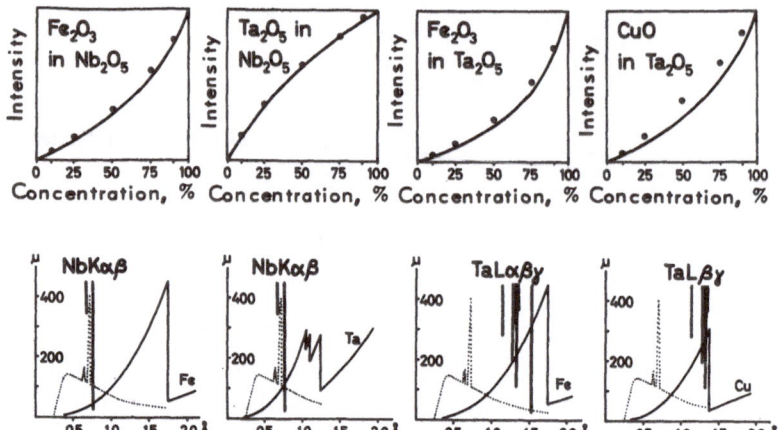

Fig. 7.3. *Upper portion*: Comparison of measured fluorescent intensities (· · ·) with calculated intensities (—). Interelemental excitation was not taken into account in the calculation. The magnitude of interelemental excitation may be estimated from the difference between the curves and the dots. *Lower portion*: Wavelength of the radiation of the particular associated element in relation to the absorption edge of the element that is to be excited. Lightly dotted curve is the tube spectrum of the molybdenum tube at an accelerating potential of 50 kV (Müller, 1962*a*).

Integration over $d\lambda$ extends from the short-wavelength limit of the continuum λ_0, to the absorption edge λ_B of the associated element and covers a smaller wavelength range than the primary radiation of the x-ray tube. In the latter case, integration is extended to the absorption edge λ_A of the element that is to be excited.

Gillam and Heal estimated that in the case of an 80 % Fe–20 % Ni alloy, the secondary radiation of nickel enhances the intensity of the iron radiation by 10 %. Laffolie (1962*b*) finds that in the case of low alloyed steels, the intensity of chromium increases by a factor of 1.39 due to fluorescence by the FeK_α line. Measurements by Kopineck and Schmitt (1961) of a mixture of 50 % Fe–50 % Mo indicate, that depending upon the accelerating potential, approximately 20 to 25 % of the fluorescent intensity of the iron is produced by secondary fluorescence of the associated element molybdenum. Müller (1962*a*) calculated the fluorescent intensity for several two-component systems as a function of the concentration without taking into account interelement excitation, and compared the calculated intensities with measured intensities in order to estimate the magnitude of interelemental excitation. The magnitude of the interelemental excitation depends upon the wavelength of the radiation of the associated elements in relationship to the absorption edge of the element that is to be excited. Interelemental excitation is negligible when the radiation of the associated element is far away from the absorption edge. When the radiation is close to the absorption edge, however, the associated element may cause an increase in the fluorescent intensity of the other element of as much as 10 to 30% (Figure 7.3). The major portion of the fluorescent radiation is, however, produced by the primary radiation of the x-ray tube. Interelemental excitation is sometimes of minor importance so that the absorption coefficients allow one to estimate the magnitude of fluorescence caused by the associated element.

Interelemental excitation has been referred to as a negative absorption by Noakes (1954) and Birks (1959, p. 59). The effect of the associated element on the fluorescent intensity is summarized in the so-called absorption-excitation coefficient. This coefficient accounts for all effects of the associated elements on the fluorescent intensity of the element that is to be analyzed.

Chapter 8

Grain Size and Surface Roughness Effects

8.1. Introduction

Derivation of the intensity formula assumes that the elements are distributed homogeneously and statistically according to their true ratios in the portion of the sample which is excited to fluorescence by the incident radiation. It is further assumed that the samples have smooth surfaces. These prerequisites are fulfilled by liquids, but powdered samples and alloys occasionally show grain-size and surface effects. The influence of grain size on the fluorescent intensity of powders was studied extensively by Haftka (1959), Claisse and Samson (1962), Müller (1963), and Kopineck and Schmitt (1965). Jenkins and Hurley (1965) investigated the influence of surface roughness in the case of alloys.* We first treat the influence of grain size of powders and then discuss the effect of surface roughness of solid samples on the fluorescent intensity. The average grain diameter in common powders ranges between 30–100 μ. Müller calculated that 50% of the fluorescent radiation in iron oxide, niobium oxide, and tantalum oxide originates from a depth of 20 to 100 μ (Table 8.1) and, hence, that the intensity originates largely from the lowest or second to lowest grain layer.

In order to study the dependence of the behavior of powders as a function of grain size, one should distinguish between homogeneous and heterogeneous powders. In the case of homogeneous powders all grains have the same chemical composition. A heterogeneous powder, on the other hand, consists of a mixture of chemically different grains. A tantalum oxide grain, for example, may lie next to a niobium oxide or an iron oxide grain (Figure 8.1). Homogeneous and heterogeneous powders behave completely differently during grinding and they are therefore discussed separately. It is often difficult, however, to distinguish *a priori* between a homogeneous and a heterogeneous powder.

---

* See also Tögel (1962*b*)

Table 8.1. Average Sample Depth from Which 50% and 90%, Respectively, of the Fluorescent Radiation Originates

Compound	Exciting radiation	Fluorescent wavelength	Penetration depth from which	
			50% originates	90% originates
Fe_2O_3	1.5—1.0 Å	1.94 Å	30—60 μ	100—200 μ
Nb_2O_5	0.63 Å	0.75 Å	100 μ	300 μ
Ta_2O_5	1.0—0.71 Å	1.54 Å	20—30 μ	70—100 μ

0.63 Å = MoK_β 0.71 Å = MoK_α

8.2. Homogeneous Powders

In a homogeneous powder chemically identical grains coexist. We introduce here the term fluorescent volume which refers to that part of a grain from which a measurable fluorescent radiation is emitted. In order to characterize the fluorescent volume individual grains are subdivided into two subareas, one of which is the area where fluorescence is excited by the incident radiation. This subarea is defined by the fact that the incident radiation, with increasing penetration depth, is more and more absorbed until total absorption occurs. We are further interested in that portion from which, in principal, fluorescence may be emitted. Analogously, this subarea is defined by the fact that the emerging fluorescent radiation decreases in intensity with increasing length of path and, therefore, below a certain depth no fluorescent radiation emerges. These two subareas are not identical and, hence, fluorescent radiation cannot emerge from all parts of the grain in which fluorescence is excited (Figure 8.2). The fluorescent volume is that part of the sample in which fluorescence is excited and from which it reaches the surface. In order to compare the fluorescent volumes of powders we normalize to irradiated areas of the same size. As was shown experimentally, the fluorescent volume and, hence, the fluorescent intensity of homogeneous powders increases with decreasing grain size (Table 8.2). If the average penetration depth is assumed to be equal to the diameter

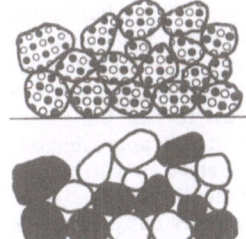

Fig. 8.1. Homogeneous and heterogeneous powders. *Upper*: homogeneous powder consisting of chemically identical grains, i.e., every individual grain consists of a mixture of A and B. *Lower*: heterogeneous powder consisting of chemically different grains, i.e., grains of the composition A are mixed with grains of the composition B.

of the spheres, then only the lower-most layer of grains fluoresces. In this case, the fluorescent volume is on the order of 80% in comparison to a closest-packed layer (pellet). In order to obtain reproducible results in practical analysis, homogeneous powders have to be packed as closely as possible. It is irrelevant whether this is achieved by grinding or by preparing a pellet.

8.3. Heterogeneous Powders

In heterogeneous powders chemically different grain coexist. A niobium oxide grain, for example, may be located next to a tantalum oxide or iron oxide grain. In order to understand the behavior of heterogeneous powders during grinding, we first investigate the dependence of fluorescent intensity on concentration of coarse-grained heterogeneous powders. In this case it is assumed that individual grains are large in comparison to the effective penetration depth of the x-ray radiation so that only the lowermost layer of grains is excited to fluorescence. Grain diameters of 100 to 300 μ fulfill this condition. Fluorescent intensity in coarse-grained powders is proportional to the number of grains of a particular type A, which are reached by the radiation and, hence, proportional to the concentration C_A

$$N_A = \text{prop } [A] = \text{prop } C_A \qquad (8.1)$$

The fluorescent intensity increases linearly with increasing concentration and a straight line is obtained as calibration curve for determination of the concentration in a two- or three-component mixture (Figure 8.3).

In contrast to coarse-grained powders, the individual grains of fine-grained heterogeneous powders are so small that x-rays pass through several grains and excite several grain layers to emit fluorescence. For this

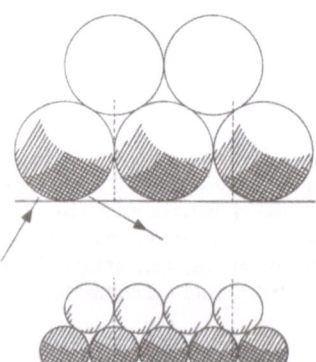

Fig. 8.2. Schematical illustration of the fluorescent volume (criss–cross hatching). The fluorescent volume covers that area in which fluorescence is excited and from which it can emerge. In the model of a coarse-grained powder (upper picture) in which the penetration depth is equal to half the sphere diameter, the fluorescent volume is approximately 35%. In the case of a fine-grained powder (lower picture), the fluorescent volume for the same penetration depth is approximately 80%. Numbers are in comparison to a closest-packed layer in a pellet.

Table 8.2. Effect of Grain Size on Fluorescence Intensity of Homogeneous Powders (counts/sec)

Grain size $\varnothing$	100% Nb$_2$O$_5$	100% Ta$_2$O$_5$	Homogeneous 25/75 mixture	
	NbK$_\alpha$	TaL$_\alpha$	NbK$_\alpha$	TaL$_\alpha$
0.25—0.10 mm	95	77	—	—
0.10—0.06 mm	98	82	92	88
<0.06 mm	100	94	100	95
4 hr of grinding	100[a]	100[a]	100[a]	100[a]

[a]Arbitrarily normalized to 100.

reason, the absorption of the incident and emerging radiation is also determined by the associated grains. The absorption coefficient is the average of the absorption coefficients of all grains through which the radiation passes. If the radiation passes through many grains, then the value of the average absorption coefficient approaches that of the coefficient for a homogeneous mixture. In this case, the relation between fluorescent intensity and concentration is derived in analogy to what was discussed in Chapter 6, and the absorption coefficient is the weighted average of the absorption coefficients of the individual components. The fluorescent intensity N_A of component A in a mixture with component B is then given as

$$N_A = \frac{q}{\sin \varphi} E_A C_A \int_{\lambda_0}^{\lambda_A} \frac{\mu_A(\lambda) N_0(\lambda) \, d\lambda}{C_A \bar{\mu}_A(\alpha) + C_B \bar{\mu}_B(\alpha)} \tag{8.2}$$

where the factor $q/\sin \varphi$ is a function of the geometry of the spectrometer; E_A is a function of the fluorescent efficiency of component A; C_A and C_B are the normalized concentrations of components A and B; $\mu_A(\lambda)N_0(\lambda)$ is the mass absorption coefficient of component A for the incident primary x-ray radiation of the wavelength λ whose intensity is $N_0(\lambda)$; and $\bar{\mu}_A(\alpha)$ and $\bar{\mu}_B(\alpha)$ are the combined mass absorption coefficients of components A and B for the incident primary radiation and the emerging fluorescent radiation of component A. The fluorescent intensity of fine-grained powders is not proportional to the concentration of a component but, depending upon the associated component, either larger or smaller than that value. The relationship between intensity and concentration is represented by a bent curve (Figure 8.3).

In order to discuss the behavior of a heterogeneous powder after grinding, the fluorescent intensities emitted by coarse- and fine-grained components are compared. When grinding coarse-grained heterogeneous powders, where the fluorescent intensity is proportional to the concentration of the respective component, the intensity departs from this proportionality and approaches the value characteristic for homogeneous mixtures. Depending upon the associated components, the fluorescent intensity either increases or decreases. Analogous to interaction with the associated component, the fluorescent volume increases and the fluorescence is enhanced. When grinding a mixture of a heavy component and a light component (for example, Ta_2O_5 in Nb_2O_5), both effects work in the same direction and the intensity increases strongly. When grinding a mixture of a light and a heavy component, on the other hand (for example, Nb_2O_5 in Ta_2O_5), both effects work in opposite directions and, due to interaction with the heavy associated component result in a decrease of the fluorescent intensity; the fluorescent volume increases, however, resulting in a more or less pronounced change in intensity.

The fluorescent intensity of a mixture of grains may be calculated as a function of the grain diameter. Such calculations, however, are only approximate. Let us assume that the grains are replaced by equally large cubes, where one face is normal to the incident and another normal to the emerging radiation, and where both rays form a 90° angle (this is almost always approximately the case). The cubes are closest-packed and completely fill the available space (Figure 8.4). The fluorescent intensity W_A of a cube of composition A is given as follows:

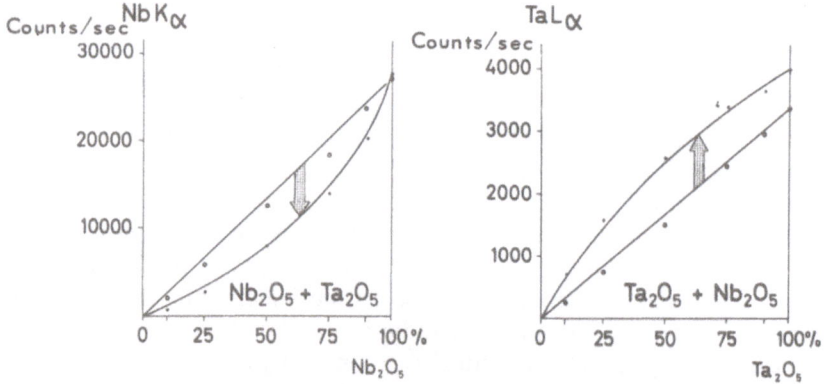

Fig. 8.3. Change of fluorescent intensity of heterogeneous powders after grinding. In general, the fluorescent intensity of the light compound (left) decreases after grinding. The fluorescent intensity of the heavier compound (right), on the other hand, increases after grinding. Straight line calibration curves are found for coarse-grained heterogeneous powders. For fine-grained powders, on the other hand, bent calibration curves originate. Comparison of experimental measurements (points) with calculated curves for a mixture of niobium oxide and tantalum oxide grains of a diameter ranging from 60 to 100 μ and for a powder after 7 hr of grinding (Müller, 1963).

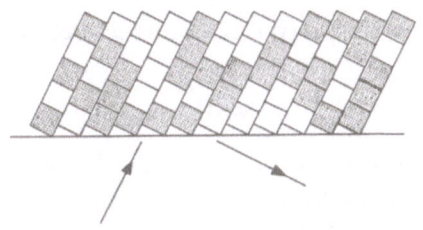

Fig. 8.4. Model for calculating the fluorescent intensity of a heterogeneous mixture consisting of equally large cubes of the composition A (black) and the composition B (white).

$$W_A = q E_A \frac{a}{\mu_A(\alpha)} \int_{\lambda_0}^{\lambda_A} N_0(\lambda) \{1 - \exp[-a\mu_A(\lambda)]\} \cdot$$

$$\cdot \{1 - \exp[-a\mu_A(\alpha)]\} \, d\lambda \qquad (8.3)$$

where a is the length of the edge of the cube; μ_A the linear absorption coefficient; and the other symbols are the ones used in Chapter 5. The relative abundance of the cubes of type B is C_B. Therefore, in the lowermost layer of N cubes there are $C_A N$ cubes of type A, whose fluorescent intensity is given as

$$N_{A1} = C_A N W_A \qquad (8.4)$$

In the second layer there are also $C_A N$ cubes of type A. Before reaching this layer, however, the incident radiation is absorbed by the preceding layer. As the first layer consists of $C_A N$ cubes of types A and $C_B N$ cubes of type B, the average absorption is

$$C_A \exp[-a\mu_A(\lambda)] + C_B \exp[-a\mu_B(\lambda)]$$

Intensity of fluorescent radiation originating in the second layer is proportional to $C_A N W_A$ and, upon emergence, is absorbed by the preceding layer in the amount

$$C_A \exp[-a\mu_A(\alpha)] + C_B \exp[-a\mu_B(\alpha)]$$

The number of cubes of type A in the third layer is again given by $C_A N$. Radiation which reaches this layer is absorbed when passing through the preceding first and second layers as follows (combination of cubes to pairs of two with corresponding relative abundance):

$$C^2_A \exp[-2a\mu_A(\lambda)] + 2\,C_A C_B \exp[-a(\mu_A(\lambda) + \mu_B(\lambda))] +$$

$$+ C^2_B \exp[-2\,a\mu_B(\lambda)] \equiv \{C_A \exp[-a\mu_A(\lambda)] + C_B \exp[-a\mu_B(\lambda)]\}^2$$

Fluorescent radiation originating in the third layer is absorbed before emergence by the average amount

$$\{C_A \exp\left[-a\mu_A(\alpha)\right] + C_B \exp\left[-a\mu_B(\alpha)\right]\}^2$$

For the intensity of fluorescent radiation originating in the fourth layer, we have to consider absorption of the incident and emerging radiation by the previous three layers. We continue to consider layer after layer until the $(i + 1)$ layer is reached. In order to calculate the absorption of the incident radiation for this particular layer, we assume (Mitra and Wilson, 1960)

$$\{C_A \exp\left[-a\mu_A(\lambda)\right] + C_B \exp\left[-a\mu_B(\lambda)\right]\}^i$$

Similarly, the absorption of the emerging radiation may be determined. In order to calculate the total fluorescence produced in all layers, the contributions by the individual layers are added and the total intensity N_A is

$$N_A = C_A N W_A \sum_{i=0}^{i=\infty} \left(\{C_A \exp\left[-a\mu_A(\lambda)\right] + C_B \exp\left[-a\mu_B(\lambda)\right]\} \cdot \right.$$
$$\left. \cdot \{C_A \exp\left[-a\mu_A(\alpha)\right] + C_B \exp\left[-a\mu_B(\alpha)\right]\}\right)^i \qquad (8.5)$$

This is the sum of a geometrical succession, and we obtain

$$N_A = \frac{C_A N W_A}{1 - \{C_A \exp\left[-a\mu_A(\lambda)\right] + C_B \exp\left[-a\mu_B(\lambda)\right]\}\{C_A \exp\left[-a\mu_A(\alpha)\right] + C_B \exp\left[-a\mu_B(\alpha)\right]\}}$$
$$(8.6)$$

In this equation the fluorescent intensity is expressed as a function of the diameter a of the cube. It is desirable to obtain the boundary conditions for very coarse and very fine grains. For very coarse grains ($a \to \infty$) all exponential terms which contain the factor a become zero, and there remains

$$N_A = C_A N W_A \qquad (8.7)$$

$C_A N$ is the number of cubes of type A present in one layer, and W_A is the fluorescent intensity of one cube. For coarse-grained mixtures, the resulting intensity is proportional to the number of grains of type A or B, respectively, present in one layer (for example, in the lowermost layer). For very small grain sizes ($a \to 0$) we develop the exponential functions of equation (8.6) in sequence and consider only the first two terms. When substituting for W_A the value given in equation (8.3) we obtain

$$N_A = C_A N q E_A \frac{a}{\mu_A(\alpha)} \cdot$$

(8.8)

$$\int_{\lambda_0}^{\lambda_A} \frac{a^2 N_0(\lambda)\, \mu_A(\lambda)\, \mu_A(\alpha)\, d\lambda}{1 - \{C_A + C_B - a C_A \mu_A(\lambda) - a C_B \mu_B(\lambda)\} \{C_A + C_B - a C_A \mu_A(\alpha) - a C_B \mu_B(\alpha)\}}$$

The integral is simplified because $C_A + C_B = 1$, and we can shorten numerator and denominator by a

$$N_A = C_A N q E_A a^2 \cdot$$

$$\cdot \int_{\lambda_0}^{\lambda_A} \frac{N_0(\lambda)\, \mu_A(\lambda)\, d\lambda}{C_A[\mu_A(\lambda) + \mu_A(\alpha)] + C_B[\mu_B(\lambda) + \mu_B(\alpha)] - a[\cdots]}$$

(8.9)

In order for the sample area not to become zero ($Na^2 \to 0$), we normalize the fluorescent intensity to a constant sample area. When $a \to 0$, all terms containing the factor a become zero, and we obtain for the fluorescent intensity of a fine-grained heterogeneous powder

$$\frac{N_A}{Na^2} = C_A q E_A \int_{\lambda_0}^{\lambda_A} \frac{N_0(\lambda)\, \mu_A(\lambda)\, d\lambda}{C_A[\mu_A(\lambda) + \mu_A(\alpha)] + C_B[\mu_B(\lambda) + \mu_B(\alpha)]}$$

(8.10)

Thus, we obtain the well-known formula for the fluorescent intensity of component A in a mixture with component B.

8.4. Effects of Surface Roughness

Massive samples may be analyzed directly provided they are prepared to the appropriate dimensions by either sawing, turning, filing, or grinding. When preparing samples, rough surfaces frequently originate which can only be made smooth by prolonged polishing. However, of importance are not only the nature of the grooves and surface profile but also the orientation of the sample and its grooves in relation to the direction of x-rays. Massive samples are therefore often rotated during measurement. However, this does not eliminate the influence of surface properties entirely but only eliminates the effects of random orientation of the samples. If samples cannot be rotated during measurement, then they should be oriented in such a way that the grooves are parallel to the plane formed by incident and emergent radiation (as is shown in the right picture of Figure 8.5). In this arrangement the effects of surface grooving on intensity are at a minimum. Following the model which we developed for heterogeneous grain mixtures, we calculate the dependence of secondary fluorescent intensity upon groove

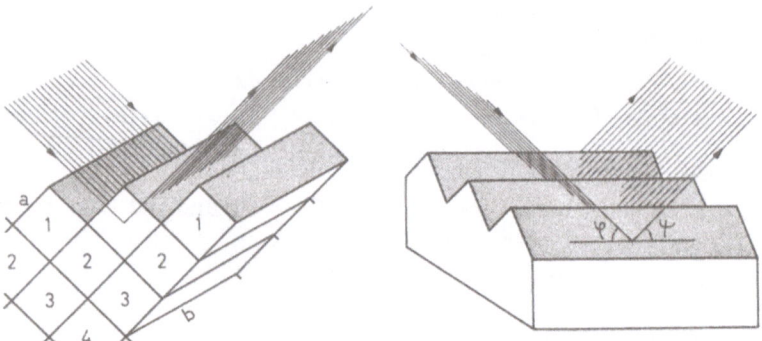

Fig. 8.5. Effect of grooving of massive samples on fluorescent intensity. *Left*: When the grooves are perpendicular to the plane formed by incident and emerging x-ray beam, then the fluorescent intensity decreases with increasing size of the grooves *a* and increasing absorption coefficient. *Right*: When the grooves are parallel to the plane formed by incident and emerging x-ray beam, then the influence of groove size *a* on the fluorescent intensity of the sample is negligible.

depth *a* and groove orientation. Instead of cubes we assume prisms of square cross section (length of edge = *a*) whose longer axes (length *b*) are parallel among themselves and also parallel to the grooves. First, the sample may be oriented with the grooves perpendicular to the plane formed by incident and emerging x-ray beam. The primary x-ray radiation is assumed to penetrate the sample in the direction of an *a*-edge, while the fluorescence is assumed to emerge in the direction of another *a*-edge (Figure 8.5).

For the intensity originating from a prism P_A we obtain

$$P_A = q\,E_A\,\frac{b}{\mu\,(\alpha)}\int_x^{\lambda_A} N_0\,(\lambda)\,\{1 - \exp\,[\,-\,a\,\mu\,(\lambda)]\}\,\{1 - \exp\,[\,-\,a\,\mu\,(\alpha)]\}\,d\,\lambda$$

$$(8.11)$$

This fluorescence corresponds to a prism in the first row which is marked by the number (1) in Figure 8.5 (left). Fluorescent intensity originating in a prism of row (2) is lower by the following factor:

$$\exp\,[-a\mu(\lambda)-a\mu(\alpha)]$$

This is because incident and emerging radiation is, in addition, absorbed by a prism of the preceding row (1). Fluorescence originating in a prism of row (3) is lower by the following factor:

$$\{\exp\left[-a\mu(\lambda)-a\mu(\alpha)\right]\}^2$$

For the total intensity we obtain

$$N(\alpha) = P_A \sum_{i=0}^{i=\infty} \{\exp\left[-a\mu(\lambda)-a\mu(\alpha)\right]\}^i$$

This corresponds to the sum of a geometrical succession. When the corresponding values are substituted for the sum and for P_A, we obtain

$$N(\alpha) = q E_A \frac{b}{\mu(\alpha)} \int_{\lambda_0}^{\lambda_A} \frac{N_0(\lambda)\{1-\exp\left[-a\mu(\lambda)\right]\}\{1-\exp\left[-a\mu(\alpha)\right]\}\,d\lambda}{1-\exp\left[-a\mu(\lambda)-a\mu(\alpha)\right]}$$

$$(8.12)$$

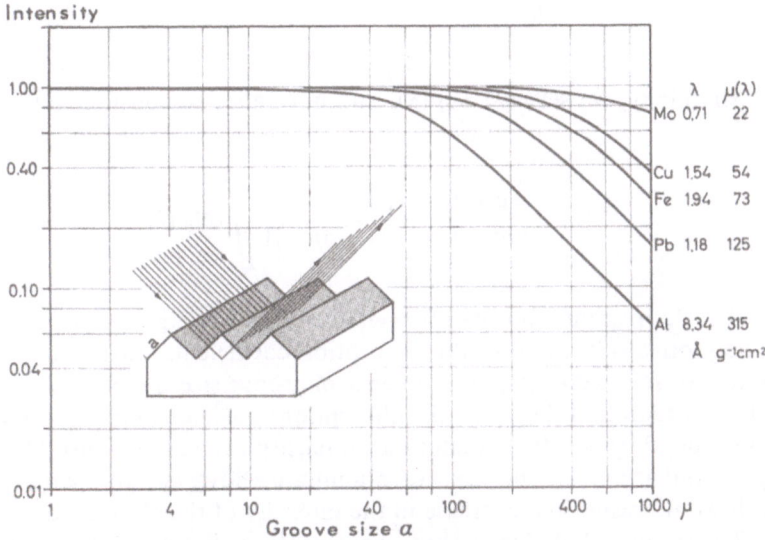

Fig. 8.6. Change in relative fluorescent intensity of massive samples as a function of groove size a for different wavelengths; calculated for pure elements according to equation (8.13). In case of large absorption coefficients, surface irregularities have a more pronounced effect on fluorescent intensities than in case of small absorption coefficients. Calculations for lead indicate that the critical parameter is not the wavelength but the absorption coefficient.

Table 8.3. Limiting Value S_{max} for Maximum Groove Depth
(Height Difference between Hill and Valley). No appreciable
intensity decrease is recorded. Jenkins and Hurley (1965).

Sample	Element	Content	Line	S_{max}, μ
1	Al	92.6%	AlK_α	70
	Si	0.16%	SiK_α	80
	Fe	0.27%	FeK_α	>180
	Cu	4.43%	CuK_α	>180 [a]
2	Cu	87.4%	CuK_α	130
	Zn	3.75%	ZnK_α	120
	Sn	5.25%	SnK_α	>180
			SnL_α	60
	Pb	3.60%	PbL_α	125 [a]
			PbM_α	60
3	Mn	0.55%	MnK_α	>180
	Ni	2.87%	NiK_α	>180
	Cr	20.4%	CrK_α	130
	Fe	62.2%	FeK_α	140

[a]Undesirable effects caused by smearing.

For simplicity, we set $\mu(\lambda) = \mu(\alpha)$, and normalize to constant sample area
ab

$$N(\alpha) \cong \frac{qE_A}{a\mu(\alpha)} \frac{1-\exp\left[-a\mu(\alpha)\right]}{1+\exp\left[-a\mu(\alpha)\right]} \int_{\lambda_0}^{\lambda_A} N_0\,(\lambda)\,d\lambda \qquad (8.13)$$

The fluorescent intensity of a massive sample is expressed as a function
of the groove size a and of the absorption coefficient, $\mu(\alpha)$; it may be calculated for selected examples. Influence of groove size is more severe for the
particular radiation the larger the absorption coefficient. Grooves in copper
of 35-μ depth ($a = 50\,\mu$) result in an intensity decrease of only 1% in the
copper radiation. In the case of aluminum, however, the same groove
depth would result in a decrease in the intensity of the aluminum radiation
of 17% (Figure 8.6). Since short-wavelength radiation is in general less
absorbed than radiation of longer wavelength, the former is less sensitive
to surface irregularities and, therefore, preferred for analysis.

In the second case, the sample may be oriented with the grooves
parallel to the plane formed by the incident and emerging x-ray beam.
Incidence and emergence angles are designated φ and ψ, respectively (Figure

8.5). When the fluorescent intensity is normalized to a constant area, we obtain

$$N(\alpha) = qE_A \int\limits_{\lambda_0}^{\lambda_A} \frac{N_0(\lambda)\,\mu(\lambda)\,d\lambda}{\mu(\lambda)/\sin\varphi + \mu(\alpha)/\sin\psi} \qquad (8.14)$$

In this case, the fluorescent intensity is independent of the groove depth.

Studies of Jenkins and Hurley (1965) indicate that for most of the samples studied, the decrease in intensity with increasing depth of grooves is analogous to Figure 8.6. Deviations from these curves indicate that in these cases additional undesirable effects are present, such as, smearing of a soft component over a large surface area during grinding and polishing of multicomponent mixtures. Considering the intensity decrease with increasing depth S of the groove, as measured by the difference in height between hill and valley, a boundary value S_{max} may be defined where no appreciable intensity decrease is recorded, although depth of the groove has already reached a certain value. This limiting value was determined by Jenkins and Hurley (1965) for several elements and found to vary between 60–180 μ (Table 8.3).

Chapter 9

Intensity Formula for a Divergent Primary Beam

So far we have assumed that the incident primary beam is parallel. In reality, however, the primary beam has lateral as well as longitudinal divergence (Figure 9.1). Let us compare the intensity formulas for divergent primary and parallel beams, respectively. We consider the crossover of the anode to be a point source of the power Q (photons/sec) located at the distance l from the sample (distance of the crossover from the sample surface is assumed to be large in comparison to the penetration depth of the x-rays; $l \gg x$). The intensity of the tube radiation which is emitted onto the sample surface under angle φ with a lateral divergence ϑ per steradian, $d\varphi \, d\vartheta$ is given as

$$\frac{Q \sin^2 \varphi \cos^2 \vartheta}{4 \, \pi \, l^2} \, d\varphi \, d\vartheta \tag{9.1}$$

Before reaching layer dx, the radiation is absorbed by the following factor:

$$\exp\left[- \frac{x \varrho \mu}{\sin \varphi \cos \vartheta} \right]$$

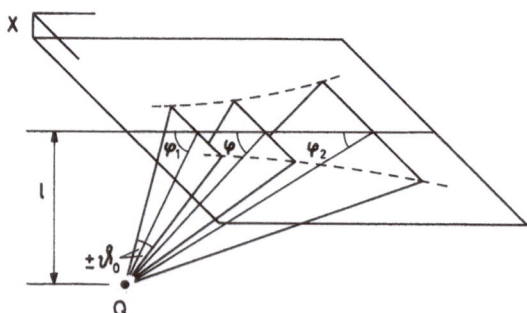

Fig. 9.1. Calculation of fluorescent intensity excited by a slightly divergent beam.

where μ is the mass absorption coefficient and ρ the average density of the sample. When passing through layer dx the following fraction is absorbed:

$$\frac{\rho \mu \, dx}{\sin \varphi \cos \vartheta}$$

Lateral divergence is limited by the angles $+\vartheta_0$ and $-\vartheta_0$, and longitudinal divergence is limited by the angles φ_1 and φ_2. The intensity absorbed in dx which is critical for the production of fluorescence is then given as

$$\frac{Q}{4\pi l^2} \, \rho \, \mu \, dx \int_{\varphi_1}^{\varphi_2} \sin \varphi \int_{-\vartheta_0}^{+\vartheta_0} \cos \vartheta \, \exp\left[-\frac{x \rho \mu}{\sin \varphi \cos \vartheta}\right] d\vartheta \, d\varphi \qquad (9.2)$$

According to the average-value theorem of the integral calculus, we obtain for the integral over $d\vartheta$ (with $0 < \vartheta' < \vartheta_0$)

$$\int_{-\vartheta_0}^{+\vartheta_0} \cos \vartheta \, \exp\left[-\frac{x \rho \mu}{\sin \varphi \cos \vartheta}\right] d\vartheta = 2\,\vartheta_0 \cos \vartheta' \, \exp\left[\frac{x \rho \mu}{\sin \varphi \cos \vartheta'}\right]$$

Analogously, the integral over $d\varphi$ is as follows (with $\varphi_1 > \varphi' > \varphi_2$):

$$\int_{\varphi_1}^{\varphi_2} \sin \varphi \, \exp\left[-\frac{x \rho \mu}{\sin \varphi \cos \vartheta'}\right] d\varphi = (\varphi_2 - \varphi_1) \sin \varphi' \cdot$$

$$\cdot \exp\left[-\frac{x \rho \mu}{\sin \varphi' \cos \vartheta'}\right]$$

For the intensity we obtain

$$\frac{Q}{4\pi l^2} \, \rho \mu \, dx \, (\varphi_2 - \varphi_1) \, 2 \, \vartheta_0 \sin \varphi' \cos \vartheta' \exp\left[\frac{x \rho \mu}{\sin \varphi' \cos \vartheta'}\right] \qquad (9.3)$$

Thus, we have calculated the intensity of the radiation absorbed in a layer dx. In order to establish a relationship to equations presented earlier, we consider the intensity $N_0(\lambda) \, d\lambda$ which impinges per unit sample area through the angles $\varphi = \varphi'$ and $\vartheta_0 = 0$:

$$N_0 \, (\lambda, \varphi', 0) \, d\lambda = \frac{Q \sin^2 \varphi'}{4 \pi l^2}$$

We substitute this value in (9.3) and obtain

$$N_0 \, (\lambda, \varphi', 0) \, \frac{\rho \mu \, dx}{\sin \varphi'} \, (\varphi_2 - \varphi_1) \, 2 \, \vartheta_0 \, \cos \vartheta' \, \exp\left[-\frac{x \rho \mu}{\sin \varphi' \cos \vartheta'}\right] d\lambda \qquad (9.4)$$

The absorbed energy of a parallel primary x-ray beam of the cross section F in a layer dx is calculated as follows (Chapter 5):

$$F N_0(\lambda) \frac{\varrho \mu \, dx}{\sin \varphi} \exp\left[-\frac{\varrho \mu x}{\sin \varphi} \right] d\lambda$$

These two values differ in the constant factor:

$$(\varphi_2 - \varphi_1) \, 2 \, \vartheta_0 \cos \vartheta' \quad \text{or} \quad F$$

In the exponential function, they also differ in the constant factor, $\cos^{-1} \vartheta'$ $\sin^{-1} \varphi'$ or $\sin^{-1} \varphi$. After integrating over all x in the numerator, we obtain

$$\left[\frac{\mu(\lambda)}{\sin \varphi' \cos \vartheta'} + \frac{\mu(\alpha)}{\sin \psi} \right]$$

instead of

$$\left[\frac{\mu(\lambda)}{\sin \varphi} + \frac{\mu(\alpha)}{\sin \psi} \right]$$

Nothing, however, changes in the basic structure of the intensity formula. Aside from certain constant factors, the formulas for a parallel primary beam may, in principle, also be applied to a slightly divergent beam.

Apparatus

10.1. Instrumentation of X-Ray Fluorescence Units

Figure 10.1 is a schematic of a commercial x-ray fluorescence unit. The primary radiation of an x-ray tube excites the elements of the sample to emit fluorescent radiation. This radiation is diffracted, according to wavelength, by an analyzing crystal and the corresponding angle of deflection is measured with a goniometer. A counter registers the intensity of the diffracted radiation. Frequently, a driving motor is attached to the spectrometer so that the complete wavelength range may be scanned automatically. Simultaneously, a recorder registers the intensity corre-

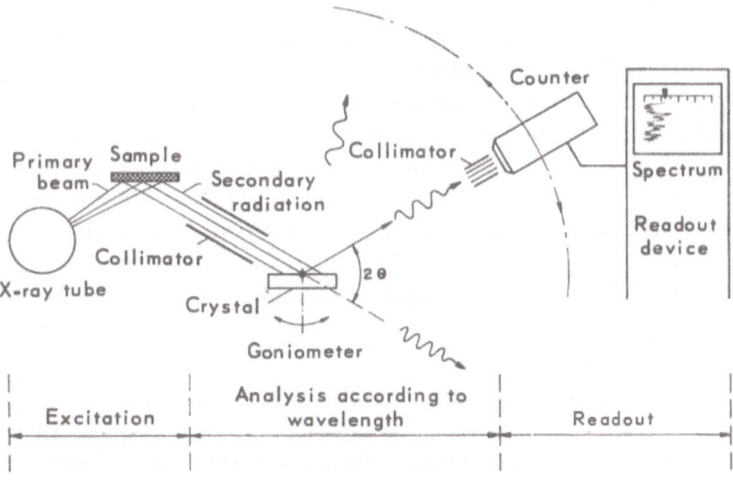

Fig. 10.1. Schematic diagram of a commercial x-ray fluorescence unit consisting of three major sections which serve, first, to excite fluorescent radiation; second, analyze the fluorescent radiation according to wavelength; and, third, read the diffraction angle and the intensity of the radiation (courtesy of Arsuffi, 1960).

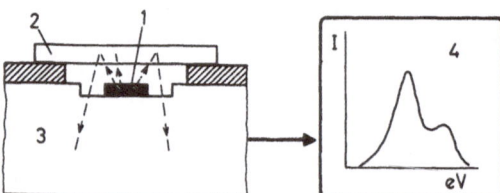

Fig. 10.2. Unit with nondispersive geometry, whereby the fluorescent radiation is excited by a radioactive sample. (1) Radioactive source, (2) sample, (3) proportional counter, (4) pulse-height analysis of quantum energy for element identification.

sponding to a certain diffraction angle. Instead of an x-ray tube a radioactive source may also be used to excite fluorescence. The resulting fluorescent radiation, however, is considerably weaker and diffraction by a crystal is therefore avoided; instead, the quantum energy of the radiation is determined with the aid of a pulse-height analyzer. In the case of direct-emission analysis, the sample is located in a high vacuum on the anticathode whereby the characteristic x-ray spectra of the elements are produced directly. X-ray fluorescence units are produced commercially by several companies; for example, Applied Research Laboratories, Glendale, Calif.; Compagnie Générale de Radiologie; General Electric; Hilger & Watts Ltd.; C. H. F. Müller GmbH; N. V. Philips Gloeilampenfabrieken; Picker X-ray Corp.; Rich. Seifert & Co.; and Siemens & Halske AG.

10.2. Generators

X-ray tubes are operated with high-voltage generators of 50, 60, or 100 kV at 1 to 2 kilowatts. High-voltage and tube current are electronically stabilized. In the case of a monitor system, part of the x-ray radiation is internally directed onto a sample whose fluorescent radiation serves as a reference signal. The voltage may be regulated in steps and, by proper choice, may serve to excite fluorescence of only selected elements of the sample, enabling spectral overlap to be avoided. The books by Glocker (1949), Neff (1959a), and Schaaffs (1957) are recommended for further reading on the technology of high-voltage generators and x-ray tubes.

10.3. X-Ray Tubes

In commercial x-ray tubes, electrons are emitted by a hot-filament cathode and are accelerated in a high-voltage field (Figure 10.4). Their kinetic energy D after acceleration by the voltage is as follows:

$$\frac{1}{2} m v^2 = e V$$

When impinging on the anode, the accelerated electrons are slowed down. The strong temperature increase of the anode indicates that most of the kinetic energy of the electrons is transformed into heat while only 0.01–1 % is transformed into radiation. The electrons release their energy in the form of radiation by one or more subprocesses. It is for this reason that the x-ray tube emits a continuous x-ray spectrum (bremsstrahlung). The energy of the most energetic quantum $h\nu$, which is emitted as radiation in the deceleration process, is equal to the kinetic energy of the electron after its passage through the voltage field V. The continuous x-ray spectrum, therefore, has a short-wavelength (high-energy) limit λ_0:

$$\lambda_0 = \frac{hc}{eV} = \frac{12'395}{V} \qquad [\lambda_0] = \text{Å}$$

In this equation h is Planck's constant, c the velocity of light, e the electrical charge of the electron, and V the accelerating potential. The highest spectral intensity of the continuous spectrum is at a wavelength which is approximately 3/2 of the short-wavelength limit λ_0. Relative intensity distribution in a tube spectrum may be determined approximately by scattering of tube radiation by a sample consisting of light elements (e.g., plexiglass, soot, starch), where the scattered radiation is measured just like the fluorescent radiation. Part of the scattered radiation, however, is modified by the amount of the Compton wavelength. Depending upon the readout system, either the spectral intensity (energy per unit time per wavelength range) or the spectral pulse rate (number of photons per unit time per wavelength range) are measured.

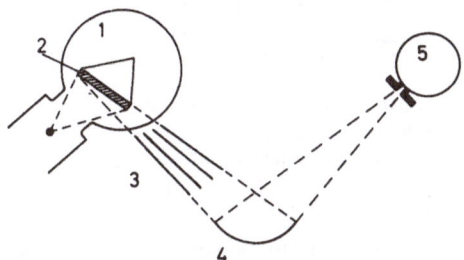

Fig. 10.3. Unit for direct emission with curved analyzing crystal. (1) X-ray tube, (2) sample, which serves as anode of the tube under direct bombardment by electrons, (3) divergent slit, (4) analyzing crystal, and (5) counter.

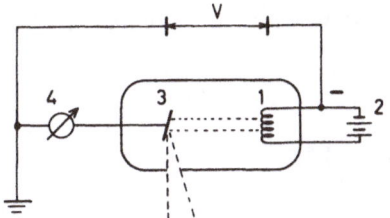

Fig. 10.4. Schematic diagram of an x-ray tube. (1) Hot-filament cathode; (2) hot-filament cathode power supply; (3) anode; and (4) milliammeter for measurement of the tube current. The high voltage V is between the hot-filament cathode and the anode.

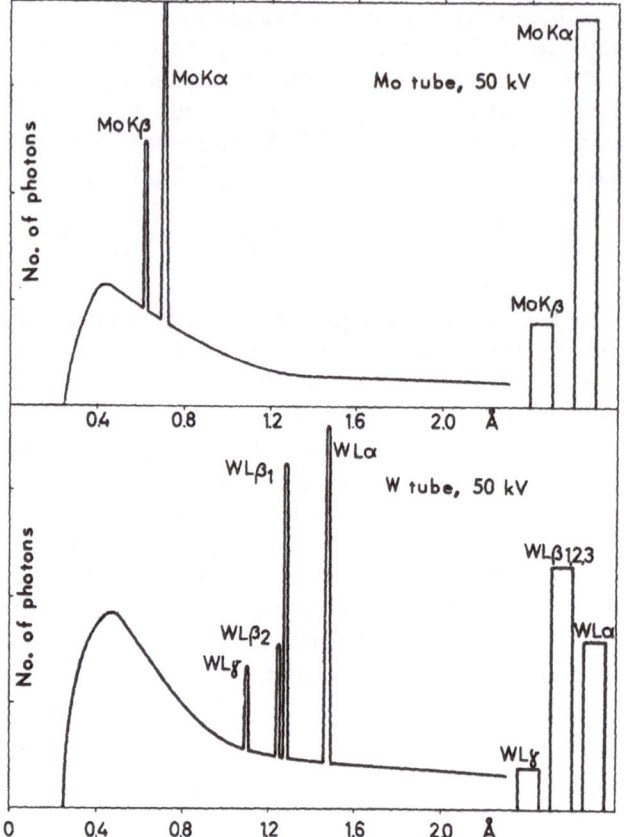

Fig. 10.5. Emission spectra, registered by the scattering method on plexiglass, of molybdenum and tungsten tubes at 50 kV, with equivalent areas for the intensity of the characteristic radiation.

The intensity of the continuous x-ray spectrum increases proportionally with the tube current i and the atomic number Z of the anode material. With increasingly higher voltage, the short-wavelength limit of the continuum shifts towards shorter wavelengths, and the total intensity of the continuous spectrum increases approximately with the square of the voltage

$$I = \int_0^\infty I(\lambda)\ d\lambda = \text{prop } i\,V^2 Z$$

The continuous x-ray spectrum was studied in detail by Ulrey (1918), and Kulenkampff and associates (1922, 1943). Further detailed discussions may be found in Stephenson (1957). Spectral intensity distribution in emission spectra of x-ray tubes, for example, was measured by Gilfrich and Birks (1968).

The characteristic radiation of the anode material is superimposed onto the continuous spectrum of the x-ray tube. Characteristic radiation occurs when the electrons have sufficient energy to ionize the K or L shells in the atoms of the anode material. The critical excitation potential for the excitation of characteristic radiation, for example, is 6 kV for a chromium tube and 20 kV for a molybdenum tube. The intensity of the characteristic radiation for the accelerating potential V is approximately proportional to the square of the overvoltage (valid to $V \leq 3V_0$, where V_0 is the critical excitation potential) and proportional to the tube current i

$$I \approx \text{prop } i \, (V - V_0)^2$$

10.4. Spectrometers

Most of the commercially available spectrometers are built according to one of the following three principles (Figures 10.7–10.9). In a spectrometer with a flat analyzing crystal, both crystal and counter are turned around a common axis and the counter travels at twice the speed of the crystal. By moving crystal and counter simultaneously, the individual lines of a spectrum may be registered. In this process, one wavelength after the other is reflected by the crystal according to Bragg's law ($2d \sin \theta = n\lambda$) and registered by the counter. Collimators serve to provide a parallel beam and thereby determine the incident angle of the radiation onto the crystal

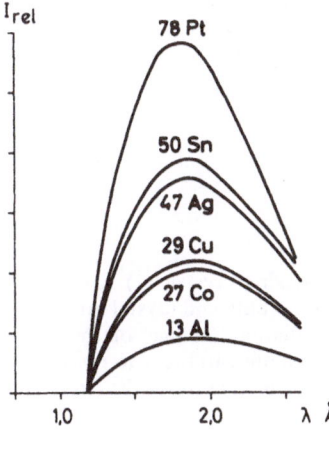

Fig. 10.6. Energy distribution of continuous spectra produced when electrons of 10.47 kV impinge upon various anode materials. Spectral and integral intensities are nearly proportional to the atomic number of the anode material (Kulenkampff, 1922).

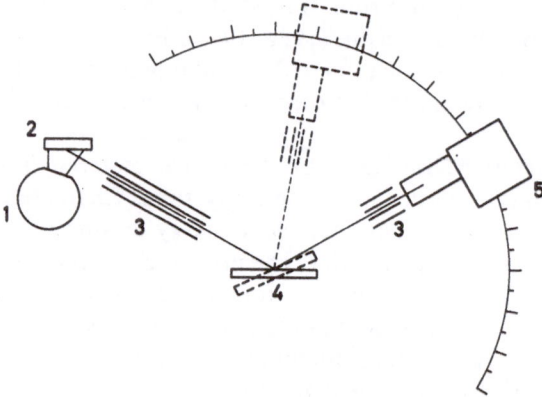

Fig. 10.7. Spectrometer with flat analyzing crystal. (1) X-ray tube; (2) sample; (3) collimator; (4) analyzing crystal; and (5) counter. Crystal and counter are turned around a
common axis, and the counter is driven at twice the speed of the crystal. A collimator
serves to provide a parallel beam of incident radiation before it reaches the crystal. In
order to register the different lines of a spectrum, crystal and counter are turned around
a common axis until Bragg's law is fulfilled for a particular line.

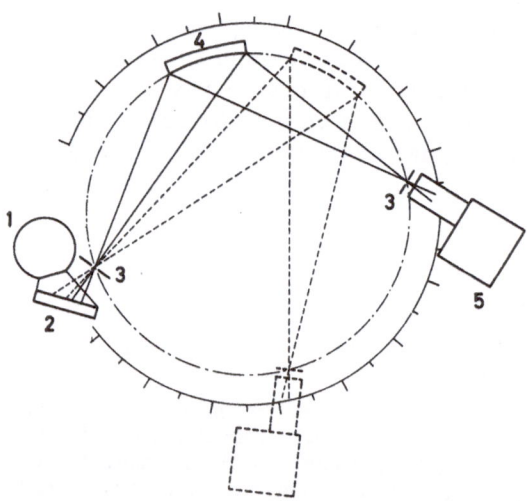

Fig. 10.8. Spectrometer with curved and polished analyzing crystal. (1) X-ray tube;
(2) sample; (3) slits; (4) analyzing crystal; and (5) counter. Slits and crystal are located
on the same circle. After passing through the slit, the radiation impinges on the curved
crystal as a divergent beam and is focused towards the counter. In order to register the
various lines of the spectrum, crystal and counter have to be moved on the circle until
the diffraction condition is fulfilled.

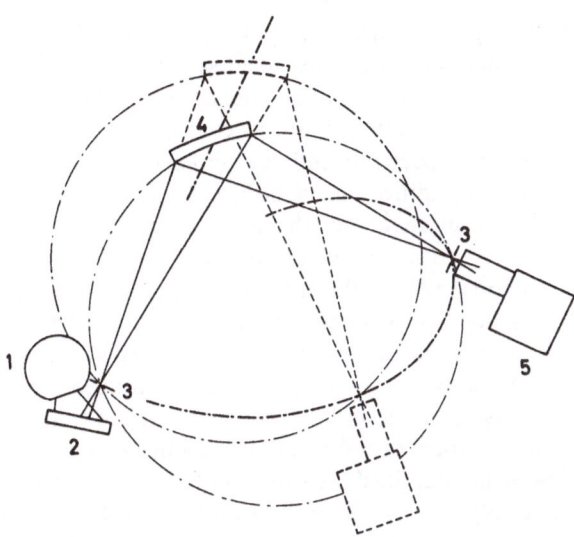

Fig. 10.9. Spectrometer with curved and ground analyzing crystal. (1) X-ray tube; (2) sample; (3) slits; (4) analyzing crystal; and (5) counter. In this construction, the crystal moves along a straight line while being turned. The counter, on the other hand, moves along a rosette-type curve (epicycloid). The center of the focusing circle also moves.

within a narrow limit. Two versions of the spectrometer with curved and ground analyzing crystal are in use. In the first type, which has a focusing circle of fixed center, both crystal and counter move on that circle, and the counter moves twice as fast as the crystal. In contrast to the spectrometer with a flat analyzing crystal, a divergent beam is provided by a slit and the curved analyzing crystal focuses that beam towards the counter. The lattice planes of the analyzing crystal are curved in such a way that the radius of curvature is exactly twice as large as the radius of the circle along which crystal and counter move; the surface of the crystal is ground until it has the same curvature as the circle. A disadvantage of this construction is that different portions of the sample surface are analyzed depending upon the position of the crystal. In the second type, the same area in the sample surface is always analyzed. In order to achieve this, the crystal is moved along a straight line while it is being turned; the counter slit and the counter, on the other hand, move along an epicycloid.

The same principles may be used when constructing units in which the complete measurement of the sample is carried out automatically. In such an instrument, fluorescent lines of several elements may be measured in one run. Two types of automatic units are distinguished, namely the multichannel and single-channel units. A multichannel unit consists of several channels which are centrally located around the sample. Each channel has its own analyzing crystal and counter. The crystal of each channel is adjusted so that only one preselected fluorescent line is reflected from the polychromatic spectrum and registered in the counter. For each channel, the counting conditions may be freely chosen. As all fluorescent lines are measured simultaneously, the total counting time is dependent upon the time required to measure the weakest line. The intensities of all other lines are measured within the same time span. In such units, a large number of samples may be measured in comparatively short time. Their disadvantage is that in order to prevent large intensity losses, no more than about nine different channels may be grouped around the centrally located sample. Although it is rarely necessary to measure more than nine elements simultaneously, a number of additional measurements of background and of lines of internal standards is often required so that the number of elements which actually may be measured simultaneously is further reduced. This disadvantage is eliminated in the single channel unit. This unit consists basically of one spectrometer whose crystal and counter are programmed for certain wavelengths in such a way that the various fluorescent lines are measured one after the other. For every crystal position, the conditions of measurement and of counting may be freely chosen. The total counting time is given by the sum of the counting times of the individual lines.

10.5. Collimators

Collimators serve to select a parallel beam of radiation and are used to accurately provide the angle which the incident fluorescent radiation forms with the analyzing crystal. Usually, collimators consist of a number of parallel lamellas. The longer and the finer the collimators, the smaller is the divergence of the selected beam and the better is the spectral resolution. The intensity of the beam, however, is also reduced. When measuring weak signals, highest possible intensity is desired. Choice of suitable collimators is therefore always a compromise between high resolution and low intensity. Resolution is also strongly affected by the choice of a proper analyzing crystal. In the case of fluorescence of heavy elements, where the intensity is usually sufficient and high resolution is most desirable, collimators with very fine lamellas may be used (for example, 160 μ). For measurement of fluorescent lines of light elements, whose intensities are usually low, collimators with comparatively large distances between individual lamellas are preferred (for example, 480 μ). After passing through the collimator, the beam is not strictly parallel but always retains a small divergence. The divergence angle 2δ of a slightly divergent beam may be calculated from the average distance b of the lamellas and the length of the collimator l:

$$2\,\delta \cong \frac{2\,b}{l}$$

For collimators of, respectively, 160 and 480 μ distance between the lamellas and an effective length of 10 cm, residual divergence of the radiation is calculated to be

$$160\ \mu:\ 2\,\delta = 0.18° \qquad 480\ \mu\ :\ 2\,\delta = 0.55°$$

10.6. Crystals

The fluorescent spectrum emitted by a sample contains lines of different wavelengths. Analysis of the spectrum according to wavelength is carried out with the aid of an analyzing crystal of known lattice parameter and the angle of deflection is measured with a goniometer. The wavelength may be determined by measuring the diffraction angle 2θ of the incident radiation. According to Bragg's law, the following simple relation exists between the angle θ, the lattice parameter d of the crystal, and the wavelength λ:

$$\sin\,\theta = n\,\frac{\lambda}{2\,d}$$

In this equation $n = 1, 2, 3, \ldots$ and represents the orders of reflections. Small values of d result in a strong dispersion of the spectrum and high resolution but are only of use for relatively short wavelengths. The largest wavelength that may be diffracted by a crystal is limited to $\lambda = 2d$ which, for technical reasons, is often reduced to $\lambda = 1.6d$. For quick identification of individual lines, it is common use to tabulate for a given analyzing

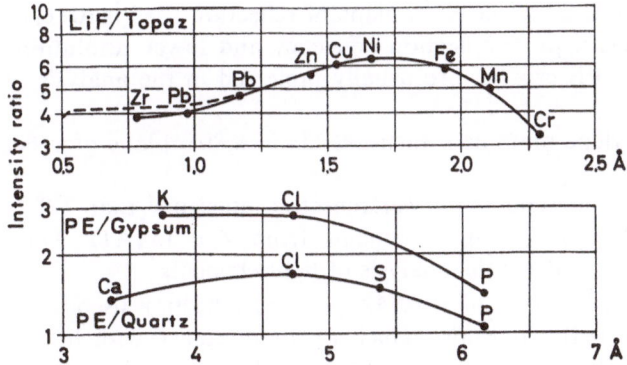

Fig. 10.10. Relative reflection efficiencies of some crystals. Intensity ratio as a function of wavelength obtained with various analyzing crystals (Laffolie, 1961; Louis, 1964a).

crystal (fixed d value) the angles of deflection of all lines in order of their appearance. The degree of reflection, i.e., the intensity of the diffracted radiation, is different from crystal to crystal. Occasionally, improved reflection efficiency is preferred over high resolution. Laffolie (1961) and Louis (1964a) have measured and compared the reflection efficiencies of various crystals (Figure 10.10).

The following analyzing crystals are commonly used:

Topaz (Al silicate): $2d = 2.712$ Å, plane of reflection used is (303).

Topaz gives high resolution with a reflection efficiency of 1/2 to 1/10 that of LiF. Used when high resolution is desirable. Undesirable spurious signals, which are due to reflections from other planes, are often observed when analyzing short wavelengths. (Spielberg and Ladell, 1960; Eckhardt and Fesser, 1961.)

Lithium Fluoride (LiF): $2d = 2.848$ Å, plane of reflection used is (220).

In this mode, LiF has similar resolution and a higher reflection efficiency than topaz. May be used instead of topaz.

Lithium Fluoride (LiF): $2d = 4.028$ Å, plane of reflection used is (200).

In this mode, LiF has very high reflection efficiency combined with reasonably good resolution. This is the most commonly used crystal for all elements from $Z = 20$ (Ca) to $Z = 92$ (U).

Sodium Chloride (NaCl): $2d = 5.640$ Å, plane of reflection used is (200).

This crystal is rarely used. The reflection efficiency for chlorine and sulfur is equal to that of EDDT.

Silicon (Si): $2d = 6.271$ Å, plane of reflection used is (111).

Silicon has an intermediate reflection efficiency and covers the spectra of light elements from $Z = 16$ (S). Second-order reflections do not occur.

Fluorite (CaF$_2$): $2d = 6.30$ Å, plane of reflection used is (111).

Fluorite covers the spectra of the light elements from $Z = 15$ (P).

Quartz (SiO$_2$): $2d = 6.686$ Å, plane of reflection used is (01$\bar{1}$1).

Quartz has a poor reflection efficiency and lower resolution than LiF. EDDT or PE crystals are usually preferred in the analysis of light elements.

EDDT (ethylene diamine-D-tartrate): $2d = 8.808$, plane of reflection used is (020).

EDDT has a good but lower reflection efficiency than LiF and covers the spectra of the light elements from $Z = 13$ (Al). This crystal is commonly used for the analysis of light elements.

PE (pentaery-thritol): $2d = 8.742$, plane of reflection used is (002).

PE has a better reflection efficiency than EDDT with the same resolution. However, it is temperature and radiation sensitive.

ADP (ammonium dihydrogen phosphate): $2d = 10.648$ Å, plane of reflection used is (011).

ADP has a poor reflection efficiency and is not particularly stable. It is mostly used in magnesium analysis.

Gypsum ($CaSO_4 \cdot 2H_2O$): $2d = 15.185$ Å, plane of reflection used is (020). Gypsum has an intermediate reflection efficiency. It covers the spectrum of the light elements from $Z = 11$ (Na).

Beryll (Be–Al silicate): $2d = 15.954$ Å, plane of reflection used is (100). Beryll has a low reflection efficiency and is used for $Z = 11$ (Na) to $Z = 30$ (Zn).

Muscovite (K–Al silicate): $2d = 19.8$ Å, plane of reflection used is (002). May be used in the analysis of fluorine.

KHP (potassium hydrophthalate): $2d = 26.4$ Å, plane of reflection used is ($10\bar{1}0$).
KHP has a good reflection efficiency suitable for the analysis of light elements, particularly Mg, Na, and F.

Ba Stearate and Pb Stearate: $2d \sim 100$ Å.

In order to register the spectra of very light elements, such as B, C, N, O, and F, gratings (600 lines/mm) or soap-type pseudo-crystals consisting of monomolecular layers are employed.

A large number of the commercially available units use flat analyzing crystals. In order to obtain higher reflection efficiencies, however, crystals should be curved, thus focusing the radiation. In order to fulfill the focusing condition the curvature of the crystal together with the incident angle or the wavelength of the radiation have to be simultaneously changed, or else the crystal as a whole must be moved in a circular path. For this reason, application of bent crystals is limited; they are predominantly used when only a particular fixed wavelength is to be analyzed.

The spectral resolution is a measure of the degree of separation of two neighboring lines of wavelength λ and $\lambda + \Delta\lambda$. The resolution A is commonly defined as the ratio of the wavelength λ of a line to its half-width $\Delta\lambda$ (line width at half height). Two lines are considered sufficiently resolved when they are separated by their half-width

$$A = \frac{\lambda}{\Delta\lambda}$$

From Bragg's equation we obtain

$$\lambda = 2\,d\,\sin\theta \quad \text{and} \quad \Delta\lambda = 2\,d\,\cos\theta\,\Delta\theta$$

where $\Delta\theta$ defines the angle at which the intensity of the line has decreased to half its peak value.

For the resolution we obtain

$$A = \frac{\lambda}{\Delta\lambda} = \frac{\tan\theta}{\Delta\theta}$$

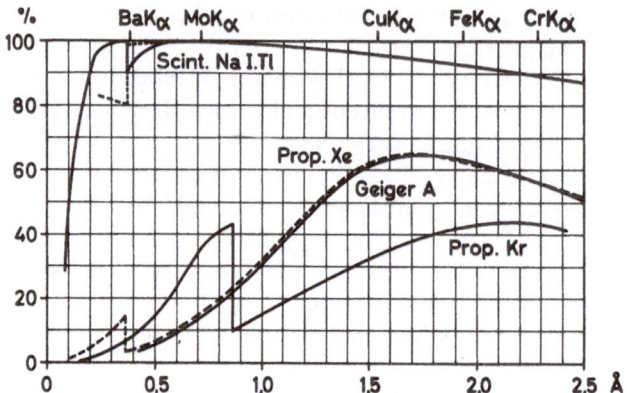

Fig. 10.11. Efficiency of various counters calculated from the absorption coefficient and plotted as a function of the wavelength. (*Geiger A*) Geiger counter with 550 torr argon, effective length of 10 cm, and a 13-μ mica window; (*Prop. K*) proportional counter with 500 torr krypton, effective length of 2.7 cm, and a side window of 130-μ Be and 13-μ mica; (*Prop. Xe*) proportional counter with 300 torr xenon, effective length of 2.7 cm, and a side window of 130-μ Be and 13-μ mica; (*Scint. NaI·Tl*) scintillation counter with Tl-activated NaI crystal of 1-mm thickness with a 130-μ Be layer (Taylor and Parrish, 1955); (– – – –) scintillation counter in the range 0.3–0.5 Å measured by the author.

With this equation, the resolution or selectivity $\Delta\lambda$ of a spectrometer may be calculated. If we assume $\Delta\theta$ to be equal to the residual divergence of the beam which passes through a collimator with a lamella distance of 160 μ, we then obtain for the selectivity of a topaz crystal near the MnK_α line $\Delta\lambda = 0.006$ Å   ($2\theta \approx 101°$; $\lambda = 2.1$ Å). The $MnK_\alpha = 2.103$ Å and $CrK_\beta = 2.085$ Å lines are separated by $\Delta\lambda = 0.018$ Å and, hence, may be resolved with a topaz crystal and a collimator of 160 μ lamella distance.

10.7. Counters

Three types of counters are employed in modern x-ray fluorescence analysis, proportional, flow proportional, and scintillation counters. Geiger counters are not commonly used anymore. Because of the low deadtime of the proportional counter, high counting rates (10^4 to 10^5 counts/sec) may be recorded. The amplitude of a pulse is proportional to the energy of the absorbed photon and, as a result, photons of different quantum energies may be separated by pulse-height analysis. The spectral sensitivity of the proportional counter depends upon the wavelength and is usually limited to a wavelength range of 1.5 to 2.3 Å (CrK_α to CuK_α and CeL_α to TaL_α).

The flow proportional counter is basically a proportional counter whose window consists of a very thin mylar film. The thin mylar film is required to lower the absorption of the incident radiation as much as possible. In order to eliminate gas loss through the thin mylar or polypropylene window, the flow proportional counter is flushed constantly by a steady stream of gas, most commonly argon. The flow proportional counter is used for spectra of the light elements in the wavelength range 1.5–12 Å (NaK_α to CuK_α and SnL_α to TaL_α). Because of the low dead time, high counting rates (10^4 to 10^5 counts/sec) may be recorded. The amplitude of a pulse is proportional to the energy of the absorbed photon and, hence, pulses of different quantum energies or different wavelengths may be separated by pulse-height analysis. Sample, analyzing crystal, and flow proportional counter are normally located in vacuum in order to reduce the strong absorption in air of the long-wavelength radiation.

For the wavelength range 0.3–2.5 Å (VK_α to SmK_α and CeL_α to UL_α), scintillation counters are most commonly used. Owing to low dead time (less than 1 μsec), counting rates of 10^4 to 10^5 counts/sec may be measured with a spectral sensitivity of nearly 100% over the whole wavelength range. The amplitude of a pulse is proportional to the energy of the absorbed photon and, hence, pulses of different quantum energies (different wavelengths) may be separated by pulse-height analysis. Laffolie (1961) and Louis (1964a) have compared efficiencies of scintillation and flow proportional counters for various wavelengths (Figure 10.13).

The Geiger counter has a large dead time (200 μsec) so that only counting rates up to 10^3 counts/sec may be counted. The spectral sensitivity is about equal to that of the proportional counter but the amplitude of the pulses is larger by about a factor of 10^3. Owing to the considerably higher

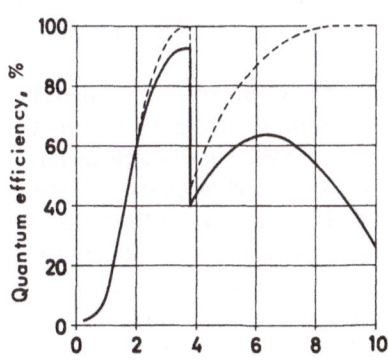

Fig. 10.12. Efficiency of an argon-filled flow proportional counter (760 torr; 90% Ar + 10% methane; effective length of 24 mm). (– – – –) Absorption by the counter gas; (——) absorption by the counter gas after absorption of the incident radiation by a 6-μ thick mylar foil. (Courtesy of Neff, 1959c.)

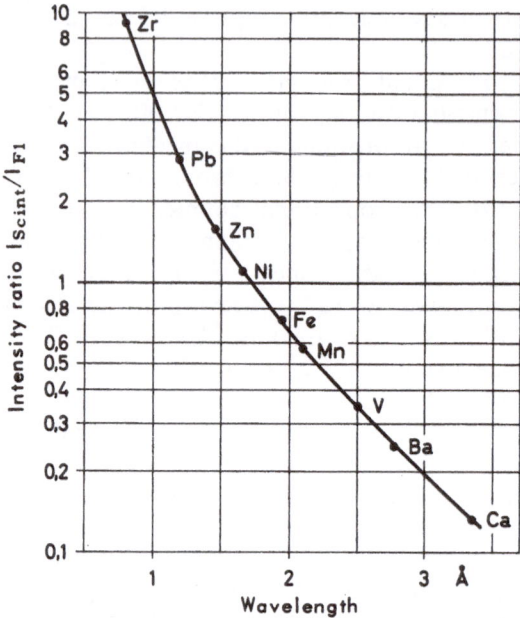

Fig. 10.13. Counters. Intensity ratios obtained with a scintillation counter and an argon-filled flow proportional counter as a function of wavelength (Laffolie, 1961; Louis, 1964*a*). It should be noted that in the case of these measurements, the scintillation counter was located outside the spectrometer while the flow proportional counter was within the spectrometer.

pulse amplitude there is no relation between output amplitude and the energy of the absorbed quantum and, furthermore, no need for very sophisticated electronics. A detailed description of counters may be found by Fünfer and Neuert (1954), Tayler and Parrish (1955), Parrish and Kohler (1956), and Neff (1959*a, c*).

Every quantum absorbed in a counter results in an electrical pulse. These pulses may be counted by common counting electronics, and the intensity of the radiation reaching the counter may be measured in terms of number of pulses. Pulses may, however, also be used to charge a capacitor for a predetermined time of measurement. The intensity of the radiation which reaches the counter, is then measured in terms of voltage. Flow proportional and scintillation counters with associated electronics can register counting rates of 10^4 to 10^5 counts/sec. When counting rates are higher than about 2×10^4 counts/sec, noticeable counting losses occur, i.e., not every photon absorbed in the counter is actually registered. This phenomenon will be discussed in more detail in Chapter 11 which deals with counting statistics and counting losses.

10.8. Pulse-Height Analysis and Discrimination

Amplitudes of the electrical pulses obtained from proportional, flow proportional, and scintillation counters are proportional to the energies of the absorbed photons. A simple relation exists between the energy $h\nu$ and the wavelength λ of the photon

$$E_{\text{Phot}} = h\nu = hc/\lambda$$

where c is the velocity of light. Radiation in x-ray fluorescence analysis is normally identified on the basis of its wavelength. The analyzing crystal, however, cannot separate two different radiations whose wavelength λ_1 and λ_2 are given as follows:

$$\lambda_1 = n \cdot \lambda_2 \qquad n = 1, 2, 3, \ldots$$

For both wavelengths, the condition for reflection

$$2 d \sin \theta = n\lambda$$

is fulfilled for the same angle θ, and only the values of n (orders of reflection) are different. Thus, the long-wavelength radiation overlaps the short-wavelength radiation in the spectrometer. Using an electronic discriminator, however, pulse-height analysis may be used to eliminate either of the two radiations. Their quantum energies and the corresponding amplitudes are given as $1 : n$. In a hafnium-containing zirconium mixture, for example, the $\text{Hf}L_\alpha$ line ($\lambda = 1.569$ Å) is overlapped by the second-order reflection of the $\text{Zr}K_{\alpha 1}$ line ($\lambda = 2 \times 0.786 = 1.572$ Å). However, the interfering line may be eliminated using pulse-height discrimination (Figure 4.7). Pulse-height discrimination may also be used to lower the general background radiation so that weak lines are often more distinctly separated from the common scattered background.

In pulse-height discriminators one usually distinguishes between a "window" or "channel," and an average or lower "window" height or "channel height." "Window" or "channel" may be changed arbitrarily in width and height in order to separate those pulses from the spectrum which are located within the chosen interval. Thus, pulse-height distribution of radiation may easily be experimentally determined. Only a narrow range of pulse heights is allowed to enter the counting electronics (registration with a narrow window) and the height of the window is varied in steps. In this way the frequency distribution of the average pulse heights is ob-

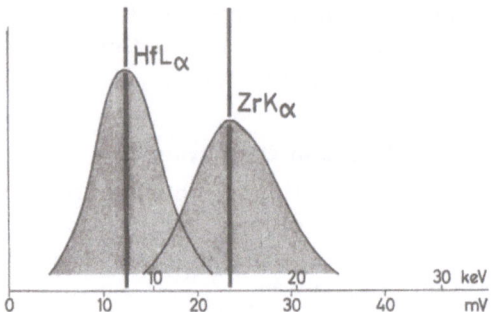

Fig. 10.14. Pulse-height distribution for the ZrK_α and HfL_α lines registered with a scintillation counter. Incident radiation originally reaching the counter is monochromatic (symbolized as a vertical line); in the counter and associated electronics, however, it is broadened to yield a normal (Gaussian) distribution. Both the HfL_α and the second-order ZrK_α line appear, in the spectrometer, to have the same wavelength and overlap each other. Using pulse-height analysis and discrimination, however, either one of the two radiations may be eliminated because the ratios of effective wavelengths and quantum energies, respectively, are 1:2.

tained in the form of a normal (Gaussian) distribution (Figure 10.14). When the pulse-height distribution of monochromatic radiation is determined with a proportional or flow proportional counter, pulses of low amplitude are detected in addition to the normal pulse heights. These additional pulses represent the so-called escape peak. In order to explain the escape peak, the process of absorption of a photon has to be considered in somewhat more detail. When a photon is absorbed, one atom of the counter gas is ionized in its K energy level. One electron of the K energy level is given the total energy $h\nu$ of the photon in form of kinetic energy, $\frac{1}{2}mv^2$, minus the ionization work W which is necessary to remove the electron from the atom

$$\frac{1}{2}mv^2 = h\nu - W$$

This photoelectron of the energy $\frac{1}{2}mv^2$ collides with other atoms and ionizes their outermost shell. The number of secondary ions formed in this process is proportional to the kinetic energy of the original photoelectron. By expelling a second electron (Auger electron) from the atom which originally was ionized in the K energy level, the atom returns within approximately 10^{-8} sec to the ground state. The kinetic energy of the Auger electron is approximately equal to the ionization work W. Just as the photoelectron, the Auger electron as well ionizes further atoms on its path. The total number N of the ions is therefore proportional to the sum

$$N = \text{prop}\left(\frac{1}{2}\,mv^2 + W\right) = \text{prop}\,h\nu$$

and, hence, proportional to the originally absorbed energy of the photon. However, it is also possible that the original atom which was ionized in the K energy level, emits a fluorescent photon. In all probability, its energy is now too low for ionization of further atoms (because the fluorescent photon escapes from the counter) and the total number of ions N' which is formed by absorption of the quantum energy $h\nu$ is only

$$N' = \text{prop}\,(h\nu - W)$$

These ions are accelerated in the electric field of the counter and, when reaching the counter wire, produce an electric pulse whose amplitude is proportional to the number of ions. At the counter exit, pulses of the following amplitudes occur:

$$A = \text{prop}\,h\nu$$
$$A' = \text{prop}\,(h\nu - W)$$

The pulses of amplitude A correspond to the energy of the absorbed photons, and the pulses of amplitude A' correspond to the escape peak. For a counter filled with argon the value for W is approximately 3 keV, and for a NaI scintillation counter it is approximately 29 keV.

Pulse-height analysis may also be employed to identify the fluorescent radiation of an unknown sample without using an analyzing crystal (socalled nondispersive geometry). Quantum energy and relative pulse height, respectively, are proportional to the reciprocal wavelength. However, energy curves of neighboring fluorescent lines overlap in pulse-height analysis because the energy resolution of common counters is rather low. Dolby (1959) and Zemany (1960) describe two methods to unfold overlapping spectra in order to obtain the contributions made by individual lines. Birks and Batt (1963), and Birks, Labrie, and Criss (1966) illustrate Dolby's method for a number of examples.

It is important to note that the average pulse height at high counting rates shifts towards smaller values; this is due to the fact that the recovery time of the counter for two successive incident photons is too low. Amplitude shift occurs at those counting rates for which appreciable counting losses are recorded.

10.9. Filters

In contrast to pulse-height discriminators, filters often allow discrimination of lines of very similar wavelength. If, for example, it is neces-

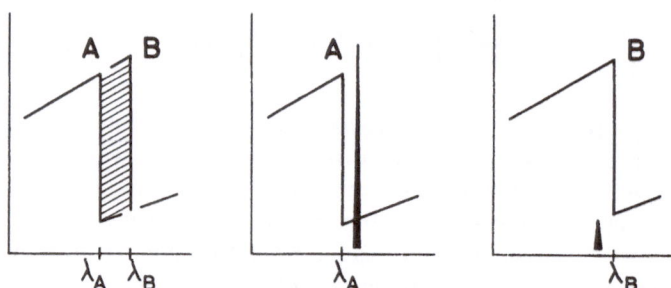

Fig. 10.15. Filter-difference method. Using two filters of proper choice it is possible to measure selectively the spectral intensity in the wavelength range between λ_A and λ_B. The approximate spectral intensity in the interval $\lambda_A - \lambda_B$ is obtained as the difference of two measurements. This difference, in addition, has to be corrected for the absorption during measurement with a filter A in order to obtain the true spectral intensity.

sary to determine small amounts of tantalum in a mixture with large amounts of tungsten, a nickel filter of approximately 20 μ in thickness reduces the WL_α line to 0.2% of its original intensity. At the same time, the intensity of the TaL_α line is reduced to only 44% of its original intensity. A nickel filter 24 μ in thickness may be employed in the analysis of brass alloys to nearly completely suppress the ZnK_α line, while the neighboring CuK_α line is reduced in intensity to only 35% of its original value (Tanemura, 1961). Analogously, a 10-μ aluminum filter suppresses the SiK_α line nearly completely, while the AlK_α line is absorbed by only about 50%.

Natelson *et al.* (1962) mounted a 0.1-mm thick titanium foil directly onto the x-ray tube in order to suppress the undesired CrK_α and WL_α lines of the primary x-ray radiation. As a result, the peak-to-background ratios for measurement of trace amounts of Cr, Mn, Fe, Co and Cu are considerably improved. Leroux and Mahmud (1967a) mounted a nickel foil of 45 μ thickness directly on a tungsten tube to suppress the characteristic tungsten lines of the tube when analyzing for tungsten and trace amounts of zinc and lead.

Filters may also be used to select a narrow wavelength range from a polychromatic spectrum and enable measurement of its intensity without taking recourse to spectral analysis by a crystal. For this method, two balanced filters will suffice [filter-difference method (Ross, 1926; Mecke, 1963)]. The absorption edge of one filter limits the wavelength range towards the short-wavelength side, and the absorption edge of the second filter limits the wavelength range towards the high-wavelength side. The thickness or mass density of the filter is chosen so that both filters absorb equally well outside the range in question (Figure 10.15). The intensity of the radiation is measured with a short wavelength filter and then with a long-wavelength filter. The difference between the two intensity measurements corresponds to the intensity within the selected wavelength range. As an example, the measurement of the intensity of the tin radiation in an ore sample is discussed. The filter pair, palladium–silver, may be used to measure selectively the intensity in the wavelength range from $\lambda = 0.486$ to 0.509 Å. This range contains the SnK_α line of wavelength $\lambda = 0.492$ Å. A palladium foil of 0.08 mm thickness absorbs radiation in this wavelength range to about 0.5% of the original value, while a silver foil of 0.0806 mm thickness absorbs the radiation in this particular wavelength range to only 41% of the original intensity. Analogously, a manganese–iron filter pair suppresses radiation in the wavelength range from 1.743 to 1.896 Å and makes it possible to selectively measure the CoK_α line ($\lambda = 1.789$ Å).

In the so-called filter scan technique, instead of two balanced filters Dothie (1962) used a whole set of tuned filters placed in sequence into the beam. If the intensity changes suddenly upon insertion of a certain filter, then a strong emission line is present in the corresponding wavelength interval which can easily be recognized and identified. In favorable cases, a combination of filters and pulse-height discriminators allows analysis without the use of an analyzing crystal.

Chapter 11

Measurement Techniques

11.1. Principles of Statistics

Just as any other technique of measurement, the determination of the elemental content of a sample is governed by the laws of statistics. It is for this reason that the principles of statistics are treated briefly in this chapter. Further discussions may be found in Kaiser and Specker (1956), Linder (1957), Davies (1957), and Neff (1961).

Every determination results in a number which is designated x. If further determinations are carried out independently from the first measurement we obtain the numbers $x_1, x_2, \ldots, x_n$ which represent a random sample of the population n. The scatter of the individual numbers are a measure of the accuracy of the method and allows conclusions to be drawn as to the confidence of the analytical result.

Individual measurement → number x
Repetition of the measurement → numbers $x_1, x_2, \ldots, x_n$
Scatter → accuracy of the result
Test procedure → confidence of the result

A random sample is best characterized by three terms: mean, scatter, and population (number of individual measurements). The sequence of the individual numbers within a random sample is of no importance. The best approximation for the desired number x is the mean $\bar{x}$:

$$E(x) = \bar{x} = \sum^{n} \frac{x_i}{n} \tag{11.1}$$

The variance and the standard deviation (the latter may be derived from the former) are a measure for the scatter

$$\text{Variance} \quad s^2 = \sum^{n} \frac{(x - \bar{x})^2}{n - 1}$$

$$\text{Standard deviation} \quad s = \sqrt{\sum^{n} \frac{(x - \bar{x})^2}{n - 1}} \tag{11.2}$$

Let us consider a second random sample of measurements. Just as the individual measurements vary, so do the means which are calculated from the individual sets of measurements; however, the variation of the latter is less than that of the former. The variance and standard deviation, respectively, of the mean are given as follows:

$$\text{Variance of the mean} \quad s_D^2 = \frac{s^2}{n}$$

$$\text{Standard deviation of the mean} \quad s_D = \frac{s}{\sqrt{n}} \tag{11.3}$$

where n is the population of the random sample from which the mean is calculated.

We must distinguish between total population and random sample. The total population contains all values which determine a result unambiguously, while the random sample contains only a limited number of individual measurements of the total population. Customarily, values of the total population are designated with Greek letters, and those of the random sample with Latin letters. The total population is described by the mean μ, the scatter σ, and the total number of measurement ν. It can be shown that the mean and the scatter, calculated from a random sample, are the best approximations for the mean and the scatter, respectively, of the total population.

$\bar{x}$ is the best approximation for μ

s^2 is the best approximation for σ^2

Normal Distribution and Statistical Test Procedures

For the following considerations, it is assumed that the total population of the individual measurements follows a Gaussian (normal) distribution. Let us consider the distribution of the individual measured numbers in the total population. The total number of measurements in the total population is infinite. The Gaussian distribution of the value x is given by the following equation:

$$\Phi(x) = \frac{1}{\sigma \sqrt{2\pi}} \cdot \exp\left[-\frac{(x-\mu)^2}{2\sigma^2}\right] \tag{11.4}$$

By coordination transformation of the type

$$u = \frac{x-\mu}{\sigma} \tag{11.5}$$

we obtain the standardized normal distribution

$$\Phi(u) = \frac{1}{\sqrt{2\pi}} \exp\left[-\frac{u^2}{2}\right] \tag{11.6}$$

The factor $1/\sqrt{2\pi}$ is chosen in such a way that the area under the curve is equal to 1

$$\int_{-\infty}^{+\infty} \Phi(u)\, du = 1$$

The value Φ as a function of u may be found in tables; it is the well-known Gaussian curve (Figure 11.1). More important, however, are tables which contain specific integrals for $\Phi(u)$. Commonly used are the following values and their corresponding tables:

$$P_D = 1 - \int_{-u}^{+u} \Phi(u)\, du \tag{11.7}$$

This corresponds to the area which is under the curve $\Phi(u)$ but outside the limits $-u$ and $+u$. The term P_D is referred to as the residual probability;

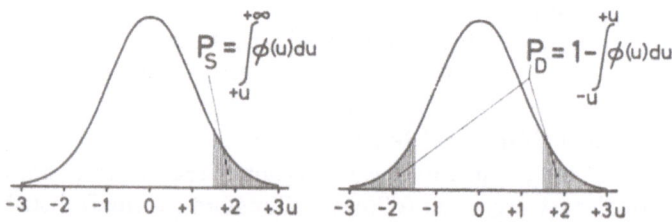

Fig. 11.1. Standardized normal distribution of measured values around the mean. The functions P_S and P_D are used to statistically test a measured value; they determine the probability of a measured value to be by chance in the cross-hatched area.

it determines the probability of a value from outside these limits to still belong to the same total population. The term P_D is the answer. Since limits exist on both sides ($-u$ and $+u$), P_D is referred to as the so-called two-sided limiting value. A further term is in common use

$$P_S = \int_{+u}^{+\infty} \Phi(u)\, du \qquad (11.8)$$

The term P_S determines the probability of a value outside the limit $+u$ to still belong to the same total population. As only one limiting value is determined, the term P_S is referred to as one-sided limiting value. It can easily be shown that

$$P_D = 2\,P_S \qquad (11.9)$$

The following question may be answered with the aid of the u-distribution: What is the range of the limits in which 95% of all individual values are located (question as to the scattering range)? Conversely, this question may be phrased as follows: A measured value x is given. What is the range of the limits which in 95% of all cases contains the mean μ? What is the probability of a value larger than the limiting value u to still belong to the same total population (question of detection limit)?

Answer to the first question: 95% of all values are to be located within the interval $-u$ to $+u$, and 5% of the values are to be located outside this limit. Since two limits are given the term P_D is chosen. For $P_D = 0.05$, the corresponding value for u ($u = 1.960$) may be taken from Table 11.1. The scattering range which, with a probability of 95%, contains all individual measured values may be determined using equation (11.5):

$$u = \frac{x - \mu}{\sigma}$$

and is found to be as follows:

$$x_{1,2} = \mu \pm 1.960\,\sigma$$

In this equation μ is the mean and σ the standard deviation of the total population. If the scattering range is to be calculated for, respectively, 90% and 99% of the values, then the corresponding u-values are found by taking $P_D = 0.10$ and 0.01, respectively.

The second problem originates from a reversion of the first: when measuring a number, the mean of the particular total population is unknown (in fact, the measurement is made to determine this value). We therefore have to consider the probability distribution of the true value or of the mean μ around a measured value x and it can be shown mathematically that the same distribution function results. If the standard deviation of the analytical technique is σ, then the desired mean μ is, with 95% confidence, within the scattering range

$$x \pm 1.960\ \sigma$$

Let us attempt to solve the third problem, which concerns the probability of a value larger than u to belong to the same total population. As there is only one limit, the term P_S applies. If, for example, a value of $u = 3.090$ is chosen, then $P_S = 0.001$. The probability, however, is so low that we consider occurrence of such an event to be most unlikely. Whenever a measurement yields a u-value larger than 3, we consider this value to be larger than the total population in question and, hence, not to belong to the same population. This limiting value may be found from equation (11.5).

$$x_G = \mu + u\sigma$$

With this equation the detection limit may be determined. The detection limit is that concentration for which a measured value results which is u times larger than the statistical variation of the background or of the dry run. The particular u-value beyond which the occurrence of a certain event is considered improbable, is chosen rather arbitrarily. The probability of 1% is often used as a boundary value in statistics. In many spectroscopic papers by Kaiser and Specker (1956) and Neff (1961), a u-value of 3 is used; this corresponds to a probability of 1.35 per thousand.

t-Distribution and Corresponding Test Procedures

So far, we have assumed that the mean μ and the scatter of the total population σ^2 are known. In most instances, however, this is not the case, and we are forced to rely on the approximate values $\bar{x}$ or s^2 which are obtained from a random sample. Furthermore, the number of individual measurements in a random sample are often limited and, due to the considerable work effort involved in most analytical procedures, 10 to 20 measurements are usually considered an adequate test for the particular analytical method. A statistical test procedure, however, may be applied in such a case and in place of the term u, the term $t(n)$ is defined

$$t(n) = \frac{x - \mu}{s} \qquad (11.10)$$

where the standard deviation s is determined from a random sample consisting of n measurements. The t-distribution is also symmetric, but some-

Table 11.1. Normal Distribution. Values of u for various probabilities P_D and P_s (P_s always equals $\frac{1}{2}P_D$).

P_D	.00	.01	.02	.03	.04	.05	.06	.07	.08	.09
.0	∞	2.576	2.326	2.170	2.054	1.960	1.881	1.812	1.751	1.695
.1	1.645	1.598	1.555	1.514	1.476	1.440	1.405	1.372	1.341	1.311
.2	1.282	1.254	1.227	1.200	1.175	1.150	1.126	1.103	1.080	1.058
.3	1.036	1.015	0.994	0.974	0.954	0.935	0.915	0.896	0.878	0.860
.4	0.842	0.824	0.806	0.789	0.772	0.755	0.739	0.722	0.706	0.690
.5	0.674	0.659	0.643	0.628	0.613	0.598	0.583	0.568	0.553	0.539
.6	0.524	0.510	0.496	0.482	0.468	0.454	0.440	0.426	0.412	0.399
.7	0.385	0.372	0.358	0.345	0.332	0.319	0.305	0.292	0.279	0.266
.8	0.253	0.240	0.228	0.215	0.202	0.189	0.176	0.164	0.151	0.138
.9	0.126	0.113	0.100	0.088	0.075	0.063	0.050	0.038	0.025	0.013

Low probability

P_s	0.000	0.001	0.002	0.003	0.004	0.005	0.006	0.007	0.008	0.009
P_D	0.000	0.002	0.004	0.006	0.008	0.010	0.012	0.014	0.016	0.018
u	∞	3.090	2.878	2.748	2.652	2.576	2.512	2.457	2.409	2.366

Very low probability

P_s	0.0000005	0.000005	0.00005	0.0005
P_D	0.000001	0.00001	0.0001	0.001
u	4.892	4.417	3.891	3.291

what more shallow than the corresponding normal distribution. The larger the individual number of measurements of the random sample from which the value for s is obtained, the closer the t-distribution approximates a normal distribution, until for $n = \infty$ it is identical to it. The magnitude of $t(n)$ which is to be chosen for a particular probability, depends upon the number of measurements from which the standard deviation is calculated. As an exercise, the same three problems are discussed again: What is the range of scatter which contains 95% of all measured values? It is assumed that the standard deviation is calculated from 10 individual measurements. Table 11.2 lists the values for P_S and P_D

$$P_S = \int_t^\infty \Phi(t)\, dt \qquad P_D = 1 - \int_{-t}^{+t} \Phi(t)\, dt \qquad (11.11)$$

For a population of the random sample of $n = 10$ and $P_D = 0.05$, a $t(n)$-value of 2.26 is found. It may, therefore, be assumed that at least 95% of all measured values are located within the range

$$x_{1,2} = \mu \pm 2.26\, s$$

Conversely, it may also be assumed that in at least 95% of all cases, the desired mean μ is within the range

$$x \pm 2.26\,s$$

where x is an arbitrarily chosen measured value. A boundary value for the detection limit is obtained as follows:

$$x_G = \mu + t(n)s \cong \bar{x} + t(n)s$$

The limited population of the random sample requires that larger ranges are considered in order to be confident of the result.

Error Propagation

Let us consider two measured values x_A and x_B as well as the respective standard deviations s_A and s_B. The relative standard deviations ϵ are

$$\varepsilon_A = \frac{s_A}{x_A} \qquad\qquad \varepsilon_B = \frac{s_B}{x_B}$$

Table 11.2. *t*-Distribution. Values of $t(n)$ as a function of the population of the random sample for various probabilities P_S and P_D.

P_S	0.1	0.05	0.025	0.01	0.005	0.0025	0.0005
n P_D	0.2	0.1	0.05	0.02	0.01	0.005	0.001
2	3.08	6.31	12.7	31.8	63.7	127.3	636.6
3	1.89	2.92	4.30	6.96	9.92	14.1	31.6
4	1.64	2.35	3.18	4.54	5.84	7.45	12.9
5	1.53	2.13	2.78	3.75	4.60	5.60	8.61
6	1.48	2.01	2.57	3.36	4.03	4.77	6.87
7	1.44	1.94	2.45	3.14	3.71	4.32	5.96
8	1.42	1.89	2.36	3.00	3.50	4.03	5.41
9	1.40	1.86	2.31	2.90	3.36	3.83	5.04
10	1.38	1.83	2.26	2.82	3.25	3.69	4.78
11	1.37	1.81	2.23	2.76	3.17	3.58	4.59
12	1.36	1.80	2.20	2.72	3.11	3.50	4.44
13	1.36	1.78	2.18	2.68	3.05	3.43	4.32
14	1.35	1.77	2.16	2.65	3.01	3.37	4.22
15	1.34	1.76	2.14	2.62	2.98	3.33	4.14
16	1.34	1.75	2.13	2.60	2.95	3.29	4.07
17	1.34	1.75	2.12	2.58	2.92	3.25	4.02
18	1.33	1.74	2.11	2.57	2.90	3.22	3.97
19	1.33	1.73	2.10	2.55	2.88	3.20	3.92
20	1.33	1.73	2.09	2.54	2.86	3.17	3.88
25	1.32	1.71	2.06	2.49	2.80	3.09	3.75
30	1.31	1.70	2.05	2.46	2.76	3.04	3.66
40	1.30	1.68	2.02	2.43	2.71	2.98	3.56
60	1.30	1.67	2.00	2.39	2.66	2.92	3.46
120	1.29	1.66	1.98	2.36	2.62	2.86	3.37
∞	1.28	1.64	1.96	2.33	2.58	2.81	3.29

and the standard deviations in $\%$ are $100\epsilon_A$ and $100\epsilon_B$. We calculate the standard deviation for a new term formed from the two measured values. If the sum or the difference is formed, then the standard deviation of the new term is

$$x_A \pm x_B: \quad s = \sqrt{s_A^2 + s_B^2} \tag{11.12}$$

The relative standard deviation is given as follows:

$$x_A \pm x_B: \quad \varepsilon = \frac{\sqrt{(x_A \varepsilon_A)^2 + (x_B \varepsilon_B)^2}}{x_A \pm x_B} \tag{11.13}$$

When forming the product or the quotient, the absolute standard deviation of the new number is obtained as follows:

$$x_A \cdot x_B^{\pm 1}: \quad s = x_A \cdot x_B^{\pm 1} \sqrt{\left(\frac{s_A}{x_A}\right)^2 + \left(\frac{s_B}{x_B}\right)^2} \tag{11.14}$$

Analogously, the relative standard deviation is

$$x_A \cdot x_B^{\pm 1}: \quad \varepsilon = \sqrt{\varepsilon_A^2 + \varepsilon_B^2} \tag{11.15}$$

If several measured values are combined, then the errors are propagated according to the above-described rules. Depending upon the type of combination it is therefore desirable to calculate with the absolute standard deviations s or with the relative standard deviations ϵ. In the case of sums or differences, the absolute standard deviations s are preferred; for product and quotient formation, on the other hand, the relative standard deviations ϵ are preferred. In order to keep the resulting total error as low as possible it is desirable that all errors of the individual measurements are approximately of the same magnitude, because the contribution of the largest error of an individual measurement to the total error is relatively high.

11.2. Counting Statistics and Counting Losses

Intensities of spectral lines are most commonly registered with counters, where the incident x-ray quanta are counted as electrical pulses. The intensity I in terms of energy transport per unit time may easily be calculated from the pulse rate or the number of pulses by multiplying the pulse rate N with the quantum energy of the radiation $h\nu$.

$$I = N h \nu \quad \text{[erg]}$$

The word "intensity" is often used to describe the pulse rate (i.e., number of pulses per unit time). In order to distinguish clearly between these two terms we designate the pulse rate (number of counts per unit time) by N and the intensity in the strict sense of the word by I.

Measurement of fluorescent intensity is only possible with a certain accuracy and various effects contribute to its limitations: variations in number of pulses per unit time as a result of counting statistics; short-term drift in the apparatus; and variability from determination to determination as a result of differences in the apparatus and sample preparation. Furthermore, long-term drift is important in cases where a longer period of time elapses between two individual measurements. In this chapter, however, only errors resulting from counting statistics and counting losses are considered. The number of counts measured in two equally long time intervals is usually not the same. This is because individual pulses do not arrive at a uniform rate; rather, their occurrence, in time, varies statistically. Hence, the number of pulses accumulated for fixed time intervals is not a constant and forms a Poisson distribution. It is for this reason that even in the case of an ideal apparatus the average pulse rate or the total number of pulses per unit time T can only be measured with a certain error. Two methods exist for the measurement of the intensity: the measurement may be carried out either on the basis of a fixed-time interval, or on the basis of a fixed-pulse number. In the fixed-time method, the time during which the incident pulses are counted, is predetermined, where either the total number of pulses accumulated in this time interval or the average pulse rate (number of pulses per unit time) is determined. In the case of the fixed pulse number method, a certain number of pulses is chosen and the time necessary to accumulate the pre-selected number of pulses is measured. The counting error in both procedures depends upon the total number of accumulated pulses, and the standard deviation is chosen as a measure of that error. If N is the total number of accumulated pulses, then the corresponding standard deviation of the Poisson distribution is $\sqrt{N}$; the desired value is then with 68 % probability within the range N $\pm$ $\sqrt{N}$; with 95 % probability in the range N $\pm$ $2\sqrt{N}$; and with 99.7 % probability in the range N $\pm$ $3\sqrt{N}$. Depending upon the technique of measurement, however, the total number of accumulated pulses N is not obtained directly; rather, the average pulse rate N during the counting time T or the counting time T itself, may be obtained. The simple proportionality N = NT exists between these three terms, so that the standard deviation corresponding to the other measured values may be calculated. Below, the measured values and the corresponding formulas for the standard deviations are given for every method of measurement.

The accuracy of a measurement increases with increasing number of pulses and increasing length of time of measurement. For 10,000 pulses, for example, the standard deviation is 100 pulses or 1 %; for 100,000 pulses it is

Method of measurement	Measured values and corresponding standard deviation	
Preselected time of measurement T	Measurement of number of pulses accumulated per preselected time interval	N (pulses)
	Corresponding standard deviation	$\pm\sqrt{N}$
	Standard deviation in percent	$\pm 100\sqrt{1/N}$
Preselected time of measurement	Measurement of average pulse rate per time interval T	N (pulses/second)
	Corresponding standard deviation	$\pm\sqrt{N/T}$
	Standard deviation in per cent	$\pm 100\sqrt{1/NT}$
Preselected number of pulses N	Measurement of time required to accumulate preselected number of pulses	T (sec/pulse)
	Corresponding standard deviation	$\pm\sqrt{T^2/N}$
	Standard deviation in %	$\pm 100\sqrt{1/N}$

316 pulses or 0.3%; and for 1,000,000 pulses it is 1000 pulses or 0.1%. However, it is usually not meaningful to attempt lowering the standard deviation resulting from counting statistics to below 0.1–0.3%; drift in the apparatus is frequently larger than 0.1–0.3% and accuracy of measurement is therefore often determined by apparatus drift rather than by counting statistics. By measuring the same fluorescent line 100 to 200 times it is possible to test experimentally whether the standard deviation approaches the theoretical value. As a rule, the value for the standard deviation of the counting statistics is somewhat larger than the theoretically expected value because short-term drift of the apparatus is often superimposed onto the counting statistics, thus increasing the counting error (the author's measurements yield numbers which are 1.25 to 1.7 times as large as the theoretically expected values).

Counting Loss in Case of High Counting Rates

It is an experimental fact that a photon which is absorbed and registered by a detector prompts this detector to be insensitive to additional photons for a very short period of time. After a counter receives a photon its counting efficiency for addition photons is essentially 0 but it then increases to its original value. Certain shielding phenomena in the counter as well as the sluggishness of the associated electronics are responsible for this effect. Let us consider in more detail the behavior of the counting apparatus after a photon has entered the counter. For this purpose the term "idealized counting apparatus" is introduced. After a photon enters the counter, the sensitivity of the idealized counting apparatus drops to zero, remains zero, and after a certain time interval t_D (dead time) increases to its original value. After incidence of a photon, a certain time span elapses until the next photon arrives. For the purpose of calculating the counting efficiency we consider two boundary cases: In the first case, a photon arriving within the dead time t_D of a previous pulse is not detected and this new photon again restricts the counting apparatus for time t_D, i.e., the dead time begins again and the time span in which no pulse can be registered is prolonged. If N is the number of incident and N' the number of registered pulses, then the counting efficiency N'/N is found as follows:

$$\frac{N'}{N} = \exp\ [-Nt_D]$$

In the second case, the photon arriving during the dead time is also not detected, but in this case the new pulse does not reinitiate the dead time. The maximum dead time is then t_D and the count yield is as follows:

$$\frac{N'}{N} = \frac{1}{1 + Nt_D} \quad \text{or} \quad N = \frac{N'}{1 - N't_D}$$

The value $(N - N')/N$ is referred to as the counting loss. In practice, scintillation counters commonly have counting losses of 15% for pulse rates of 10^5 counts/sec. In Figure 11.2, N'/N is plotted as a function of N'.

Additivity of Correction Factors

The magnitude of the counting loss may be determined experimentally by several methods (Lonsdale, 1948; Short, 1960; Sawatzky and Jones, 1967). Other than the multifilter technique of Lonsdale (1948), the following procedures are to be considered:

1. *Method employing a filter or aperture.* A thin metal foil (filter) or an aperture are required and the intensity of a line is measured with and without filter for different counting rates (the filter is not mounted directly in front of the counter entrance slit in order to avoid simultaneous recording of the fluorescent radiation originating from the metal of the filter). The "absorption factor" and the ratios of the two intensities, respectively, remain constant as long as no counting losses occur. When plotting this ratio as a function of the unfiltered counting rates, a straight line originates as long as the counter operates without counting losses; this plot becomes nonlinear, however, when counting losses occur. Let us designate N_1 and N_1' as the true and measured counting rates, respectively, for the nonfiltered

radiation, and N_2 and N'_2 as the true and measured counts, respectively, for the radiation after passing through the filter. Following the counting loss-free range in the plot is a range for which

$$N'_1 < N_1 \quad \text{and} \quad N'_2 = N_2$$

In order to determine the true value it is necessary to calculate the correction factor F by which the measured value N'_1 has to be multiplied to obtain N_1, and to plot F as a function of N'_1. With (N_2/N_1) as the intensity ratio in the counting loss-free range, we obtain

$$N_1 = N'_1 \underbrace{(N'_2/N_1)/(N_2/N_1)}_{\text{correction factor}} = N'_1 \times F(N'_1)$$

The following relation exists when the filtered radiation approaches high counting rates where counting losses occur:

$$N'_1 < N_1 \quad \text{and} \quad N'_2 < N_2$$

Under these circumstances, the absorption effectiveness of filter and aperture, respectively, has to be increased so that the intensity of the filtered radiation is once again registered in the linear counting range. If this is not the case, the observed counting rate N'_2 has to be corrected using the previously obtained data, and the following relation has to be considered to derive the true counting rate N_1:

$$N_1 = N'_1 \times \frac{(N'_2/N)'_1 \times F(N'_2)}{(N_2/N_1)} = N'_1 \times F(N'_1)$$

2. *Method employing first and second order reflections.* The intensity ratio of first- and second-order reflections of a particular line is a constant. This ratio is determined for different counting rates and constant values are obtained as long as there is no counting loss. The ratio changes, however, when counting losses occur, and the numerical determination of the counting loss is then analogous to the previously described filter method.

3. *Method employing two samples.* Two identical samples, but masked to expose surface areas of different sizes, are employed resulting in emission of different fluorescent intensities. The determination of the counting loss is again analogous to the previously described filter method.

II.3. Background Correction

In order to obtain the true intensity of the fluorescent radiation, the measured intensity value has to be corrected for the background radiation. The background is superimposed onto the fluorescent line and consists largely of diffusely scattered primary tube radiation. The intensity

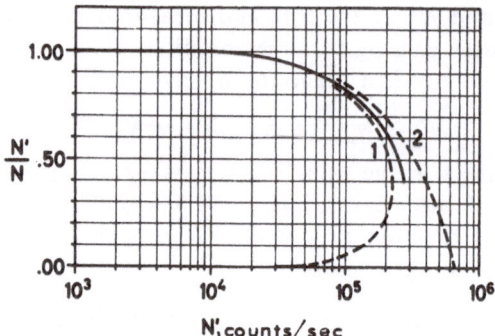

Fig. 11.2. Counting loss of a scintillation counter at high counting rates, presented as the experimentally determined counting efficiency N'/N (measured pulse rate N' over true pulse rate N) as a function of N'. Curves 1 and 2 represent the counting efficiencies calculated according to the two models discussed in the text.

of the background, however, can not be measured directly at the wavelength of the fluorescent line in question, but has to be determined in its neighborhood. Extrapolations and interpolations are necessary to derive the intensity at the desired wavelength under the peak. The type of extrapolation or interpolation to be applied in a particular case depends largely on the nature of the background in the area of the fluorescent line which, in turn, determines the error in the background measurement. Four typical cases are presented in Figure 11.3.

Case (A): The background radiation in the area of the fluorescent line is constant. It is sufficient to measure the background somewhere in the neighborhood of the line. For calculation of the error in the determination of the true intensity of the line, its total intensity is designated P (peak), the intensity of the background at i is designated U_i, and the value of the background at the desired wavelength is designated U_0. The error in the determination of the value P is S_P, while the individual values U_i all have the same error S_U. The desired value U_0 and the net fluorescent intensity N for case (A) are given as follows:

$$U_0 = U \pm S_U; \quad N = (P - U_0) \pm \sqrt{S_P^2 + S_U^2}$$

Case (B): The intensity of the background radiation changes linearly with the wavelength. Its intensity may be determined by measuring the intensities at two wavelengths, one on the right and one on the left of the

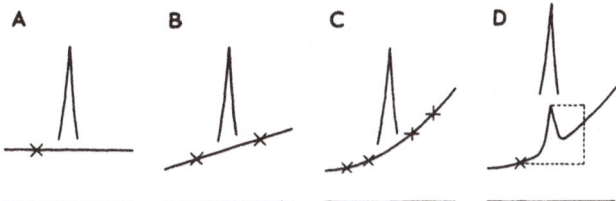

Fig. 11.3. Schematic presentation of the nature of the background in the area of the fluorescent line. The background, in the wavelength range in question, may either be constant (A), may change linearly (B) or nonlinearly (C) with wavelength, or the fluorescent line may be overlapped by a spurious line originating in the apparatus (D). Errors in the determination of the true intensity of a line are different from case to case.

peak, where both positions are equidistant from the peak. The desired value of the background may then be determined by forming the arithmetic average of the two measurements. The error in the determination of the net fluorescent intensity may be calculated from the errors of the individual measurements according to the equation

$$U_0 = \frac{1}{2}(U_1 + U_2) \pm \sqrt{2}\,S_U; \qquad N = (P - U_0) \pm \sqrt{S_P^2 + 2\,S_U^2}$$

Case (C): The background changes nonlinearly in the neighborhood of the characteristic fluorescent line. The shape of the background curve and, hence, the intensity of the background at the desired wavelength under the peak of the characteristic line may be determined by measuring the background at several positions on the right and on the left of the line. The error in the determination of the net fluorescent intensity is found to be approximately

$$U_0 = U_0\,(U_1, U_2, U_3, U_4) \pm 2\,S_U; \qquad N = (P - U_0) \pm \sqrt{S_P^2 + 4\,S_U^2}$$

Case (D): A spurious line originating in the apparatus itself (Fe-, Ni-, Cu-, Zn-, Pb-, W-lines from steel and brass parts of the unit, the lead shielding, and the hot-filament cathode) may sometimes be visible in the background. When the desired fluorescent line is overlapped by such a spurious line, a numerical relationship between the intensity of the background radiation to that of the spurious line has to be established with the aid of ultrapure samples. Only then is it possible to correct the measured fluorescent intensity for the contribution by the spurious line. The error in the determination of U_0 and of the net fluorescent intensity, in such a case, is largely determined by the accuracy of the correlation

$$U_0 = kU \pm S_k U; \qquad N = (P - U_0) \pm \sqrt{S_P^2 + S_k^2\,U^2}$$

In this equation, k is the factor correlating the intensity of the background to that of the contamination line, and S_k is the error in the value of k (S_k is usually 5 to 10% relative). Occasionally, background corrections may be neglected, particularly in cases where the intensities of the fluorescent lines are very large in comparison to those of the background radiation; errors are then largely due to errors in the measurement of the fluorescent lines themselves.

11.4. Optimal Conditions of Analysis

The purpose of optimizing conditions of analysis is to determine the proper parameters required for most successful analytical work. A number of independently variable parameters have to be considered:

Anode material of the x-ray tube

Excitation potential and current of the x-ray tube

Path of the x-ray beam in the spectrometer (air, helium, vacuum)

Slit-width of collimator

Analyzing crystal (dispersion and reflection efficiencies)

Counter

Setting of pulse-height discriminator

Counting technique

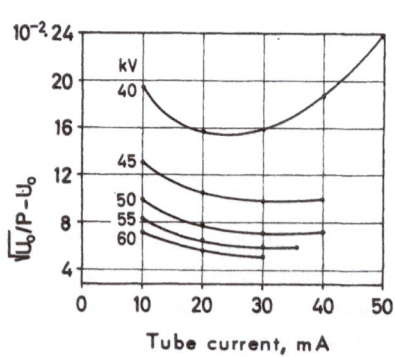 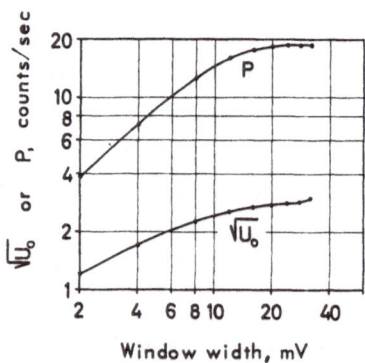

Fig. 11.4. *Left*: Optimal anode voltages and tube currents for the determination of antimony in galena. Optimal setting for detection limit is at the lowest point of the curve, i.e., at minimum error (Siemens, 1962). *Right*: Optimal discriminator setting for determination of arsenic in pharmaceuticals. Upper Curve: change in fluorescent intensity of arsenic as a function of the window width; lower curve: change in background variation as a function of the window width. Plotted on a logarithmic scale. The ratio of arsenic fluorescence to the variations in the background radiation is at a maximum at the position of the largest distance between the two curves.

Optimal analytical conditions are those which yield a net intensity of lowest possible relative error. The net intensity N and the corresponding error are given as follows:

$$N = (P - U_0) \pm \sqrt{S_P^2 + n S_U^2}$$

where P and U_0 are the counting rates for, respectively, the uncorrected fluorescence intensity and the intensity of the background radiation; S_P and $n S_U$ are the corresponding errors. The factor n is a term which depends upon the particular extrapolation procedures employed in the determination of the background U_0 (see the chapter on background correction). When assuming that the error is due only to statistical variations in the counting rate and every measurement is made for the same time T, the relative error ϵ is

$$\varepsilon = \frac{1}{P - U_0} \sqrt{\frac{P}{T} + n \frac{U_0}{T}}$$

The most favorable parameters are those for which the relative error is at a minimum:

$$\frac{\sqrt{P + n U_0}}{P - U_0} = \min \quad \text{or} \quad \frac{P - U_0}{\sqrt{P + n U_0}} = \max \qquad (11.16)$$

Figures 5.3 and 5.4 illustrate changes in the fluorescent intensity as a function of the anode material of the x-ray tube. Choice of collimator and analyzing crystal is determined by the required dispersion and reflection efficiencies (Chapter 10.6; *Crystals*). The counter is chosen according to the particular wavelength in question (Chapter 10.7; *Counters*). Figure 11.4 illustrates the selection of the most favorable conditions for x-ray tube voltages and currents as well as for pulse-height discriminator settings.

Optimal Counting Time

Special consideration has to be given to proper choice of optimal counting time, i.e., to the choice of counting times for individual lines which account for the fact that there is a practical limit as to the total time available for a particular analytical problem.

Measurement of Differences. Let us consider the case where the difference between two intensities is to be determined as accurately as possible in a given time period. The average pulse rate of the one intensity measurement is assumed to be P counts/sec and is measured in the time interval of T_1 sec, and the average pulse rate of the other measurement is U_0 counts/sec and is obtained by extrapolation from several measurements

U_1, U_2, etc. For simplicity, the individual values U are counted for the same time interval T_2. The error for the difference $(P - U_0)$ is given as follows:

$$S = \sqrt{\frac{P}{T_1} + n\frac{U_0}{T_2}}$$

where the factor n depends upon the particular extrapolation procedures applied for the determination of the background U_0. The value n, however, does not have to be equal to the number m of the measured background positions. The total required counting time T is $T_1 + mT_2$ (counting for every one of the m background positions is for T_2 seconds). Optimal selection of counting times is achieved when the following equations are fulfilled:

$$\frac{\partial S}{\partial T_1}dT_1 + \frac{\partial S}{\partial T_2}dT_2 = 0 \quad \text{with} \quad dT_1 = -mdT_2$$

or

$$m\frac{\partial S}{\partial T_1} = \frac{\partial S}{\partial T_2} \tag{11.17}$$

The ratio for the counting times may be derived from this equation as follows:

$$\frac{T_1}{T_2} = \sqrt{\frac{mP}{nU_0}} \tag{11.18}$$

The smallest error in the difference $(P - U_0)$ is obtained when the counting times for the individual intensities are selected according to the equation presented above. The total available counting time is T and, hence, we find for T_1 and T_2

$$T_1 = \frac{T\sqrt{\dfrac{mP}{nU_0}}}{m + \sqrt{\dfrac{mP}{nU_0}}} \quad ; \quad T_2 = \frac{T}{m + \sqrt{\dfrac{mP}{nU_0}}} \tag{11.19}$$

Measurement of Intensity Ratios. Let us consider the case where the ratio of two intensities is to be determined as accurately as possible in a given time interval. The relative error ϵ of the ratio P/U, with P and U as the pulse rates, is as follows:

$$\varepsilon = \sqrt{\frac{1}{PT_1} + \frac{1}{UT_2}}$$

The total counting time T required is $T_1 + T_2$. The ratio of the two counting times is given in analogy to equation (11.17):

$$\frac{T_1}{T_2} = \sqrt{\frac{U}{P}} \qquad (11.20)$$

The smallest error for the ratio P/U is obtained when the counting times for measurement of the individual intensities are selected according to the above equation. The total available counting time is T and, hence, we find for T_1 and T_2

$$T_1 = \frac{T\sqrt{\dfrac{U}{P}}}{1 + \sqrt{\dfrac{U}{P}}} \quad ; \quad T_2 = \frac{T}{1 + \sqrt{\dfrac{U}{P}}} \qquad (11.21)$$

Counting Techniques

Two techniques are employed for measurement of intensities, namely measurement with reference to fixed number of pulses (fixed-pulse measurement) and measurement with reference to a fixed counting time (fixed-time measurement). Depending upon the nature of the problem, apparatus, and readout device, either one of the two procedures may be employed. Let us consider both techniques with particular emphasis on optimal counting times. For this purpose, we consider two sets of four examples each and calculate for every example the resulting errors for either fixed-time or fixed-pulse measurement; these errors are compared to the errors for optimal settings. The first four examples serve to find the most advantageous technique of measurement for the determination of differences, while the second four examples serve to find the most suitable technique of measurement for the ratios of two intensities. It is assumed that the ratios are 1:1, 10:1, 100:1, and 1000:1 (Table 11.3). In a difference determination, the higher intensity has to be measured for a longer period of time, because variations in the larger value contribute more to the resulting error than variations in the smaller value. For fixed-time measurement, where both intensities are measured in equal time intervals, no effort is made to achieve optimal time settings; thus, the resulting error is larger than the optimal error. For fixed-pulse measurement, where the same number of counts is accumulated for both intensities, measurement procedures are opposite to what is required for optimal settings: the higher intensity is measured for a shorter period of time than the lower intensity and, hence, the resulting error is larger than the optimal error. Measurement with fixed-time intervals is therefore preferred for difference measure-

Table 11.3. Procedures for Selecting Optimal Techniques of Measurement. [For Case (A) p.118.]

Formation of differences

Intensity ratio	1 : 1	10 : 1	100 : 1	1000 : 1
Optimal counting times	$T_1 = 0.5$	0.76	0.91	0.97
	$T_2 = 0.5$	0.24	0.09	0.03
Counting times for	$T_1 = 0.5$	0.5	0.5	0.5
fixed-time measurements	$T_2 = 0.5$	0.5	0.5	0.5
Counting times for fixed-pulse	$T_1 = 0.5$	0.09	0.01	0.001
measurements	$T_2 = 0.5$	0.91	0.99	0.999
Error for fixed-time measurement expressed as multiple of optimal error	1	1.13	1.29	1.37
Error for fixed-pulse measurement expressed as multiple of optimal error	1	2.55	9.09	30.7

Formation of ratios

Intensity ratio	1 : 1	10 : 1	100 : 1	1000 : 1
Optimal counting times	$T_1 = 0.5$	0.24	0.09	0.03
	$T_2 = 0.5$	0.76	0.91	0.97
Counting times for fixed-time	$T_1 = 0.5$	0.5	0.5	0.5
measurements	$T_2 = 0.5$	0.5	0.5	0.5
Counting times for fixed-pulse	$T_1 = 0.5$	0.09	0.01	0.001
measurements	$T_2 = 0.5$	0.91	0.99	0.999
Error for fixed-time measurement expressed as multiple of optimal error	1	1.13	1.29	1.37
Error for fixed-pulse measurement expressed as multiple of optimal error	1	1.13	1.29	1.37

ments. When determining intensity ratios, the lower intensity has to be measured for a longer period of time than the higher intensity because variations of the lower value are relatively large. For fixed-time measurements, where both intensities are measured for the same time interval, optimal settings are not achieved and, hence, the resulting error is larger than the optimal error. In the case of fixed-pulse measurement, where the same number of pulses are accumulated for both intensities, i.e., where long counting times for low pulse rates and short counting times for high pulse rates are employed, "counting time" selection is basically correct. However, the resulting error is nevertheless larger than the optimal error because "counting time" selection for the two intensities goes beyond the optimal setting. Both fixed-pulse and fixed-time measurements, however, result in the same error for ratio measurements (Table 11.3). On the basis of these results, the most suitable counting techniques may be selected:

1. Expected counting rates are unknown or highly variable.
   Goal: Formation of difference
   Technique: Fixed-time measurement
   Goal: Formation of ratio
   Technique: Either fixed-time or fixed-pulse measurement
2. The magnitude of the expected pulse rate is known (e.g., routine analysis in production control).
   Goal: Formation of difference or of ratios
   Technique: Measurement with fixed-time setting and variable counting intervals. Preselected counting intervals are adjusted according to the individual pulse rates so that they come as close as possible to the optimal setting. Measurement with fixed-pulse setting and variable total number of pulses. The total number of pulses for every set of measurements is chosen so that the resulting counting intervals correspond as closely as possible to the optimal time settings.

Studies of optimal recording conditions and counting times were made by Loevinger and Berman (1951), Parrish (1956a), Mack and Spielberg (1958), and Neff (1963).

11.5. Short-Term Drift and Variations From Measurement to Measurement

Accuracy of a measurement, in many cases, is not only limited by counting statistics but also by short-term drift and variations from measurement to measurement which are caused by the apparatus. Drift arises from both the primary x-ray tube radiation as well as from variations in the readout system employed to record the secondary fluorescent radiation. Variations from measurement to measurement are largely due to problems in sample positioning and spectrometer settings. Commercial companies claim stabilities for high voltage and current of x-ray tubes to be on the order of 0.02%. Similar values apply to high-voltage power supplies of counters. Counting rates of flow proportional counters are temperature and pressure dependent, and the multiplication factor of the secondary-electron multiplier in scintillation counters is also temperature sensitive. The electronics of the counting and readout unit and, in particular, of the discriminator are temperature insensitive and constant only within certain limits. The lattice parameter of the analyzing crystal and, hence, the glancing angles for the diffracted radiation are temperature dependent. Setting of individual sample holders in spectrometer chambers equipped for multiple sample positioning are only reproducible within 0.2 to 0.3%.

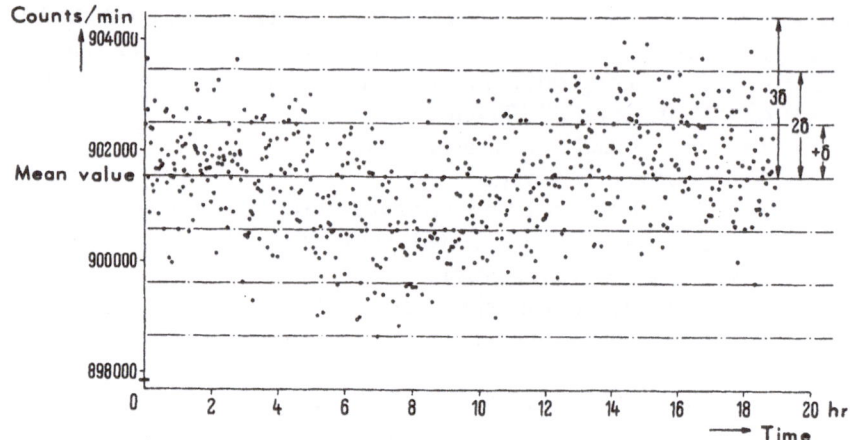

Fig. 11.5. Data scatter due to drift in an x-ray fluorescence unit over a time interval of 20 hr. (Courtesy of Neff, 1961.)

If samples are separated from the x-ray tube by foils, then the intensity of the soft radiation is affected by deviations in the foil thickness and by the quality of the vacuum (Alley and Higgins, 1963). The combined standard deviation, accounting for drift and variations from measurement to measurement, commonly ranges between 0.2 to 1%.

11.6. Long-Term Drift

The efficiency of an x-ray fluorescence unit changes over longer periods of time. High-voltage generators, for example, occasionally produce higher than anticipated effective voltages due to the aging of resistors. After prolonged use, anodes of x-ray tubes tend to develop surface roughness which results in emission of increasingly lower intensities. Old counters usually have poor counting efficiencies, and aged counting and readout electronics tend to become inaccurate. Long-term drift is different from unit to unit; however, an annual decrease in efficiency of 2 to 10% appears to be normal. In order to compare measurements taken at widely separated dates, it is recommended to simultaneously measure comparative standards (external standards) and to correct the measured fluorescence intensities in relation to the standards. It is also common practice to periodically remeasure calibration curves, either in part or in total, in order to account for long-term drift.

11.7. Errors Due to Sample Preparation and Calibration

Sample preparation may also be the cause of errors in an analysis. However, these errors largely depend upon the particular procedure employed and more general suggestions can only be made with reservations. The fluorescent intensity of a powdered sample is affected by grain size and grain-size distribution. If two chemically identical powders are different in grain size and grain-size distribution, then fluorescent intensities may differ by several %. Decomposition of samples in a borax melt, admixing of an internal standard, and grinding and polishing of massive samples may result in errors. Errors due to sample preparation are superimposed onto other errors of measurement and, hence, increase the overall error according to the following equation:

$$S_{total} = \sqrt{S^2_{Meas} + S^2_{Prep}}$$

Standard deviation of the total error is usually on the order of 0.3 to 3 %.

In quantitative analysis, it is necessary to transform the measured fluorescent intensity of the particular element in question to concentration. For this purpose, a relation between concentration and fluorescent intensity is established with the aid of standards. It is desirable to use several standards so that the error due to calibration is held small; however, the transformation is usually subject to a small uncertainty. This uncertainty may be relatively large in the case of massive samples where the synthetic standards deviate systematically in grain size, texture, formation of solid solution series, contamination, etc. from the samples that are to be analyzed. Although the standards may well be reproducible among themselves, they nevertheless may be the cause of systematically erroneous analyses. Accuracy of the results may be tested by comparing the analytical data to data obtained by other techniques. Comparison of measurements obtained by x-ray emission spectroscopy with those of other analytical methods has been made by numerous authors. Deviations are usually within 1 to 4 % relative; however, it has to be kept in mind that the comparative analytical methods are subject to errors just as is x-ray emission spectroscopy.

Reproducibility (precision) of an analytical technique describes the magnitude of deviation which occurs after repeatedly measuring the same sample by the same technique. Repeated measurement of one and the same sample and repeated sample preparation allow experimental estimation of the reproducibility. Depending upon the particular apparatus, technique of measurement, and sample preparation the errors encountered in practical analysis may actually be larger or smaller than the values given in this book. Accuracy includes the reproducibility and describes the deviation between the true and the analytical value. Accuracy not only depends upon errors which arise from transformation of intensity measurements into concentrations, but also from statistical scattering of the measurements.

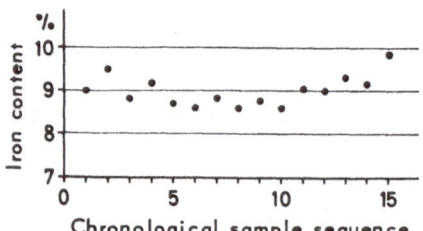

Fig. 11.6. Inhomogeneity of Thomas slags. Individual random samples deviate in their iron contents by approximately $\pm 5\%$ relative (Kopineck, 1956).

11.8. Detection Limits

The detection limit was defined in the section on *Principles of Statistics* (Section 11.1) as that concentration for which the net fluorescent intensity, with a certain confidence, is larger than the statistical variations of the background or of the test sample. It is a matter of agreement to define for what probability a result is considered to be reliable or unreliable. One usually distinguishes between a detection limit resulting from the particular technique of measurement, and the detection limit resulting from sample preparation. The former determines the lowest possible content which may.be measured with this particular technique, while the latter involves reproducibility in sample preparation and the magnitude of scattering from random sample to random sample for the same concentration. Let us assume an experimental arrangement for which the standard deviation of the measurement of the iron fluorescence is less than 0.5% relative. Due to problems in the sample preparation, the standard deviation of the iron intensity measurement, in spite of all precaution, may not be better than 1% relative. For example, the composition of slags may vary greatly on a small scale, and scatter in the iron values from random sample to random sample may be as high as $\pm 5\%$. The iron determination is made, however, not to obtain the average iron content of a random sample but of the slag as whole; hence, the detection limits due to the method and sample preparation, respectively, differ considerably.

An analytical value is considered to be realistic when the difference between the total intensity and the background $(P - U_0)$ is larger than a multiple of the standard deviation. The detection limit is that content for which a multiple of the standard deviation is equally as large as the difference $(P - U_0)$.

If the standard deviation of a measurement of a line is designated σ, then the standard deviation of the difference $(P - U_0)$ is equal to $\sigma\sqrt{(n + 1)}$, where n depends upon the particular extrapolation procedure used in the determination of U_0 from U, and which may have values between $n = 1$ to 4 (Section 11.3, Cases (A) to (C); with $\sigma_P = \sigma_U$). It is required that the difference $(P - U_0)$ is larger, by a multiple u, than the corresponding standard deviation

$$(P - U_0) = u \sigma \sqrt{n + 1} \qquad (11.22)$$

The detection limit C_{AG} of the technique for component A may be calculated with a standard

$$C_{AG} = u \sigma \sqrt{n + 1} \left(\frac{C_A}{N_A} \right)_{\text{Standard}} \qquad (11.23)$$

The value of u for 99 % confidence is 2.33, while $\sqrt{(n + 1)}$, depending upon the background extrapolation, ranges between 1.42 to 2.42. We then obtain

$$C_{AG} = (3.3 - 5.2) \sigma \left(\frac{C_A}{N_A} \right)_{\text{Standard}} \qquad (11.24)$$

The magnitude of the standard deviation σ is usually calculated on the basis of the pulse rate and the counting time (Section 11.2).

Detection limits depend upon the fluorescence efficiency of the desired element, the apparatus, and the matrix in which the particular element is to be determined. Detection limits are therefore different for different elements and matrices and range from 0.1 % for very light elements in heavy matrices to 0.00001 % or 0.1 ppm for heavy elements in light matrices. Figures 19.1 to 25.1 illustrate the dependence of the detection limits of elements upon the matrices.

PART II
QUANTITATIVE ANALYSIS

Chapter 12

Calibration Curves and Regression Coefficients

The relationship between fluorescent intensity and concentration in a two-component mixture may be illustrated by a curve. Every value of concentration has a corresponding value for the fluorescent intensity (Figure 12.1). Depending upon the nature of the mixture, the curves are bent to a different degree and may be determined experimentally with the aid of standard samples of known composition. Determination of concentrations in two-component mixtures is commonly carried out with calibration curves. The relationship between fluorescent intensity and concentration may also be expressed algebraically in the form of an equation where the regression function is determined instead of the calibration curve. Abstract formulation, as a regression function, is of importance in the analysis of multi-component mixtures where graphic illustration of the relationship between intensity and concentration is impossible. The concentration is then determined by calculation.

The regression function is first derived for the fluorescent intensity of component A in a mixture with component B, where interelemental excitation may be neglected. The intensity of element A as pure A and as a mixture with component B may be calculated as follows:

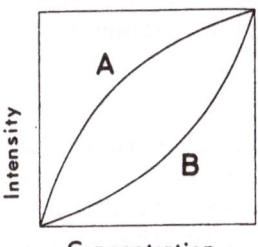

Concentration

Fig. 12.1. Calibration curves. Relationship between fluorescent intensity and concentration in a two-component mixture is illustrated by a curve. Curve A corresponds to the fluorescent intensity of iron in a mixture of Fe_2O_3–SiO_2; curve B corresponds to the fluorescent intensity of iron in a mixture of Fe_2O_3–Ta_2O_5.

$$N_A = \frac{q}{\sin \varphi} E_A \int_{\lambda_0}^{\lambda_A} \frac{C_A \mu_A(\lambda) N_0(\lambda) d\lambda}{C_A \bar{\mu}_A(\alpha) + C_B \bar{\mu}_B(\alpha)}$$

$$N_{A100} = \frac{q}{\sin \varphi} E_A \int_{\lambda_0}^{\lambda_A} \frac{\mu_A(\lambda) N_0(\lambda) d\lambda}{\bar{\mu}_A(\alpha)}$$

$$C_A + C_B = 1$$

The intensity in a mixture is expressed as a fraction of the fluorescent intensity in the pure element:

$$\frac{N_A}{N_{A100}} = \frac{\displaystyle\int \frac{C_A \mu_A(\lambda) N_0(\lambda) d\lambda}{C_A \bar{\mu}_A(\alpha) + C_B \bar{\mu}_B(\alpha)}}{\displaystyle\int \frac{\mu_A(\lambda) N_0(\lambda) d\lambda}{\bar{\mu}_A(\alpha)}} \rightarrow \frac{\dfrac{C_A \mu_A(\bar{\lambda}) N_0(\bar{\lambda})}{C_A \bar{\mu}_A(\alpha) + C_B \bar{\mu}_B(\alpha)} (\lambda_0 - \lambda_K)}{\dfrac{\mu_A(\bar{\lambda}) N_0(\bar{\lambda})}{\bar{\mu}_A(\alpha)} (\lambda_0 - \lambda_K)}$$

$$(12.1)$$

If we assume that the radiation, in both cases, is in general produced by the same wavelength range of the primary x-ray spectrum and that the weighted-average wavelength λ is the same in both cases, then the intensity ratio for the first term on the right side of the expression is simplified. If one considers that $C_B = 1 - C_A$, the so-called regression function $R(C_A)$ is obtained (Guinier, 1961; Müller, 1962a; Kopineck, 1962):

$$R(C_A) = \frac{N_A}{N_{A100}} = \frac{C_A}{C_A + (1 - C_A) r_{AB}}$$

$$(12.2)$$

$$r_{AB} = \bar{\mu}_B(\alpha) / \bar{\mu}_A(\alpha) \qquad \text{regression coefficient}$$

The relation between the fluorescent intensity N_A and the concentration C_A of element A is therefore clearly expressed by the regression coefficient r_{AB}; for every value of N_A there is a value of C_A, and vice versa. Analogously to a calibration curve, the regression function relates the fluorescent intensity to the concentration; the relation however, is not displayed graphically but algebraically. The regression coefficient

$$r_{AB} = \bar{\mu}_B(\alpha) / \bar{\mu}_A(\alpha)$$

is the ratio of the average combined mass absorption coefficient of associated component B to the average combined mass absorption coefficient of component A. In analogy to calibration curves, the regression coefficient is determined with the aid of standard samples. Every standard sample yields one value for r_{AB} (Figure 12.2)

$$r_{AB} = \frac{C_A}{1 - C_A} \frac{N_{A100} - N_A}{N_A}$$

$$(12.3)$$

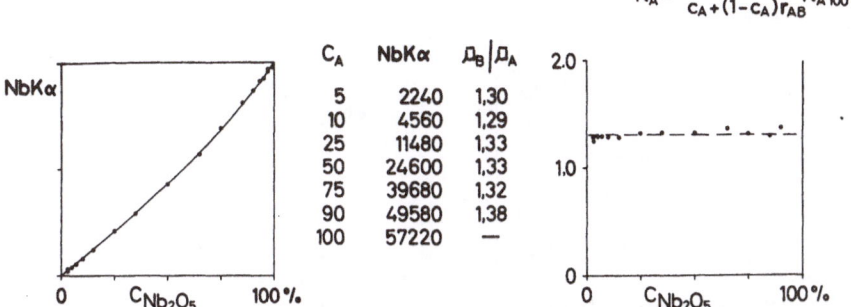

C_A	NbKα	μ_B / μ_A
5	2240	1,30
10	4560	1,29
25	11480	1,33
50	24600	1,33
75	39680	1,32
90	49580	1,38
100	57220	—

$$N_A = \frac{C_A}{C_A + (1 - C_A) r_{AB}} N_{A\,100}$$

Fig. 12.2. Data obtained with standard samples (middle column) may either be used to construct a calibration curve (*left*) or to calculate the regression coefficient which is constant over the entire concentration range (*right*). Determination of niobium oxide in a mixture of Nb_2O_5–Ta_2O_5; borax pellets.

The value of r_{AB} is constant for the entire range of concentrations. The assumption that the weighted-average wavelength is indeed the same for all concentrations may thus be tested.

Occasionally, the value of N_{A100} of the fluorescent intensity of pure component A is unknown. In this case, the regression coefficient r_{AB} is calculated with the aid of the fluorescent intensity of a second standard sample

$$r_{AB} = \frac{N_{A2} - N_{A1}}{N_{A2} - N_{A1} + \dfrac{N_{A1}}{C_{A1}} - \dfrac{N_{A2}}{C_{A2}}} \qquad (12.4)$$

The regression coefficient expresses all interactions between the two components in a single number, for example, 1.7, 0.8, or 1.1, where the last number indicates that the average combined mass absorption coefficient of the associated component is 1.1 times as large as the average mass absorption coefficient of component A. The value of r_{AB} allows one to determine the curvature of the corresponding calibration curve. Three cases are possible for the ratios which determine the curvatures:

$$\bar{\mu}_B(\alpha) / \bar{\mu}_A(\alpha) < 1 \qquad N_A = \frac{C_A N_{A100}}{C_A + (1 - C_A)\,\bar{\mu}_B(\alpha) / \bar{\mu}_A(\alpha)} > C_A N_{A100}$$

$$\bar{\mu}_B(\alpha) / \bar{\mu}_A(\alpha) = 1 \qquad N_A = \frac{C_A N_{A100}}{C_A + (1 - C_A)\,\bar{\mu}_B(\alpha) / \bar{\mu}_A(\alpha)} = C_A N_{A100}$$

$$\bar{\mu}_B(\alpha) / \bar{\mu}_A(\alpha) > 1 \qquad N_A = \frac{C_A N_{A100}}{C_A + (1 - C_A)\,\bar{\mu}_B(\alpha) / \bar{\mu}_A(\alpha)} < C_A N_{A100}$$

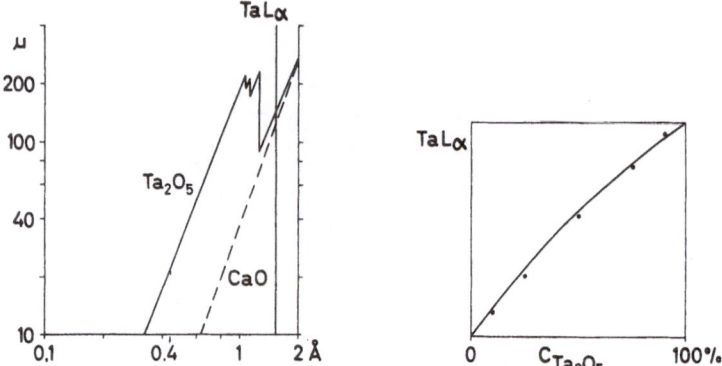

Fig. 12.3. Mass absorption coefficients and calibration curve for the determination of Ta_2O_5 in a mixture of Ta_2O_5–CaO. The average absorption
coefficient of the associated component CaO is smaller than that of the
desired component Ta_2O_5; hence, the regression coefficient is less than one
and the calibration curve is located above the diagonal (case 1).

In the first case, the calibration curve is located above the diagonal (as
determined by C_AN_{A100}); in the second case, it coincides with the diagonal;
and in the third case, it is located below the diagonal. Frequently, the value
for r_{AB} and, hence, the curvature of the calibration curve, may be predicted
from the mass absorption coefficients of the two components. The average
combined mass absorption coefficient is composed of the average absorption
coefficients for, respectively, the incident polychromatic radiation and the
emerging monochromatic fluorescent radiation.

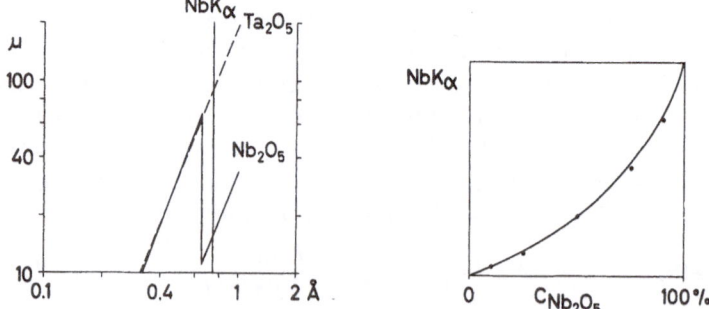

Fig. 12.4. Mass absorption coefficient and calibration curve for the measurement of Nb_2O_5 in a mixture of Nb_2O_5–Ta_2O_5. The average absorption
coefficient of the associated component Ta_2O_5 is larger than that of the
desired component Nb_2O_5; hence, the regression coefficient is larger
than 1 and the calibration curve is located below the diagonal (case 3).

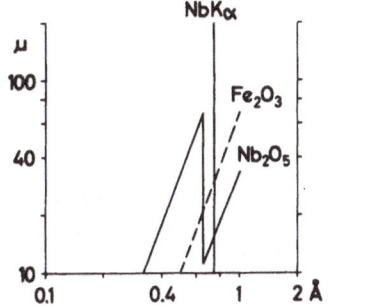 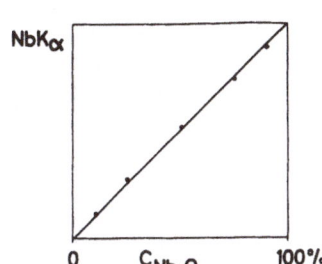

Fig. 12.5. Mass absorption coefficient and calibration curve for the measurement of Nb_2O_5 in a mixture of Nb_2O_5-Fe_2O_3. The average absorption coefficients of both components are about equally large; hence, the regression coefficient equals 1 and the calibration curve coincides with the diagonal (case 2).

$$\bar{\mu}(\alpha) \equiv \frac{\mu(\lambda)}{\sin \varphi} + \frac{\mu(\alpha)}{\sin \psi}$$

The ratio of the average absorption coefficients $\bar{\mu}_B(\alpha)/\bar{\mu}_A(\alpha)$ may easily be determined from a graphical illustration, such as that in Figure 12.3 illustrating the determination of tantalum oxide in a mixture of tantalum oxide and calcium oxide. The relation between mass absorption coefficient and wavelength is illustrated for the two components on the left of the figure, and the position of the fluorescent line of tantalum is indicated. It is evident that the average absorption coefficient of the associated component, calcium oxide, is smaller than that of tantalum oxide and, hence, the regression coefficient is less than one. The calibration curve, therefore, is located above the diagonal. Further examples for the prediction of regression coefficients and calibration curves are given in Figures 12.4 and 12.5. Calibration curve and regression function reflect the relationship between fluorescent intensity and concentration of a component and are the basis for quantitative analysis. The regression functions may be used to calculate the corresponding curve for any values of r_{AB} without having to take recourse to measurements. A single standard sample determines unequivocally the value of r_{AB} as well as the calibration curve and, hence, the number of required standard samples is reduced (Figure 12.6).

The regression coefficient depends upon the associated component, which is reflected by the fact that the mass absorption coefficient of the associated component is in the denominator. As the mass absorption coefficients are numbers which depend upon the wavelength, the regression coefficient also depends upon the spectral composition of the primary x-ray

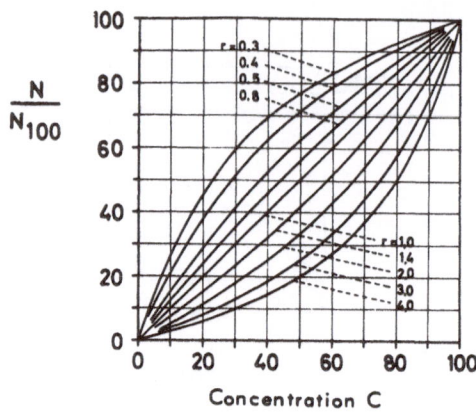

Fig. 12.6. Calibration curves for different values of r_{AB}. A single standard sample determines unequivocally the value of r_{AB} and the complete calibration curve.

radiation used for excitation. The calibration curve for the determination of tantalum oxide in a mixture of tantalum oxide and niobium oxide is different when tantalum is excited to fluorescence by a tungsten instead of molybdenum tube; hence, the regression coefficient also has a different value (Figure 12.7).

The calibration curve for the concentration $C_A = 0$ passes through zero, and the tangent at zero is given as follows:

$$\frac{dR(C_A)}{dC_A} = \frac{1}{r_{AB}} \tag{12.5}$$

The slope of the tangent at zero is identical to the regression coefficient, and the tangent corresponds to the calibration curve for the determination of traces of A in mixture with A + B and in the matrix B, respectively.

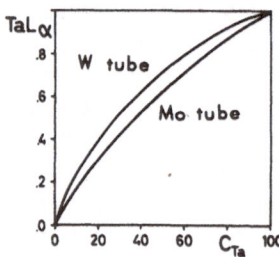

Fig. 12.7. Calibration curves for the determination of tantalum oxide in a mixture of Ta_2O_5–Nb_2O_5. Tantalum was excited by a tungsten tube and a molybdenum tube, respectively. All other conditions are identical. Two different calibration curves and regression coefficients result.

Determination of Low Concentrations

The fluorescent intensity of an element present in low concentrations $(10^{-3}$ to $10^{-6})$ is proportional to the concentration of the element in question, and the calibration curve is a straight line. The nature of the associated components determines the slope of the calibration curve because the magnitude of absorption of the x-ray radiation is different for different associated components. The fluorescent intensity of niobium in a mixture $Nb_2O_5-TiO_2$, for example, is approximately four times as large as in a mixture $Nb_2O_5-Ta_2O_5$ (Figure 13.1). For measurement of low concentrations, standard samples of known trace element contents and of matrix compositions similar to those of the samples are used. The calibration curve may be calculated directly, provided the regression coefficient is known. For low concentrations, the regression function is simplified as follows:

$$N_A = \frac{C_A N_{A100}}{C_A + (1 - C_A)\, r_{AB}} \xrightarrow{C_A \to 0} \frac{C_A\, N_{A100}}{r_{AB}} \qquad (13.1)$$

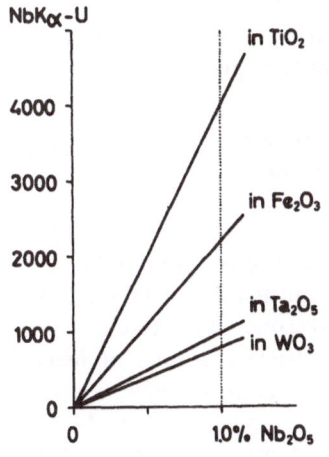

Fig. 13.1. Calibration curves for the determination of small amounts of Nb_2O_5 in various two-component mixtures. The slope of the calibration curve is dependent upon the nature of the associated component.

13.1. Effects of the Matrix on Fluorescent Intensity

Effects of the matrix on the fluorescent intensity were investigated by Hower (1959), Dwiggins and Dunning (1960), and Müller (1964). If the desired component is present in low concentrations, then the formula for the fluorescent intensity is simplified. We designate the desired component by A and summarize all associated components in the term "matrix." The fluorescent intensity of A is then

$$N_A = \frac{q}{\sin \varphi} E_A \int_{\lambda_0}^{\lambda_A} \frac{C_A \mu_A(\lambda) N_0(\lambda) d\lambda}{C_A \bar{\mu}_A(\alpha) + C_M \bar{\mu}_M(\alpha)}, \qquad C_M \cong 1 \quad (13.2)$$

The contribution of a component to the absorption of x-ray radiation is proportional to its concentration. As the concentration of A varies in the range from 10^{-3} to 10^{-6}, the contribution of A to the absorption, in comparison to that of the matrix, may be neglected. The formula for the fluorescent intensity is then simplified to

$$N_A = \frac{q}{\sin \varphi} E_A C_A \int_{\lambda_0}^{\lambda_A} \frac{\mu_A(\lambda) N_0(\lambda) d\lambda}{\bar{\mu}_M(\alpha)}$$

$$\bar{\mu}_M(\alpha) \equiv \left\{ \frac{\mu(\lambda)}{\sin \varphi} + \frac{\mu(\alpha)}{\sin \psi} \right\}_M$$

$$(13.3)$$

Only terms which are independent of the concentration of A and which have a constant value for a given matrix M are in the integral. In case the element A is in addition excited to fluorescence by the associated element B of the matrix, then the following term has to be added to the above equation to account for interelemental excitation (Chapter 7):

$$+ N_A' = \frac{q}{2 \sin \varphi} E_A C_A \mu_A(\beta) \int_{\lambda_0}^{\lambda_B} \frac{E_B C_B \mu_B(\lambda) N_0(\lambda) d\lambda}{\bar{\mu}_M(\alpha)} \cdot L \quad (13.4)$$

All terms in the integral have a constant value for a given matrix M.

In the following we limit our discussion to multicomponent mixtures where the component A is not excited to fluorescence by interelemental excitation. This is the case when analyzing for trace amounts of heavy elements in light matrices (organic substances, petroleum, silicates, cement, light-metal alloys), and for certain elements in ores and steel (e.g., Mo or Nb in steel). When comparing the fluorescent intensity of element A in matrix M1 to the intensity of the same amount of A in matrix M2, one obtains

$$\frac{N_{A\ M1}}{N_{A\ M2}} = \frac{\displaystyle\int_{\lambda_0}^{\lambda_A} \frac{\mu_A(\lambda)\,N_0(\lambda)\,d\lambda}{\{\mu(\lambda)/\sin\varphi + \mu(\alpha)/\sin\psi\}_{M1}}}{\displaystyle\int_{\lambda_0}^{\lambda_A} \frac{\mu_A(\lambda)\,N_0(\lambda)\,d\lambda}{\{\mu(\lambda)/\sin\varphi + \mu(\alpha)/\sin\psi\}_{M2}}} \qquad (13.5)$$

The mass absorption coefficient μ_i of an element i is wavelength dependent. It is given by the following empirical formula, where Z_i is the atomic number of the element i:

$$\mu_i(\lambda) = \text{const } \lambda^n Z_i{}^m$$

The mass absorption coefficient $\mu_M(\lambda)$ of the matrix M for wavelength λ is the weighted sum of the absorption coefficients of the individual elements (C_i = concentration of the element i; $\Sigma C_i = 1$):

$$\mu_M(\lambda) = \overset{i}{\Sigma}\, C_i \mu_i(\lambda) = \text{const } \lambda^n \overset{i}{\Sigma}\, C_i Z_i{}^m$$

An analogous expression may be written for the value of $\mu_M(\alpha)$.

Matrix M1 is composed of elements i of concentrations C_i and the matrix M2 of elements j of concentrations C_j. The two expressions may then be written:

$$\{\mu(\lambda)/\sin\varphi + \mu(\alpha)/\sin\psi\}_{M1} = \text{const}\left(\frac{\lambda^n}{\sin\varphi} + \frac{\alpha^n}{\sin\psi}\right) \cdot \overset{i}{\Sigma}\, C_i Z_i{}^m$$

$$\{\mu(\lambda)/\sin\varphi + \mu(\alpha)/\sin\psi\}_{M2} = \text{const}\left(\frac{\lambda^n}{\sin\varphi} + \frac{\alpha^n}{\sin\psi}\right) \cdot \overset{j}{\Sigma}\, C_j Z_j{}^m$$

and

$$\{\mu(\lambda)/\sin\varphi + \mu(\alpha)/\sin\psi\}_{M2} = K \cdot \{\mu(\lambda)/\sin\varphi + \mu(\alpha)/\sin\psi\}_{M1}$$

with

$$K = \overset{j}{\Sigma}\, C_j Z_j{}^m \Big/ \overset{i}{\Sigma}\, C_i Z_i{}^m$$

where K is dependent upon the composition of the respective matrices. If this is substituted into the expression for the ratio of the fluorescent intensities of element A in matrices M1 and M2 [equation (13.5)] then this expression may be simplified and the integrals disappear

$$\frac{N_{A\ M1}}{N_{A\ M2}} = \frac{\displaystyle\int_0^{\lambda_A} \frac{\mu_A(\lambda)\, N_0(\lambda)\, d\lambda}{\{\mu(\lambda)/\sin\varphi + \mu(\alpha)/\sin\psi\}_{M1}}}{\displaystyle\int_{\lambda_0}^{\lambda_A} \frac{\mu_A(\lambda)\, N_0(\lambda)\, d\lambda}{K\cdot\{\mu(\lambda)/\sin\varphi + \mu(\alpha)/\sin\psi\}_{M1}}} = K \qquad (13.6)$$

The simplest method for determining the constant K is that of calculating the ratio of the mass absorption coefficient of matrix M2 to the mass absorption coefficient of matrix M1 for a given wavelength λ

$$\frac{\mu(\lambda)_{M2}}{\mu(\lambda)_{M1}} = \frac{\text{const}\cdot\lambda^n\, \Sigma\, C_j Z_j{}^m}{\text{const}\cdot\lambda^n\, \Sigma\, C_i Z_i{}^m} = K$$

The ratio of the fluorescent intensity of element A in matrix M1 to the intensity in matrix M2 is equal to the reciprocal ratio of the mass absorption coefficients in the two matrices

$$\frac{N_{A\ M1}}{N_{A\ M2}} = \frac{\mu(\lambda)_{M2}}{\mu(\lambda)_{M1}} \quad \text{or} \quad N_A = \text{prop}\,\frac{1}{\mu(\lambda)_M} \qquad (13.7)$$

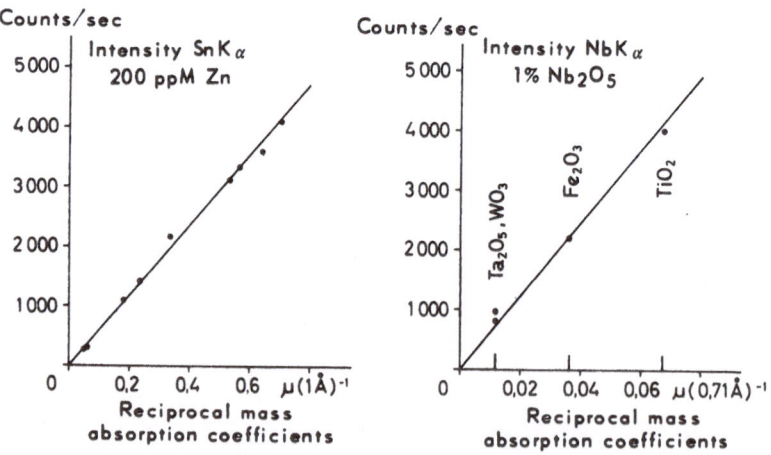

Fig. 13.2. For elements present in minor amounts, the fluorescent intensity is inversely proportional to the mass absorption coefficient of the matrix. *Left*: Fluorescent intensity of 200 ppm Zn in solvents of different mass absorption coefficients. *Right*: Fluorescent intensity of 1% Nb_2O_5 in a mixture with different metal oxides. The fluorescent intensities of niobium are taken from Figure 13.1.

Table 13.1. Test of the Empirical Formula $\mu(\lambda)M_1/\mu(\lambda)M_2$ = Constant

Wavelength (Å)	0.6	0.8	1.0	1.2	1.6	2.0	2.5
μ Alcohol C_2H_5OH	0.59	1.09	1.93	3.17	7.17	13.7	26.3
μ Talcum							
$Mg_3Si_4O_{10}(OH)_2$	2.11	4.70	8.92	15.1	34.7	66.1	125.3
μ Water H_2O	0.81	1.63	2.98	4.97	11.4	21.9	42.1
μ Alcohol $/\mu H_2O$	0.73	0.67	0.65	0.64	0.63	0.63	0.63
μ Talcum $/\mu H_2O$	2.61	2.88	2.99	3.04	3.04	3.02	2.98
μ Alcohol $/\mu$ Talcum	0.28	0.23	0.22	0.21	0.21	0.21	0.21

The intensity of element A, when A is present only in trace amounts, is inversely proportional to the mass absorption coefficient of the matrix.

This relation may be tested experimentally, for example, by dissolving 200 ppm of a heavy element in solvents of different mass absorption coefficients, and plotting the resulting fluorescent intensities for a given wavelength vs the reciprocal mass absorption coefficients of the solvents. The intensity increases reversely with the mass absorption coefficients (Figure 13.2). In this derivation, an empirical formula was used for the dependence of the mass absorption coefficients on the wavelength λ

$$\mu(\lambda) = \text{const } \lambda^n \overset{i}{\Sigma} C_i Z_i{}^m$$

For every wavelength, the ratio of the mass absorption coefficients of the two matrices M1 and M2 is therefore constant

$$\frac{\mu(\lambda)_{M2}}{\mu(\lambda)_{M1}} = \frac{\text{const } \lambda^n \Sigma C_j Z_j{}^m}{\text{const } \lambda^n \Sigma C_i Z_i{}^m} = K$$

In order to test this equation, the mass absorption coefficients of alcohol, talc, and water for various wavelengths are listed in Table 13.1 and the ratios of the absorption coefficients for various wavelengths are calculated. For wavelengths longer than 0.8 Å, the ratios of the mass absorption coefficients remain constant and satisfy the above equation. For wavelengths shorter than 0.8 Å, however, the ratios of the mass absorption coefficients change with the wavelengths, because the scattered portion in comparison to the atomic absorption cannot be neglected for the shorter wavelengths. For elements which are largely excited to fluorescence by wavelengths shorter than 0.8 Å, the resulting fluorescent intensity is approximately reversely proportional to the mass absorption coefficients of the matrix for the weighted-average wavelength $\bar{\lambda}$

$$\frac{N_{A\,M1}}{N_{A\,M2}} \cong \frac{\mu(\bar{\lambda})_{M2}}{\mu(\bar{\lambda})_{M1}} \tag{13.8}$$

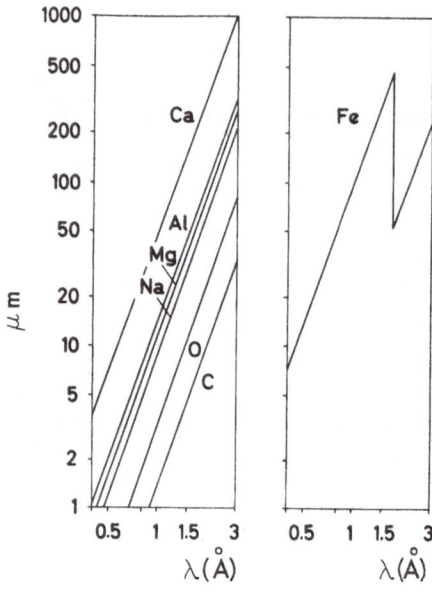

Fig. 13.3. Graphical presentation of mass absorption coefficients for the most important rock- and mineral-forming elements (Hower, 1959).

The particular wavelength λ is usually found empirically by testing the above relation for the best fit at several wavelengths. For elements bromine ($Z = 35$) and niobium ($Z = 41$), the wavelength $\lambda = 0.71$ Å (MoK_α) was found to be most suitable. For the most important rock-forming elements, Hower (1959) has shown the existence of a constant factor between the mass absorption coefficients of the individual elements. When plotted on a logarithmic scale, the curves for the mass absorption coefficients plotted as a function of the wavelength are parallel (Figure 13.3).

13.2. Quantitative Trace Element Determination

The derived relation provides for simple determination of the trace element content, and a reference sample of known trace element composition is used for calibration. The matrix of the standard sample does not have to be identical to the matrix of the sample which is to be analyzed. For convenience, a standard sample which can easily be prepared, such as an aqueous solution, is often chosen. The mass absorption coefficients for a given wavelength may be calculated from the bulk chemical compositions of the unknown and the standard sample, respectively.

The mass absorption coefficient is the weighted sum of the mass absorption coefficients of the individual elements (Table 13.2)

$$\mu(\lambda) = \frac{\Sigma\, C_i \mu_i(\lambda)}{\Sigma\, C_i}$$

For an unknown sample 1 and a standard sample 2 the concentration C_1 of the desired element may be calculated according to the following formula:

$$C_1 = \frac{N_1}{N_2} \frac{\bar{\mu}_1(\lambda)}{\bar{\mu}_2(\lambda)} C_2 \tag{13.9}$$

where C_1 and C_2 refer to the concentrations of the particular elements in the unknown and the standard, respectively; N_1 and N_2 to the fluorescent intensities of the particular elements in the unknown and the standard, respectively; and $\bar{\mu}_1(\lambda)$ and $\bar{\mu}_2(\lambda)$ to the mass absorption coefficients of the matrices of the unknown and the standard, respectively, for the same wavelength. In Table 13.3, concentrations determined with this technique are compared to chemically determined values.

13.3. Use of Diffusely Scattered X-Ray Radiation

The effect of the matrix on the fluorescent intensity may be compensated for by making use of the diffusely scattered x-ray radiation (background). Intensity of the diffusely scattered background also depends upon the absorption coefficient of the matrix (Andermann and Kemp, 1958).

The scattering efficiency of diffusely scattered radiation per unit mass σ is composed of two portions, namely the coherently scattered and the incoherently scattered radiation, respectively. The wavelength of the incoherently scattered radiation is modified in the scattering process and shifted slightly towards longer wavelengths (so-called Compton scattering)

$$\sigma = (\sigma)_{coh} + (\sigma)_{incoh} \tag{13.10}$$

The total intensity U_λ of radiation which after scattering has the wavelength λ is composed of the radiation of the wavelength λ (coherently scattered) and the radiation of the wavelength $(\lambda - \Delta\lambda)$ (incoherently scattered) which in the scattering process shifted by $+\Delta\lambda$. Analogous to the fluorescent intensity, the absorption of the incident as well as of the emerging radiation

Table 13.2. Calculation of the Mass Absorption Coefficient $\mu(1 \text{ Å})$ of $C_{15}H_{14}N_4O_2S$

	Weight %	$\mu(1 \text{ Å})$ of the elements		Proportion
C_{15}	57.3	1.40	=	80.22
H_{14}	4.5	0.40	=	1.80
N_4	17.8	2.21	=	39.34
O_2	10.2	3.31	=	33.76
S	10.2	25.6	=	261.12
	100%			416.24 $\mu(1 \text{ Å}) = 4.16$

Table 13.3. Comparison of Contents Determined by X-Ray and Chemical Techniques

	Sample	Determined x-ray fluorescence	Determined chemically
Fe determination in hydroxyphenanthroline; $\mu(1\ \text{Å})$ of hydroxyphenanthroline $=1.63$; standard solution of 200 ppm Fe in $H_2O.^a$	1	497 ± 50	503
	2	192 ± 20	215
	3	1043 ± 100	1192
	4	1090 ± 110	1208
	5	246 ± 25	260
Cu determination in a dye; $\mu(1\ \text{Å})$ of the dye $= 1.68$; standard solution of 0.5% Cu in $H_2O.^a$	1	480 ± 50	500
	2	590 ± 60	650
V determination in anthraquinone; $\mu(1\ \text{Å})$ of anthraquinone $= 1.65$; standard solution of 200 ppm V in H_2O^a	1	998 ± 100	1000
	2	4550 ± 450	5000
Pb determination in Mg-stearate; $\mu(1\ \text{Å})$ of Mg-stearate $= 1.92$ standard solution of 200 ppm Pb in $H_2O.^a$	1	58 ± 6	78

$^a\mu(1\ \text{Å})$ of $H_2O = 2.98$.

has to be considered. The intensity of the scattered radiation is then

$$U_\lambda = \frac{N_0(\lambda)\,(\sigma)_{\text{coh}}}{\mu(\lambda)/\sin\varphi + \mu(\lambda)/\sin\psi} + + \frac{N_0(\lambda-\Delta\lambda)\,(\sigma)_{\text{incoh}}}{\mu(\lambda-\Delta\lambda)/\sin\varphi + \mu(\lambda)/\sin\psi} \qquad (13.11)$$

We simplify and assume $N_0(\lambda) \approx N_0(\lambda - \Delta\lambda)$, except for wavelengths close to the characteristic line of the tube; furthermore, $\mu(\lambda) \approx \mu(\lambda - \Delta\lambda)$. The intensity of the scattered radiation is then given as follows:

$$U_\lambda \approx \frac{N_0(\lambda)\,\sigma}{\mu(\lambda)\left(\dfrac{1}{\sin\varphi} + \dfrac{1}{\sin\psi}\right)} \qquad (13.12)$$

The scattering efficiency per unit mass is approximately constant ($\sigma \approx 0.20$ cm^2/g). The intensity of the diffusely scattered background is proportional to the intensity of the primary x-ray radiation $N_0(\lambda)$ (which is a constant for a fixed accelerating potential), and inversely proportional to the mass absorption coefficient of the scattering matrix (Figure 13.4).

We assume two matrices M1 and M2 with different mass absorption coefficients, $\mu(\lambda)_{M1}$ and $\mu(\lambda)_{M2}$. The intensity ratio of the diffusely scattered radiation is then given as follows:

$$\frac{U_{\lambda\ M1}}{U_{\lambda\ M2}} \approx \frac{\mu(\lambda)_{M2}}{\mu(\lambda)_{M1}} \tag{13.13}$$

The ratio of the fluorescent intensities for element A in the two matrices, on the other hand, is given by equation (13.7):

$$\frac{N_{A\ M1}}{N_{A\ M2}} = \frac{\mu(\lambda)_{M2}}{\mu(\lambda)_{M1}}$$

When using the intensity of the diffusely scattered background instead of the mass absorption coefficients, the ratio of the fluorescent intensities of element A in the two matrices is given as follows:

$$\frac{N_{A\ M1}}{N_{A\ M2}} \approx \frac{U_{\lambda\ M1}}{U_{\lambda\ M2}} \quad \text{or} \quad \left(\frac{N_A}{U_\lambda}\right)_{M1} \approx \left(\frac{N_A}{U_\lambda}\right)_{M2} \tag{13.14}$$

The ratio of fluorescent intensity to intensity of the background is constant for a fixed concentration C_A and is independent of the matrix (Figure 13.5).

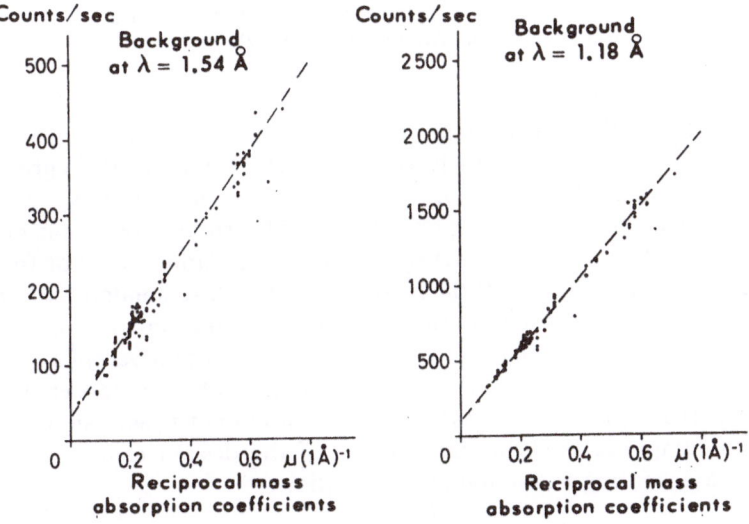

Fig. 13.4. Intensity of the diffusely scattered tube radiation as a function of the mass absorption coefficients of the scattering matrices. The intensity is approximately proportional to the reciprocal absorption coefficient.

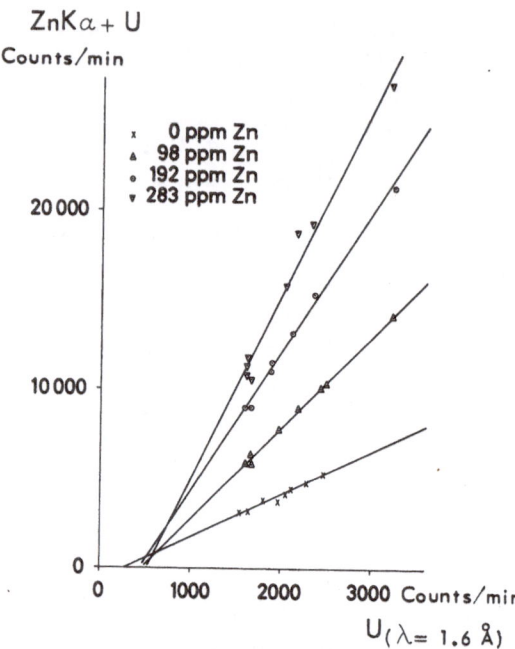

Fig. 13.5. Relation between fluorescent intensity and intensity of the background radiation for specific concentrations of zinc in Al_2O_3–$CaCO_3$ mixtures. Background was measured for a fixed reference wavelength (Kalman and Heller, 1962).

The effect of the matrix on the fluorescent intensity may be compensated for with the aid of the diffusely scattered tube radiation (background), because the intensity of the fluorescence and of the incoherently scattered background likewise depend upon the mass absorption coefficient of the matrix. From the intensity of the scattered tube radiation, a factor for the mass absorption coefficient of the matrix is obtained. A standard of known trace-element content is used in the analysis, where the matrix composition of the standard does not have to be identical to that of the sample. For the purpose of correcting for the matrix effect, the diffusely scattered tube radiation (background) is measured. For an unknown sample 1 and a standard sample 2, the concentration C_1 of the desired element may be calculated according to the following equation:

$$C_1 = \frac{N_{A1}}{N_{A2}} \frac{U_{\lambda 2}}{U_{\lambda 1}} C_2 \tag{13.15}$$

where C_1 and C_2 are the concentrations of the particular elements in the sample and standard; N_{A1} and N_{A2} are the fluorescent intensities of the particular elements in the sample and standard; and $U_{\lambda 1}$ and $U_{\lambda 2}$ are the intensities of the diffusely scattered tube radiations (background) for the sample and standard, respectively.

Handley (1960) used the diffusely scattered background to compensate for differences in matrix compositions in the determination of selenium in plants, and Kalman and Heller (1962) analyzed rocks and soil samples for heavy elements and used the intensity of the diffusely scattered background to compensate for varying sample compositions. Cullen (1962) and Strasheim and Wybenga (1964) investigated solutions and used the intensity of the background to compensate for differences in the absorption efficiencies of the solutions.

A more detailed consideration of diffuse scattering shows, however, that coherent and incoherent scattering are not equally dependent upon the matrix. Only the incoherently scattered portion of the background strictly fulfills the derived relation. In order to explain this fact, scattering of photons from a cloud of electrons consisting of Z individual electrons is considered [Compton and Allison (1935), p. 136]. When designating the phases of the waves scattered by the individual electrons as $\delta_1, \delta_2, \ldots, \delta_z$, the intensity of the resulting wave is given as follows:

$$I = I_e \left[(\cos \delta_1 + \cdots + \cos \delta_z)^2 + (\sin \delta_1 + \cdots + \sin \delta_z)^2 \right]$$

In this equation, I_e is the scattering efficiency of an individual electron. By rewriting and partial summation one obtains

$$I = I_e \left[Z + \sum_1^Z \sum_1^Z {}_{(m \neq n)} \cos (\delta_m - \delta_n) \right]$$

The expression $\cos(\delta_m - \delta_n)$ describes the phase difference between the waves scattered by the m and n electrons, respectively. Depending upon the residence probability of the electrons, different values result. When taking for the average value $\cos(\delta_m - \delta_n) = f^2$ we obtain

$$I = I_e \left[Z + f^2 (Z^2 - Z) \right]$$

The term f is the structure factor of the electrons. The coherently scattered portion has the intensity

$$I_{\text{coh}} = I_e Z^2 f^2$$

The remainder of the scattered radiation corresponds to the incoherently scattered portion. For this reason, the intensity is appropriately divided into two parts

$$I = I_e\,[Z + f^2\,(Z^2 - Z)] = \underbrace{I_e Z^2 f^2}_{\text{coherent}} + \underbrace{I_e Z\,(1 - f^2)}_{\text{incoherent}}$$

In the scattering process, the intensity of the radiation scattered incoherently by free electrons is reduced by the factor R [$R = R(\lambda$, scattering angle)]. For incoherent scattering by bound electrons, the same intensity decrease is assumed, and the intensity of the radiation scattered by an electron cloud is obtained as follows:

$$I = I_e Z^2 f^2 + I_e Z R\,(1 - f^2)$$

For a mass of $\sim 2Zm$, the scattering efficiency per unit mass is given as follows:

$$\sigma = \underbrace{\frac{I_e}{2\,m}\,Z f^2}_{(\sigma)_{\text{coh}}} + \underbrace{\frac{I_e}{2\,m}\,R\,(1 - f^2)}_{(\sigma)_{\text{incoh}}}$$

We find that only the scattering efficiency of the incoherently scattered portion is independent of the matrix and of Z, while the scattering efficiency of the coherently scattered portion has a certain dependence upon the matrix (Z-dependence) (Figure 13.6).

For trace element analysis in rocks, Reynolds (1963) used coherent scattering to compensate for the differences in sample compositions and to determine indirectly the mass absorption coefficients of the matrices.

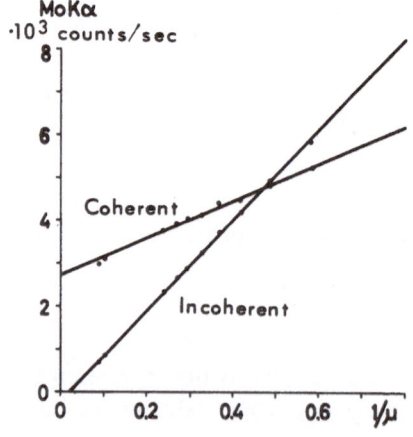

Fig. 13.6. Intensity of incoherently and coherently scattered tube radiation as a function of the mass absorption coefficients of the scattering matrices; measured for the K_α line of a molybdenum tube.

The matrix may be analyzed "indirectly" by making use of the fact that coherent and incoherent radiation depend, to a different degree, upon the mass absorption coefficients and atomic numbers of the scattering matrices. The ratio of coherently-to-incoherently scattered radiation varies as a function of the composition. For the intensity ratio we obtain

$$\frac{N_{coh}}{N_{incoh}} = \frac{f^2}{R\,(1-f^2)}\,Z \qquad (13.16)$$

A certain dependence of this ratio upon Z exists. Using this method, Dwiggins (1961) determined the C/H-ratio of liquid hydrocarbons with carbon contents between 84 to 93 %, and Toussaint and Vos (1964) used this method for the analysis of solid hydrocarbons with carbon contents ranging from 90 to 97 %. They measured the intensity of the coherently and incoherently scattered portion of a characteristic line emitted by the tube, such as $(MoK_\alpha)_{coh}$ and $(MoK_\alpha)_{incoh}$ and, with the aid of standard samples determined the dependence of this ratio upon composition.

Chapter 14

Determination of Thin Film Thicknesses

In the analysis of a thin film the total amount of the element which is to be analyzed is concentrated on the surface. The determination of the thickness, therefore, involves measurement of the mass density per unit area (grams per square centimeter). A calibration curve of fluorescent intensity *vs* mass density is determined for quantitative analysis by using standard samples of known mass density. The fluorescent intensity of an element in the case of sufficiently thin films is proportional to the mass density and, hence, the resulting calibration curves are straight lines. However, the calibration curves become more shallow for larger mass densities and film thicknesses and, at a certain film thickness, no change in the fluorescent intensity is observed. This limiting value corresponds to the maximum sample thickness which can be determined by x-ray fluorescence. Koh and Caugherty (1952) experimentally determined these thicknesses for nickel, iron, and chromium films and found values ranging from 0.003 to 0.004 cm (30 to 40 μ) (Figure 14.1).

14.1. Determination of Mass Density

Let us consider the fluorescent intensity. A thin surface layer of height h contains the element A in a concentration of C_A; provided there is no secondary excitation by the substrate, the resulting fluorescent intensity of element A is given as follows:

$$N_A = \frac{q}{\sin \varphi} E_A C_A \int_{\lambda_0}^{\lambda_A} \frac{\mu_A(\lambda) N_0(\lambda) d\lambda}{C_A \bar{\mu}_A(\alpha) + C_B \bar{\mu}_B(\alpha)} \cdot$$
$$\cdot \{1 - \exp[-h\varrho(C_A \bar{\mu}_A(\alpha) + C_B \bar{\mu}_B(\alpha))]\} \tag{14.1}$$

Associated elements are designated as B and the average density of the sample is designated as ρ. For thin films (i.e., small values of h), the

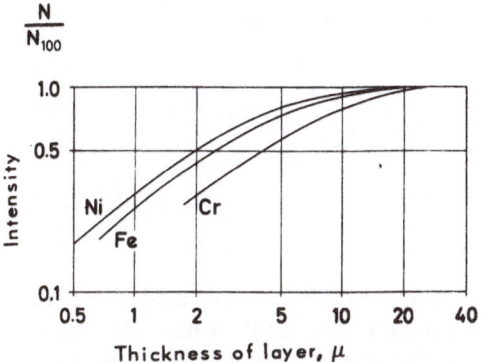

Fig. 14.1. Relative fluorescent intensity of nickel, iron, and chromium films in comparison to the fluorescent intensity of massive samples, plotted as a function of film thickness (Koh and Caugherty, 1952).

exponential function can be developed to the second term in the sequence, and the formula for the fluorescent intensity is simplified as follows:

$$N_A \cong \frac{q}{\sin \varphi} E_A C_A \varrho h \int_{\lambda_0}^{\lambda_A} \mu_A (\lambda) N_0 (\lambda) \, d\lambda \qquad (14.2)$$

The value

$$C_A \varrho h = m_A \; [\text{g/cm}^2]$$

corresponds to the mass density per unit area and is designated as m_A. If we compare the fluorescent intensity N_A of a thin film with the intensity N_{A100} of a massive sample consisting of pure A,

$$N_{A100} = \frac{q}{\sin \varphi} E_A \int_{\lambda_0}^{\lambda_A} \frac{\mu_A (\lambda) N_0 (\lambda) \, d\lambda}{\{\mu (\lambda) / \sin \varphi + \mu (\alpha) / \sin \psi\}_A}$$

we obtain

$$\frac{N_A}{N_{A100}} = \frac{m_A \int_{\lambda_0}^{\lambda_A} \mu_A (\lambda) N_0 (\lambda) \, d\lambda}{\int_{\lambda_0}^{\lambda_A} \frac{\mu_A (\lambda) N_0 (\lambda) \, d\lambda}{\{\mu (\lambda) / \sin \varphi + \mu (\alpha) / \sin \psi\}_A}} \qquad (14.3)$$

The mass density m_A is then given by

$$m_A = \frac{N_A}{N_{A100}} \cdot \frac{\displaystyle\int_{\lambda_0}^{\lambda_A} \frac{\mu_A(\lambda)\, N_0(\lambda)\, d\lambda}{\{\mu(\lambda)/\sin\varphi + \mu(\alpha)/\sin\psi\}_A}}{\displaystyle\int_{\lambda_0}^{\lambda_A} \mu_A(\lambda)\, N_0(\lambda)\, d\lambda} \tag{14.4}$$

Using this formula, Weyl (1961) determined the mass densities and film thicknesses of nickel and iron films, where the fluorescent intensities of massive samples consisting of pure nickel and iron were used for calibration. Weyl calculated the values of the two integrals from the mass absorption coefficients and the spectral intensity distribution of the particular tungsten tube employed in the experiments.

14.2. Semiquantitative Determination of Mass Density

By substituting for the integrands the value of the function at the weighted-average wavelength, the expression for the mass density may be further simplified

$$m_A = \frac{N_A}{N_{A100}} \frac{\dfrac{\mu_A(\bar{\lambda})\, N_0(\bar{\lambda})}{\{\mu(\bar{\lambda})/\sin\varphi + \mu(\alpha)/\sin\psi\}_A}}{\mu_A(\bar{\lambda}')\, N_0(\bar{\lambda}')} \frac{(\lambda_A - \lambda_0)}{(\lambda_A - \lambda_0)} \tag{14.5}$$

The weighted-averages of the two integrals are, however, not at exactly the same wavelength so that $\bar{\lambda}' \cong \bar{\lambda}$. Therefore $\mu(\lambda') \cong \mu(\lambda)$, $N_0(\lambda') \cong N_0(\lambda)$, and according to Pluchery (1963)

$$m_A \cong \frac{N_A}{N_{A100}} \frac{1}{\{\mu(\bar{\lambda})/\sin\varphi + \mu(\alpha)/\sin\psi\}_A} \tag{14.6}$$

Although the value for the weighted-average wavelength in this expression is unknown, its upper and lower limits may be given. The weighted-average wavelength is shorter than the wavelength of the absorption edge of the element that is to be excited and larger than the short-wavelength limit of the continuum of the x-ray spectrum employed for excitation. In this interval the mass absorption coefficient $\mu(\lambda)$ has a largest and a smallest value, and the value of the expression

$$\frac{1}{\mu(\bar{\lambda})/\sin\varphi + \mu(\alpha)/\sin\psi}$$

is calculated for these two extreme values of $\mu(\bar{\lambda})$.

The mass density of a thin film may therefore be determined semi-quantitatively from the fluorescent intensity emitted by the film and by a massive sample of the pure element. The accelerating potential of the x-ray tube is chosen so that the particular short-wavelength limit λ_0 is not too far away from λ_A. The limits for the expression in the determination of the mass density of thin copper films are calculated as follows (24-kV accelerating potential; $\varphi = 67°$; $\psi = 35°$):

$$\frac{1}{\mu(\bar{\lambda})/\sin\,\varphi + \mu(\alpha)/\sin\,\psi\}_{Cu}} = (2-7.4) \cdot 10^{-3} \text{ g/cm}^2$$

For a molybdenum tube, a value of 4.2×10^{-3} g/cm² is found experimentally.

14.3. Determination of Thin Film Thicknesses via Absorption

The film thickness may also be determined on the basis of the absorption efficiency of the film, namely by measuring the intensity of the fluorescent radiation emitted by the substrate: this intensity decreases with increasing thickness of the thin film until a certain thickness is reached where no fluorescence is measurable. Measurement of the film thickness using absorption is common in routine production control of tinplate. Let us consider the fluorescent intensity of the substrate. Element B of the substrate is excited to fluorescence by radiation passing through the thin cover film. Before reaching the substrate, the primary tube radiation of the intensity $N_0(\lambda)$ is decreased by absorption to the amount $N(\lambda)$. The spectral distribution of the primary tube radiation changes, however, when the radiation passes through the cover layer. This is due to the fact that absorption of the long-wavelength radiation in the cover layer is stronger than that of the shorter-wavelength radiation; hence, the substrate is excited to fluorescence by the short-wavelength radiation. $N(\lambda)$ is

$$N(\lambda) = N_0(\lambda) \exp\left[-\mu_A(\lambda)\,\varrho h/\sin\,\varphi\right] \qquad (14.7)$$

where A is the cover layer, h is its thickness, ρ is its density, and φ is the average incident angle of the tube radiation. The fluorescent intensity, which originates in the substrate and emerges towards the analyzing crystal, is designated N'_B. The fluorescent intensity of the wavelength β is absorbed to the amount N_B when passing through the cover layer. It is

$$N_B = N'_B \exp\left[-\mu_A(\beta)\,\varrho h/\sin\,\psi\right] \qquad (14.8)$$

N'_B is given as follows:

$$N'_B = \frac{q}{\sin\,\varphi}\, E_B \int_{\lambda_0}^{\lambda_B} \frac{\mu_B(\lambda)\,N_0(\lambda)}{\bar{\mu}_B(\beta)} \exp\left[-\mu_A(\lambda)\,\varrho h \sin\,\varphi\right] d\lambda$$

and, hence, we obtain for the fluorescent intensity N_B

$$N_B = \frac{q}{\sin \varphi} \, E_B \exp\,[-\mu_A\,(\beta)\,\varrho h\,/\,\sin \psi] \int\limits_{\lambda_0}^{\lambda_B} \frac{\mu_B\,(\lambda)\,N_0\,(\lambda)}{\bar{\mu}_B\,(\beta)} \, .$$
$$\cdot \exp\,[-\mu_A\,(\lambda)\,\varrho h\,/\,\sin \varphi]\,d\lambda \qquad\qquad (14.9)$$

The fluorescent intensity of the substrate decreases exponentially with increasing thickness of the cover layer (Figure 14.2). Measurement of the film thickness by absorption is best carried out by use of comparative standards of known mass density and film thickness.

The effect of surface roughness of the substrate on the analytical results was investigated for the absorption technique by Cass and Kelly (1963). They assume a rough surface to consist of rectangular cavities and elevated areas, whereby both deviate from the average film thickness h by the value Δh. (This model is not limited to rough substrate surfaces but is valid in general for cover films of irregular thicknesses. Furthermore, a cover film may differentiate into areas of various thicknesses after the layer

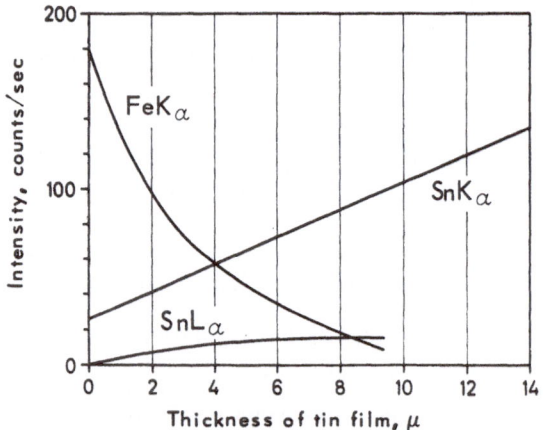

Fig. 14.2. Comparison of various methods for determination of thin film thicknesses. The mass density of a film may be determined directly from its fluorescent intensity ($SnK\alpha$ and $SnL\alpha$) or from its absorption efficiency. In the case of the absorption method, the magnitude of absorption by the cover layer of the fluorescent radiation emitted by the substrate ($FeK\alpha$) is measured. [Example: The determination of tin in tinplate. Excitation by radioactive americium-241, and measurement with a scintillation counter ($SnK\alpha$) and a flow proportional counter ($SnL\alpha$, $FeK\alpha$). (Cook, Mellish, and Payne, 1960).]

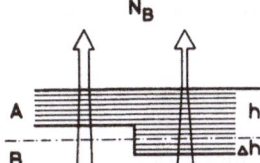

Fig. 14.3. Calculation of the effects of irregular cover film thicknesses on fluorescent intensity.

has been deposited.) For simplicity, only absorption of the fluorescent radiation emerging perpendicularly to the surface is considered (Figure 14.3). The fluorescent intensity of the substrate covered by too thin a film is given as follows:

$$N_B = N'_B \exp\left[-\mu_A(\beta)\,\varrho\,(h-\Delta h)\right] \tag{14.10}$$

The intensity of the radiation emerging from an area covered by too thick a cover film, on the other hand, is given as

$$N_B = N'_B \exp\left[-\mu_A(\beta)\,\varrho\,(h+\Delta h)\right] \tag{14.11}$$

The total fluorescent intensity of the substrate for a rough surface is

$$N_{B1} = N'_B \left\{\exp\left[-\mu_A(\beta)\,\varrho\,(h-\Delta h)\right] + \exp\left[-\mu_A(\beta)\,\varrho\,(h+\Delta h)\right]\right\} \tag{14.12}$$

The total intensity emitted by a substrate with smooth surface and uniform film thickness h is

$$N_{B2} = 2\,N'_B \exp\left[-\mu_A(\beta)\,\varrho h\right] \tag{14.13}$$

When introducing the term $u = \mu_A(\beta)\rho$, the ratio of the two intensities is calculated to be

$$\frac{N_{B1}}{N_{B2}} = \frac{\exp\left[-u\,(h-\Delta h)\right] + \exp\left[-u\,(h+\Delta h)\right]}{2 \exp\left[-uh\right]} \tag{14.14}$$

Shortened by $\exp(-uh)$, we obtain for the intensity ratio

$$\frac{N_{B1}}{N_{B2}} = \frac{\exp\left[u\Delta h\right] + e\left[-u\Delta h\right]}{2} = \cosh\left(u\Delta h\right) =$$

$$= 1 + \frac{(u\Delta h)^2}{2!} + \frac{(u\Delta h)^4}{4!} + \cdots \tag{14.15}$$

When measuring the film thickness by absorption, the fluorescent intensity of the substrate with irregular film thickness is larger than in the case of uniform thickness, i.e., the method is relatively sensitive to irregularities in the film thickness. Let us compare this to the sensitivity of the direct method. Applying the same model, the intensities from too thin and too thick a cover film [equation (14.2)] are, respectively,

$$N_A = \text{prop } C_A \varrho \ (h - \varDelta h) \quad \text{and} \quad N_A = \text{prop } C_A \varrho \ (h + \varDelta h) \quad (14.16)$$

The intensity N_{A1} of the radiation for irregular film thickness is

$$N_{A1} = 2 \text{ prop } C_A \varrho h \qquad (14.17)$$

and is equally as large as the fluorescent intensity N_{A2} of a uniformly thick cover film. The method of direct measurement of the thickness is relatively insensitive to irregularities in the film thickness.

Chapter 15

Determination of High Concentrations with Calibration Curves

The fluorescent intensity of a component is, in a first approximation, proportional to the concentration. The precise composition is determined with the aid of standard samples. For mixtures consisting of two and three components, the graphical method of determining the concentration is commonly used. Calibration curves are obtained by measuring samples of known composition for each component and by presenting graphically the relationship between intensity and concentration. The effect of associated components on the fluorescent intensity is evident from the fact that curves instead of straight lines are obtained, where the calibration curves for one and the same component may have different curvatures depending upon the associated components. The calibration curve used for determining niobium oxide in a mixture with tantalum oxide, for example, has a different curvature than the calibration curve used for determining niobium oxide in a mixture with titanium oxide (Figure 15.1). In order to balance as much as possible the differences in curvature between individual calibration points, the regression coefficient may be calculated first and the corresponding regression curve may then be plotted. The calibration curve for a two-component mixture is given approximately by the following function (see

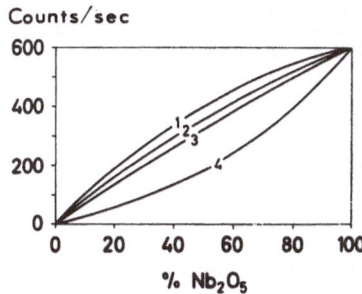

Counts/sec

% Nb_2O_5

Fig. 15.1. Calibration curves for the determination of niobium oxide. The calibration curves have different curvatures depending upon the associated components. Niobium oxide in mixture with (1) titanium oxide, (2) zirconium oxide, (3) iron oxide, and (4) tantalum oxide. (Mitchell, 1958.)

Chapter 12):

$$\frac{N_A}{N_{A100}} = \frac{C_A}{C_A + C_B r_{AB}}.$$

where r_{AB} is the regression coefficient. The complete curve is determined by one standard sample. The best value for the regression coefficient may be found mathematically when several standards are available (Chapter 16).

Only one calibration curve exists for the determination of A in a two-component mixture consisting of the two components A and B. A multitude of calibration curves is obtained, however, for the determination of the content of A in a three-component mixture, ABC. The calibration curve for C-poor mixtures is located close to the calibration curve for the two-component mixture AB. With increasing content of C, the curve departs more and more from the calibration curve for AB until for C-rich mixtures, it approaches the calibration curve for the two-component mixture AC. Let us investigate more closely the shift of calibration curves as a function of the amounts of the associated components. We first assume the three-component mixture ABC to be a two-component mixture AB where part of component B is replaced by a third component C, and where either a fixed amount of B is replaced by C (for example, always 20 wt. %), or where the replaced amount is always a certain fixed proportion of B (for example, always 1/3 of B is replaced by C). Different groups of calibration curves result.

15.1. Curves for Constant Admixture of a Third Component

Based on our previous experience with the fluorescent intensity of component A for a fixed concentration C_A, let us now consider the change in intensity when substituting part of the associated component B by component C. For the fluorescent intensity of the component A in a mixture AB or ABC, we write (Chapter 6):

$$N_{AB} = \frac{q}{\sin \varphi} E_A \int_{\lambda_0}^{\lambda_A} \frac{C_A \mu_A(\lambda) N_0(\lambda) \, d\lambda}{C_A \bar{\mu}_A(\alpha) + C_B \bar{\mu}_B(\alpha)} \tag{15.1}$$

or

$$N_{ABC} = \frac{q}{\sin \varphi} E_A \int_{\lambda_0}^{\lambda_A} \frac{C_A \mu_A(\lambda) N_0(\lambda) \, d\lambda}{C_A \bar{\mu}_A(\alpha) + C'\bar{\mu}_B(\alpha) + C_C \bar{\mu}_C(\alpha)}. \tag{15.2}$$

where C'_B is the concentration of B after part of it was replaced by C. From

$$C_A + C_B = 1 \quad \text{or} \quad C_A + C'_B + C_C = 1$$

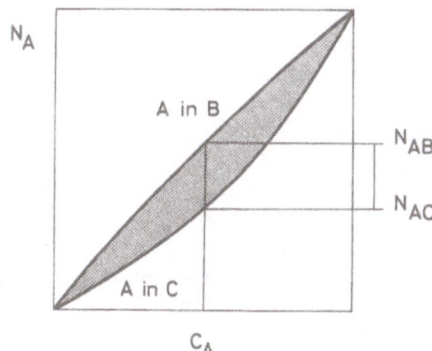

Fig. 15.2. The fluorescent intensity of component A, for a fixed concentration C_A in a three-component mixture ABC varies between the values N_{AB} to N_{AC}, depending upon the residual composition of the sample. The relationship between intensity and concentration is not given by a line but by a field.

follows

$$C'_B = C_B - C_C = 1 - C_A - C_C \qquad (15.3)$$

C_A, C'_B, and C_C are the concentrations of the components ABC in the three-component mixture; C_A and C_B are the concentrations of the components in a two-component mixture. The ratio of the two intensities is approximately

$$\frac{N_{AB}}{N_{ABC}} \cong \frac{C_A \bar{\mu}_A(\alpha) + (C_B - C_C)\, \bar{\mu}_B(\alpha) + C_C \bar{\mu}_C(\alpha)}{C_A \bar{\mu}_A(\alpha) + C_B \bar{\mu}_B(\alpha)} =$$

$$= 1 + \frac{\bar{\mu}_C(\alpha) - \bar{\mu}_B(\alpha)}{C_A \bar{\mu}_A(\alpha) + C_B \bar{\mu}_B(\alpha)}\, C_C \qquad (15.4)$$

In order to obtain the relative intensity change, the intensity ratio is written

$$\frac{N_{AB}}{N_{ABC}} = \frac{N_{AB}}{N_{AB} + \Delta N_A} = 1 - \frac{\Delta N_A}{N_{AB}} + \left(\frac{\Delta N_A}{N_{AB}}\right)^2 - \cdots$$

We consider only the first two terms in the sequence and, for the relative change in intensity, obtain approximately

$$\frac{\Delta N_A}{N_A} \approx \frac{\bar{\mu}_B(\alpha) - \bar{\mu}_C(\alpha)}{C_A \bar{\mu}_A(\alpha) + C_B \bar{\mu}_B(\alpha)}\, C_C \approx \text{prop}\, C_C \qquad (15.5)$$

When partially replacing one associated component by another, the fluorescent intensity changes approximately proportionally to the magnitude of the replacement. When determining experimental calibration curves for measurement of A in a mixture (B + C), where the mixture contains a fixed amount of C [for example, calibration curve of A in (B + 10 wt. % C); A in (B + 20 wt. % C); A in (B + 30 wt. % C), etc.], linear interpolation is

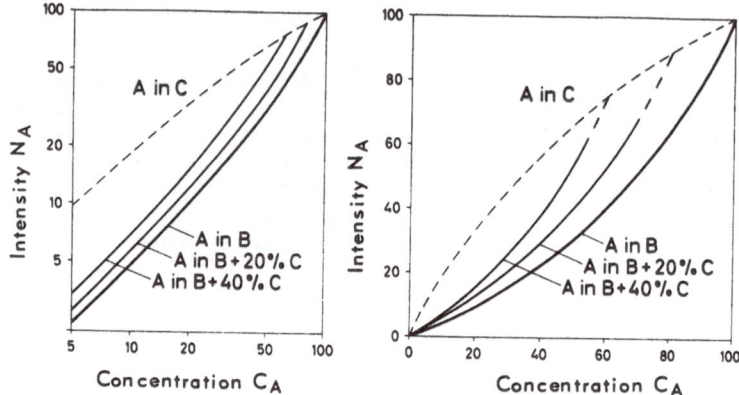

Fig. 15.3. Calibration curve for the determination of component A in a three-component mixture ABC. Calibration curves for constant admixtures of C. Linear interpolation between the individual curves is largely possible. Plot on a log–log scale (*left*), and on a linear scale (*right*). (Niobium oxide determination in a mixture of tantalum oxide and solid admixtures of titanium oxide.)

possible to a large extent between the individual calibration curves. When plotting these calibration curves on log–log paper (logarithm of intensity *vs* logarithm of concentration), approximately parallel curves are obtained which are separated by constant distances; linear interpolation is then particularly simple. When plotting the resulting calibration curves on a linear scale, however, fan-shaped groups of curves result (Figure 15.3).

15.2. Curves for Constant Mixing Ratios of Associated Components

Extension of a two-component mixture AB to a three-component mixture ABC can be said to consist of replacement of a fixed proportion of B by C, and a well-defined relationship exists between the concentration C_B and C_C of the two associated components. The fluorescent intensities N_{AB} and N_{ABC} are expressed as previously [equations (15.1) and (15.2), respectively]. After normalizing, we obtain equation (15.3):

$$C'_B = C_B - C_C$$

The mixing ratio of the associated components is defined as the ratio of the concentration of component C to the concentration of the residual amount of B

$$\frac{C_C}{C'_B} = \frac{C_C}{C_B - C_C} = v, \quad C_C = \frac{v}{1 + v} C_B = v C'_B$$

Again, we have to keep in mind that C_B and C_B' are the concentrations of B in a two- and three-component mixture, respectively. We obtain for the fluorescent intensity of A

$$N_{ABC} = \frac{q}{\sin \varphi} E_A \int_{\lambda_0}^{\lambda_A} \frac{C_A \mu_A (\lambda) N_0 (\lambda) \, d\lambda}{C_A \bar{\mu}_A (\alpha) + C_B \left[\dfrac{1}{1+v} \bar{\mu}_B (\alpha) + \dfrac{v}{1+v} \bar{\mu}_C (\alpha) \right]} \tag{15.6}$$

For the determination of the concentration of component A in a three-component mixture ABC, a calibration curve similar to the one for the determination of A in a two-component mixture is obtained. The two associated components B and C can be thought to be replaced by a single associated component whose mass absorption coefficient $\bar{\mu}_{BC}(\alpha)$ is equally as large as that of the mixture (B + C):

$$\bar{\mu}_{BC} (\alpha) = \left[\frac{1}{1+v} \bar{\mu}_B (\alpha) + \frac{v}{1+v} \bar{\mu}_C (\alpha) \right]$$

When comparing the fluorescent intensity of component A in a mixture with (B + C) to the fluorescent intensity N_{A100} of pure component A, we obtain an approximate intensity ratio

$$\frac{N_{ABC}}{N_{A100}} \cong \frac{C_A \bar{\mu}_A (\alpha)}{C_A \bar{\mu}_A (\alpha) + C_B \left[\dfrac{1}{1+v} \bar{\mu}_B (\alpha) + \dfrac{v}{1+v} \bar{\mu}_C (\alpha) \right]} \tag{15.7}$$

Upon introducing the regression coefficients r_{AB} and r_{AC},

$$r_{AB} = \frac{\bar{\mu}_B (\alpha)}{\bar{\mu}_A (\alpha)} \quad \text{and} \quad r_{AC} = \frac{\bar{\mu}_C (\alpha)}{\bar{\mu}_A (\alpha)}$$

we obtain

$$\frac{N_A}{N_{A100}} \cong \frac{C_A}{C_A + C_B \left[\dfrac{1}{1+v} r_{AB} + \dfrac{v}{1+v} r_{AC} \right]} \tag{15.8}$$

The regression coefficient of the new calibration curve for the determination of A in a mixture with (B + C) can easily be calculated from the regression coefficients r_{AB} and r_{AC} as follows:

$$r_{ABC} = \left(\frac{1}{1+v} r_{AB} + \frac{v}{1+v} r_{AC} \right), \quad v = \frac{C_C}{C_B'}$$

For numerical examples see Table 15.1.

Table 15.1. Determination of Niobium Oxide in a Mixture with Tantalum Oxide and Titanium Oxide. Numerical example for the calculation of the regression coefficients for several mixing ratios of titanium oxide–tantalum oxide (see Figure 15.4).

Mixing ratio of associated components	Calculation of regression coefficients	
TiO_2 pure	experimental	$r = 0.49$
$TiO_2/Ta_2O_5 = 3/1$	$\dfrac{3 \cdot 0.49 + 2.26}{4}$	$r = 0.93$
$TiO_2/Ta_2O_5 = 1/1$	$\dfrac{0.49 + 2.26}{2}$	$r = 1.38$
$TiO_2/Ta_2O_5 = 1/3$	$\dfrac{0.49 + 3 \cdot 2.26}{4}$	$r = 1.82$
Ta_2O_5 pure	experimental	$r = 2.26$

A group of curves of type AB of different curvature is obtained for the determination of component A in a mixture with (B + C) (Figure 15.4). A particularly simple case is the determination of small amounts of A in a mixture with (B + C). The calibration curves become straight lines for low concentrations and the regression coefficients correspond to the slopes of the straight lines (Figure 15.5).

15.3. Presentation in a Concentration Triangle

The chemical composition of a three-component mixture is determined unequivocally by a point in a concentration triangle. Every point in the concentration triangle that corresponds to three values for the concentration is also representative for the corresponding three intensities N_A, N_B, and N_C of the components A, B, and C; these intensities also determine the point and, hence, the sample unequivocally. Curves of constant fluorescent intensity may be plotted in a concentration triangle, such as curves for 1000, 2000, 3000, . . . counts/sec. A network of intensity curves is obtained when plotting the corresponding groups of curves for the three fluorescent intensities, where linear interpolation is largely possible within each sector of the network. In order to relate fluorescent intensity of a sample to concentration, the point in the intensity network, whose intensities correspond to the measured values, is determined and the corresponding composition is given by the coordinates of the concentration triangle.

N_A

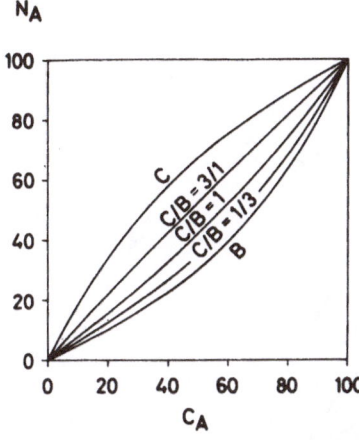

Fig. 15.4. Calibration curves for the determination of component A in a three-component mixture ABC, and for constant mixing ratios of the associated components B and C. Curves are similar to the ones for two-component mixtures. (Niobium oxide determination in mixture with tantalum oxide–titanium oxide.)

For construction of the intensity network, one appropriately starts out with the calibration curves. Groups of calibration curves for constant admixtures or for constant mixing ratios maybe used. When the calibration curve for A in B differs little from the calibration curve of A in C, the resulting curves of constant fluorescent intensity in the concentration triangle are approximately straight lines. However, if this condition is not fulfilled, slightly bent curves originate (Figure 15.6).

15.4. Correction for Effects of a Third Component by Correction Factors

The effects of a third component on the fluorescent intensity may be adjusted with the aid of correction factors. The three-component mixture

N_A

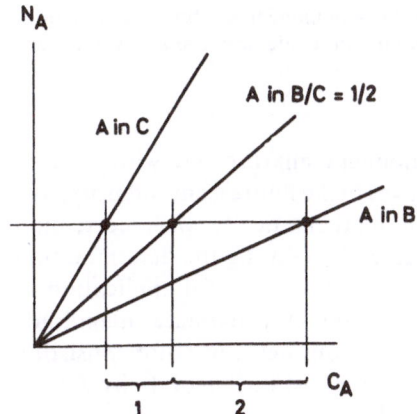

Fig. 15.5. Construction of a calibration curve for the determination of minor amounts of A in a mixture with (B + C), where the mixing ratio of associated components is fixed. The distance between the straight lines (A in B) and (A in C) is divided by the calibration curve (A in B/C) in reverse proportion to the mixing ratio of the associated components (niobium oxide determination in mixture with tantalum oxide and titanium oxide).

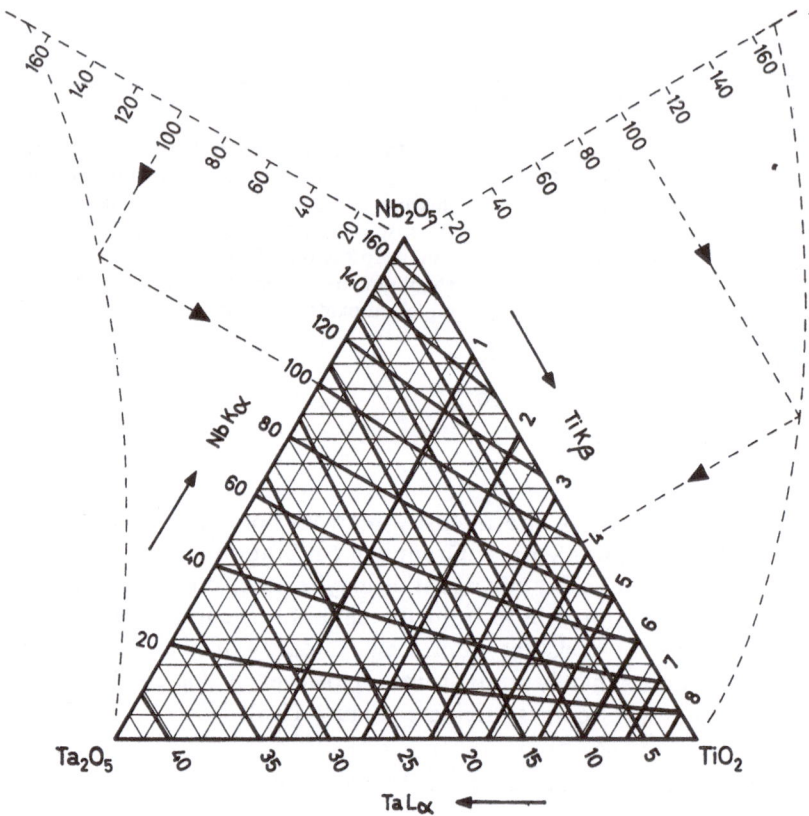

Fig. 15.6. Concentration triangle for samples of the three-component mixture niobium oxide–tantalum oxide–titanium oxide; also plotted are curves of constant fluorescent intensity of NbKα, TaLα, and TiKβ. Every point in the concentration triangle is determined unequivocally by the corresponding three fluorescent intensities. The corresponding calibration curves for the two-component mixtures of niobium oxide–tantalum oxide and niobium oxide–titanium oxide are presented on the left and right, respectively (weakly dashed).

ABC is again considered as a two-component mixture AB with varying admixtures of a third component C. Changes in the fluorescent intensity, due to the substitution, may be eliminated mathematically with correction factors and the composition may be obtained by relating the determination to the curves of the two-component mixture AB. Successful application of correction factors requires knowledge of the approximate magnitude of admixture of the third component. Let us consider a mixture consisting of three components ABC. With increasing substitution of B by C, the fluorescent intensity of the component A departs more and more from the original intensity in the two-component mixture AB, and the correction factors become larger. When applying the correction factors according to Mitchell (1958)

$$\frac{N_{A \text{ in } ABC}}{N_{A \text{ in } AB}} \qquad (15.9)$$

the intensity ratios are formed for every concentration of C, and the intensity ratios are plotted as a function of the amount of C. These correction factors, as a function of C_C, form hyperbolic curves (Figure 15.7). For the intensity ratio in the equation given above we obtain

$$\frac{N_{ABC}}{N_{AB}} \cong \frac{C_A \bar{\mu}_A(\alpha) + C_B \bar{\mu}_B(\alpha)}{C_A \bar{\mu}_A(\alpha) + (C_B - C_C) \bar{\mu}_B(\alpha) + C_C \bar{\mu}_C(\alpha)}$$

This expression is the equation of a hyperbola. In order to obtain the corrected intensity, the measured intensity is divided by the correction factor according to Mitchell.

The correction factor may better be defined as the ratio N_{AB}/N_{ABC}, and we obtain

$$\frac{N_{AB}}{N_{ABC}} \cong \frac{C_A \bar{\mu}_A(\alpha) + (C_B - C_C) \bar{\mu}_B(\alpha) + C_C \bar{\mu}_C(\alpha)}{C_A \bar{\mu}_A(\alpha) + C_B \bar{\mu}_B(\alpha)} =$$

$$= 1 + \frac{\bar{\mu}_C(\alpha) - \bar{\mu}_B(\alpha)}{C_A \bar{\mu}_A(\alpha) + C_B \bar{\mu}_B(\alpha)} C_C \qquad (15.10)$$

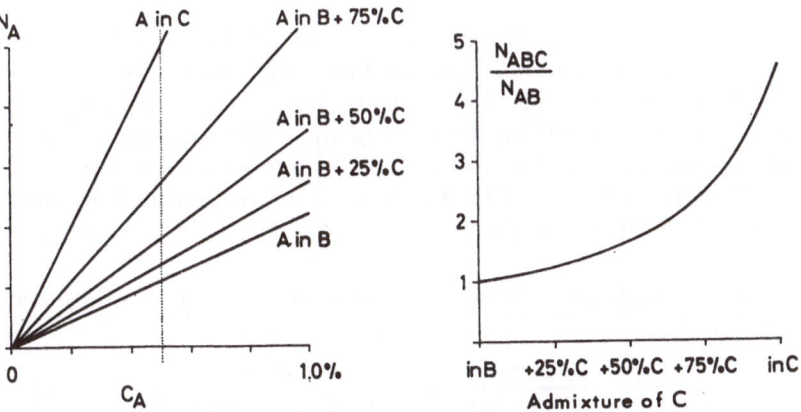

Fig. 15.7. Correction factors. *Left*: Calibration curves for the determination of concentration of minor amounts of A in a mixture with (B + C), where fixed amounts of C are added to the mixture. *Right*: Resulting correction factors according to Mitchell as a function of the content of C; hyperbolic curve.

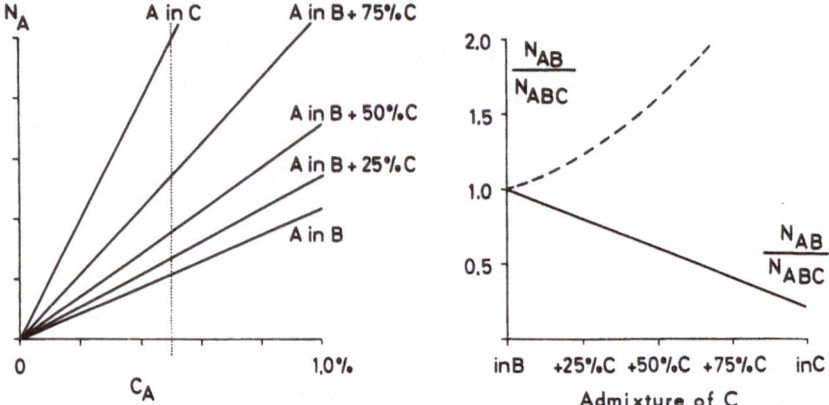

Fig. 15.8. Straight-line correction factor. *Left*: Groups of calibration curves for the determination of the concentration of small amounts of A in a mixture with (B + C), where fixed amounts of C are added to the mixture. *Right*: Correction factor N_{AB}/N_{ABC} as a function of C. The correction factor changes proportionally to the amount of C. Weakly dashed: Correction factor according to Mitchell.

This expression is the equation of a straight line (Figure 15.8). In the particular case in question, the magnitude of the correction factor is linearly dependent on the amount of admixed C. In order to obtain the corrected intensity, the measured intensity is multiplied by this correction factor. This definition of the correction factor has the advantage of including the so-called additivity of the correction factors.

The correction factors of the type N_{AB}/N_{ABC} are additive. Let us consider a four-component mixture ABCD where the effects of the associated components C and D on the fluorescent intensity of A are to be corrected in such a way that the determination of the concentration of A again may be carried out on the basis of the calibration curve of a two-component mixture AB. For the ratio of the fluorescent intensity of component A in mixture with B to the intensity of component A in mixture with [B + C + D], we obtain

$$\frac{N_{AB}}{N_{ABCD}} \cong \frac{C_A \bar{\mu}_A(\alpha) + (C_B - C_C - C_D)\bar{\mu}_B(\alpha) + C_C \bar{\mu}_C(\alpha) + C_D \bar{\mu}_D(\alpha)}{C_A \bar{\mu}_A(\alpha) + C_B \bar{\mu}_B(\alpha)}$$

$$\cong 1 + \frac{\bar{\mu}_C(\alpha) - \bar{\mu}_B(\alpha)}{C_A \bar{\mu}_A(\alpha) + C_B \bar{\mu}_B(\alpha)} C_C + \frac{\bar{\mu}_D(\alpha) - \bar{\mu}_B(\alpha)}{C_A \bar{\mu}_A(\alpha) + C_B \bar{\mu}_B(\alpha)} C_D \quad (15.11)$$

The effects of the associated components are subdivided into the effects of the component C and the effects of the component D. The magnitude of the respective effects can easily be calculated from the concentrations C_C and C_D. Knowing the admixed amounts of C and D, we find for the correction factor

$$\frac{N_{AB}}{N_{ABCD}} \cong 1 + C_C \Xi_C + C_D \Xi_D$$

where Ξ_C and Ξ_D describe the effects of the particular third components. As a result, the relative intensity change may be considered instead of the intensity ratio

$$\frac{N_{ABCD} - N_{AB}}{N_{ABCD}} = \frac{\bar{\mu}_B(\alpha) - \bar{\mu}_C(\alpha)}{C_A \bar{\mu}_A(\alpha) + C_B \bar{\mu}_B(\alpha)} C_C + \frac{\bar{\mu}_B(\alpha) - \bar{\mu}_D(\alpha)}{C_A \bar{\mu}_A(\alpha) + C_B \bar{\mu}_B(\alpha)} C_D$$

$$(15.12)$$

The relative intensity change is approximately proportional to the amount of the admixed third component. The effect of a third component on the intensity may also be corrected for by determining the relative intensity change which would occur upon removal of the interfering third component. The change in intensity, as a result of a change in the concentration of the associated components, may be calculated approximately from the mass absorption coefficients (Neff 1959b; Laffolie 1962a). Instead of the combined mass absorption coefficients $\bar{\mu}(\alpha)$ we only consider the mass absorption

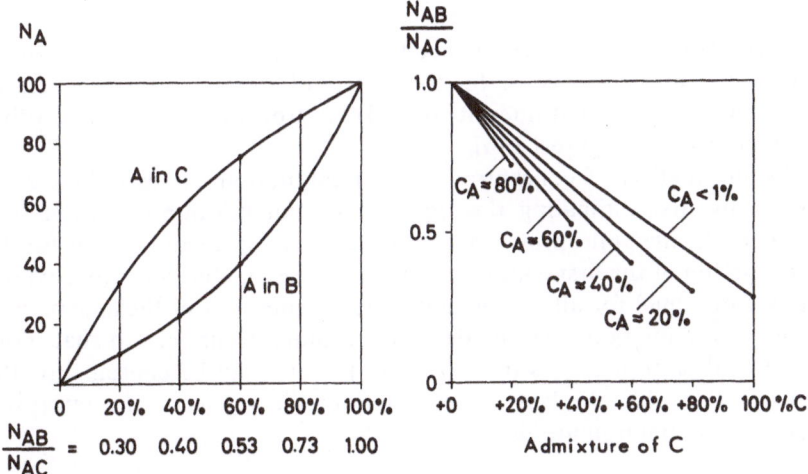

Fig. 15.9. Calibration curves (*left*) and correction factors (*right*) for the determination of component A in mixture with B and C, respectively. Correction factors for A-poor mixtures are different than factors for A-rich mixtures.

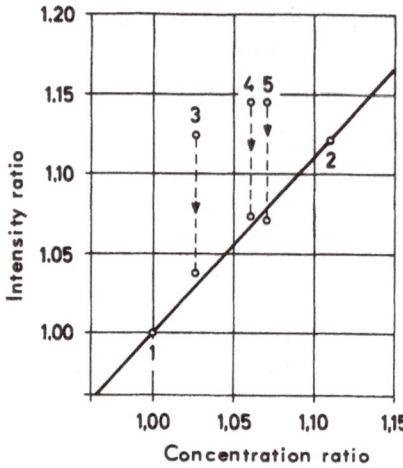

Fig. 15.10. Fluorescent intensity of nickel–iron alloys. Mathematical correction for the effects of small amounts of molybdenum on the nickel fluorescence. After correction, points 3, 4, and 5 are located on the straight-line calibration curve for pure nickel–iron alloys (Laffolie, 1962).

coefficient for the emerging fluorescent radiation of wavelength α which simplifies the expression for the relative intensity change as follows:

$$\frac{N_{ABC} - N_{AB}}{N_{ABC}} = \frac{\bar{\mu}_B(\alpha) - \bar{\mu}_C(\alpha)}{C_A \bar{\mu}_A(\alpha) + C_B \bar{\mu}_B(\alpha)} C_C \approx$$

$$\approx \frac{\mu_B(\alpha) - \mu_C(\alpha)}{C_A \mu_A(\alpha) + C_B \mu_B(\alpha)} C_C = \frac{\Delta \mu(\alpha)}{\mu(\alpha)}$$

where $\mu(\alpha)$ is the mass absorption coefficient for the fluorescent radiation of element A. Laffolie corrected for the effects of small amounts of molybdenum on the fluorescent intensity of nickel in iron–nickel alloys according to this procedure (Figure 15.10).

Determination of concentration with calibration curves is best done in steps by first estimating the approximate composition of the matrix. On this basis, the appropriate calibration curve or correction factor for the determination of the first component may be chosen. Once first approximations are obtained for all components, the composition of the matrix, in a second approximation, may be determined more accurately so that more accurate calibration curves or correction factors may be chosen. For the third (and possibly the last) approximation, the values of the improved second approximation are used, etc.

Determination of Concentration, Formulated as a Linear System of Equations

Determination of concentration in three- and multicomponent mixtures is complex because of interelement effects and is therefore best formulated mathematically as a linear system of equations. In this way relationships may be recognized more clearly. It is apparent that the determination of concentration in a multicomponent mixture is equivalent to the solution of a linear system of equations consisting of several equations with several unknowns. This analogy illustrates why the determination of concentration often can not be carried out directly but requires an appropriate technique of sample preparation by which the determination of concentration as well as the corresponding system of equations become solvable.

16.1. Derivation of the Linear System of Equations

The linear system of equations is derived with the aid of the regression equation, where a unique equation is formulated for every component. The system of equations, therefore, consists of as many equations as the compound contains components. The concentrations of the individual components are found by searching for such values and concentrations, respectively, for which the total system of equations is fulfilled simultaneously. Let us consider the fluorescent intensity N_A of the component A in a mixture with the associated components B, C, D, ... and compare these intensities to the intensity N_{A100} of the pure component A. The concentrations of the individual components are C_A, C_B, C_C, C_D, ..., and the combined mass absorption coefficients for the incident primary tube radiation and the emerging fluorescent radiation of the component A are $\bar{\mu}_A(\alpha)$, $\bar{\mu}_B(\alpha)$, $\bar{\mu}_C(\alpha)$, $\bar{\mu}_D(\alpha)$, etc. By introducing the values for N_A and N_{A100}, the ratio of the two intensities for the case where component A is not excited to fluorescence by secondary excitation, is given as follows:

$$N_A = \frac{q}{\sin \varphi} E_A \int_{\lambda_0}^{\lambda_A} \frac{C_A \mu_A(\lambda) N_0(\lambda) d\lambda}{C_A \bar{\mu}_A(\alpha) + C_B \bar{\mu}_B(\alpha) + C_C \bar{\mu}_C(\alpha) + C_D \bar{\mu}_D(\alpha) + \cdots}$$

(16.1)

$$N_{A100} = \frac{q}{\sin \varphi} E_A \int_{\lambda_0}^{\lambda_A} \frac{\mu_A(\lambda) N_0(\lambda) d\lambda}{\bar{\mu}_A(\alpha)}$$

(16.2)

$$\frac{N_A}{N_{A100}} \cong \frac{C_A \bar{\mu}_A(\alpha)}{C_A \bar{\mu}_A(\alpha) + C_B \bar{\mu}_B(\alpha) + C_C \bar{\mu}_C(\alpha) + C_D \bar{\mu}_D(\alpha) + \cdots}$$

(16.3)

This approximation assumes that the fluorescent intensity in the mixture and in the pure sample is largely produced by the same wavelength range of the primary tube spectrum and, hence, the weighted-average wavelength is the same in both cases. We obtain from this equation for the concentration C_A of the component A

$$C_A = \frac{N_A}{N_{A100} - N_A} \left(C_B \frac{\bar{\mu}_B(\alpha)}{\bar{\mu}_A(\alpha)} + C_C \frac{\bar{\mu}_C(\alpha)}{\bar{\mu}_A(\alpha)} + C_D \frac{\bar{\mu}_D(\alpha)}{\bar{\mu}_A(\alpha)} + \cdots \right)$$

(16.4)

The term outside the parentheses accounts for the dependence of the concentration on the measured fluorescent intensity: the concentration increases with increasing intensity. Terms in parentheses account for interactions of the associated components and contain the concentration of the associated component B multiplied by its "interaction coefficient" $\bar{\mu}_B(\alpha)/\bar{\mu}_A(\alpha)$; a term accounting for the interaction with the next associated component then follows, and its concentration is multiplied by the "interaction coefficient" $\bar{\mu}_C(\alpha)/\bar{\mu}_A(\alpha)$. This continues to the last component whose concentration is also multiplied by its interaction coefficient. The fluorescent intensity is modified by the associated components, and the effect of an associated component on the intensity of the radiation is proportional to its concentration and interaction coefficient.

The concentration of a particular component may be calculated from its fluorescent intensity, provided not only the individual interaction coefficients but also the concentrations of the associated components are known which, for the latter, is not normally the case. An equation analogous to that for the concentration of component A, however, can be written for component B

$$C_B = \frac{N_B}{N_{B100} - N_B} \left(C_A \frac{\bar{\mu}_A(\beta)}{\bar{\mu}_B(\beta)} + C_C \frac{\bar{\mu}_C(\beta)}{\bar{\mu}_B(\beta)} + C_D \frac{\bar{\mu}_D(\beta)}{\bar{\mu}_B(\beta)} + \cdots \right)$$

(16.5)

Similar equations may also be written for the concentrations of all other associated components

$$C_C = \frac{N_C}{N_{C100} - N_C} \left(C_A \frac{\bar{\mu}_A(\gamma)}{\bar{\mu}_C(\gamma)} + C_B \frac{\bar{\mu}_B(\gamma)}{\bar{\mu}_C(\gamma)} + C_D \frac{\bar{\mu}_D(\gamma)}{\bar{\mu}_C(\gamma)} + \cdots \right), \text{etc.}$$ (16.6)

The concentration of a component may be determined from its fluorescent intensity using one of these equations, provided the concentrations of the other components are already known. This difficulty is not eliminated by considering all equations simultaneously but the problem becomes mathematically solvable. When rewriting every equation [transferring all terms on one side and multiplying every equation by $(N_{100} - N)/N$], the particular system of equations for the determination of concentration in a four-component mixture is as follows:

$$-C_A \left(\frac{N_{A100} - N_A}{N_A} \right) + C_B \left(\frac{\bar{\mu}_B(\alpha)}{\bar{\mu}_A(\alpha)} \right) + C_C \left(\frac{\bar{\mu}_C(\alpha)}{\bar{\mu}_A(\alpha)} \right) + C_D \left(\frac{\bar{\mu}_D(\alpha)}{\bar{\mu}_A(\alpha)} \right) = 0$$

$$C_A \left(\frac{\bar{\mu}_A(\beta)}{\bar{\mu}_B(\beta)} \right) - C_B \left(\frac{N_{B100} - N_B}{N_B} \right) + C_C \left(\frac{\bar{\mu}_C(\beta)}{\bar{\mu}_B(\beta)} \right) + C_D \left(\frac{\bar{\mu}_D(\beta)}{\bar{\mu}_B(\beta)} \right) = 0$$

$$C_A \left(\frac{\bar{\mu}_A(\gamma)}{\bar{\mu}_C(\gamma)} \right) + C_B \left(\frac{\bar{\mu}_B(\gamma)}{\bar{\mu}_C(\gamma)} \right) - C_C \left(\frac{N_{C100} - N_C}{N_C} \right) + C_D \left(\frac{\bar{\mu}_D(\gamma)}{\bar{\mu}_C(\gamma)} \right) = 0$$

$$C_A \left(\frac{\bar{\mu}_A(\delta)}{\bar{\mu}_D(\delta)} \right) + C_B \left(\frac{\bar{\mu}_B(\delta)}{\bar{\mu}_D(\delta)} \right) + C_C \left(\frac{\bar{\mu}_C(\delta)}{\bar{\mu}_D(\delta)} \right) - C_D \left(\frac{N_{D100} - N_D}{N_D} \right) = 0$$ (16.7)

We are dealing here with a homogeneous linear system of equations with n equations and n unknowns, where the equations contain the measured values N_A, N_B, N_C, and N_D, the experimentally determined interaction coefficients $(\bar{\mu}_i/\bar{\mu}_j)$, and the intensities of the pure components N_{A100}, N_{B100}, N_{C100}, and N_{D100}; the desired concentrations are designated C_A, C_B, C_C, and C_D. The determination of the concentration in a multi-component mixture consisting of n components is identical to the problem of solving a system of equations with n equations and n unknowns. When deriving the system of equations, it is assumed that the equation for the concentration of component A is formally valid for the concentration of the other components as well. In the derivation of the equation for A it was assumed, however, that component A is not, in addition, excited to fluorescence by the associated elements. This restriction may be made for one component of the mixture but cannot arbitrarily be extended to fluorescent intensities of all other components; in a multicomponent system there are always certain components which are excited to fluorescence by other associated components. This uncertainty, however, may be reduced in magnitude when substituting the so-called absorption excitation coefficients according to Noakes (1954) and Birks (1959) (Chapter 7) in place of the mass absorption coefficients, and when the secondary excitation is formally treated as a "negative absorption."

16.2. Formulation According to Sherman

The linear system of equations for determination of the concentration in a multicomponent mixture was apparently first presented in a paper by Sherman (1954) and later adapted by Burnham, Hower, and Jones (1957) and by Preis and Esenwein (1959). Sherman's system of equations is as follows:

$$
\begin{aligned}
(a_{11} - t_1)\, c_1 + a_{21} c_2 \quad\quad + a_{31} c_3 \quad\;\; &= 0 \\
a_{12} c_1 \quad\quad + (a_{22} - t_2)\, c_2 + a_{32} c_3 \quad\;\; &= 0 \\
a_{13} c_1 \quad\quad + a_{23} c_2 \quad\quad + (a_{33} - t_3)\, c_3 &= 0
\end{aligned}
$$

where c_1, c_2, and c_3 are the concentrations of the individual components, and t_1, t_2, and t_3 are the counting times required to accumulate a predetermined number of counts for a particular fluorescent line. When introducing the values T_{A100}, T_{B100}, and T_{C100} as the counting times for the fluorescent lines of the pure components, and T_A, T_B, and T_C as the counting times for the fluorescent lines in a mixture, the first and, analogously, the other lines in Sherman's equation may be written as follows:

$$
\underbrace{\left(T_{A100} - T_A \right)}_{a_{11}} C_A + \underbrace{\left(\frac{\bar\mu_B(\alpha)}{\bar\mu_A(\alpha)}\, T_{A100} \right)}_{a_{21}} C_B + \underbrace{\left(\frac{\bar\mu_C(\alpha)}{\bar\mu_A(\alpha)}\, T_{A100} \right)}_{a_{31}} C_C = 0
$$

By dividing every line by the intensity T_{100} of the pure component, introducing the intensity N in place of the counting time T, and postulating

$$
T = \frac{\text{constant}}{N_A} \quad \text{and} \quad \frac{T_{100} - T}{T_{100}} = -\frac{N_{100} - N}{N}
$$

we obtain an equation of well-known form

$$
-C_A \left(\frac{N_{A100} - N_A}{N_A} \right) + C_B \left(\frac{\bar\mu_B(\alpha)}{\bar\mu_A(\alpha)} \right) + C_C \left(\frac{\bar\mu_C(\alpha)}{\bar\mu_A(\alpha)} \right) = 0 \;, \quad \text{etc.}
$$

16.3. Formulation According to Beattie and Brissey

A different formulation for the system of equations may be found in a paper by Beattie and Brissey (1954)

$$(R_a-1)\ W_a + A_{ab}\ W_b \quad\quad + A_{ac}\ W_c \quad\quad +\cdots = 0$$
$$A_{ba}\ W_a \quad\quad -(R_b-1)\ W_b + A_{bc}\ W_c \quad\quad +\cdots = 0$$
$$A_{ca}\ W_a \quad\quad + A_{cb}\ W_b \quad\quad -(R_c-1)\ W_c + \cdots = 0$$

where A_{ab} is the quotient of the combined mass absorption coefficients for the incident primary tube radiation and the emerging fluorescent radiation

$$A_{ab} = \frac{M_{ab}}{M_{aa}}, \quad M_{ab} = \left\{ \frac{\mu(\lambda)}{\sin\varphi} + \frac{\mu(\alpha)}{\sin\psi} \right\}_b$$

and $R_a = I_{aa}/I_{ab}$ corresponds to the quotient of the intensity of the pure component to the intensity of the same component in a mixture. By replacing the concentrations W_a, W_b, and W_c in a mixture by C_A, C_B, and C_C, introducing the term $\bar\mu$ instead of M for the combined mass absorption coefficients, and designating the intensities of the pure components and of the mixture with N_{A100} and N_A, respectively, we obtain the well-known formulation

$$-\left(\frac{N_{A100}}{N_A} - 1 \right) C_A + \left(\frac{\bar\mu_B(\alpha)}{\bar\mu_A(\alpha)} \right) C_B + \left(\frac{\bar\mu_C}{\bar\mu_A} \right) C_B + \cdots = 0, \text{ etc.}$$

16.4. System of Equations According to Marti

Still another linear system of equations is presented in a paper by Marti (1962)

$$I_{Cr} = i_{Cr}\ (\alpha_{Cr} C_{Cr} + \alpha_{Ni} C_{Ni} + \alpha_{Mo} C_{Mo} + \alpha_{Fe} C_{Fe} + \cdots)$$
$$I_{Ni} = i_{Ni}\ (\beta_{Cr} C_{Cr} + \beta_{Ni} C_{Ni} + \beta_{Mo} C_{Mo} + \beta_{Fe} C_{Fe} + \cdots)$$
$$I_{Mo} = i_{Mo}\ (\gamma_{Cr} C_{Cr} + \gamma_{Ni} C_{Ni} + \gamma_{Mo} C_{Mo} + \gamma_{Fe} C_{Fe} + \cdots)$$

where I is the corrected intensity, i the measured intensity, and α, β, and γ are the interaction coefficients of the elements for the particular fluorescent intensity in question. The fluorescent intensity I corrected for the effects of the associated elements, corresponds to the intensity that would result for strict proportionality between concentration and intensity, $C \times N_{100}$. For the measured intensity we substitute N for i and designate the interaction coefficients α, β, and γ with $\bar\mu_i/\bar\mu_j$. The first and, analogously, the other lines in the system of equations according to Marti may then be written as follows:

$$C_{Cr} N_{Cr100} = N_{Cr}\left[\left(\frac{\bar\mu_{Cr}}{\bar\mu_{Cr}} \right) C_{Cr} + \left(\frac{\bar\mu_{Ni}}{\bar\mu_{Cr}} \right) C_{Ni} + \left(\frac{\bar\mu_{Mo}}{\bar\mu_{Cr}} \right) C_{Mo} + \left(\frac{\bar\mu_{Fe}}{\bar\mu_{Cr}} \right) C_{Fe} + \cdots \right]$$

The concentration of the desired component appears in every line on both sides of the equation. Both sides of the equation are divided by the value N

and the concentration of the desired component on the right side of the equation is combined. When taking the value of the interaction coefficient of chromium on the fluorescent intensity of chromium ($\bar{\mu}_{Cr}/\bar{\mu}_{Cr}$), that of nickel on nickel ($\bar{\mu}_{Ni}/\bar{\mu}_{Ni}$), etc. to be equal to 1, and with

$$\left(1 - \frac{N_{Cr100}}{N_{Cr}}\right) = -\frac{N_{Cr100} - N_{Cr}}{N_{Cr}}$$

we obtain for the first line

$$0 = -\left(\frac{N_{Cr100} - N_{Cr}}{N_{Cr}}\right)C_{Cr} + \left(\frac{\bar{\mu}_{Ni}}{\bar{\mu}_{Cr}}\right)C_{Ni} + \left(\frac{\bar{\mu}_{Mo}}{\bar{\mu}_{Cr}}\right)C_{Mo} + \left(\frac{\bar{\mu}_{Fe}}{\bar{\mu}_{Cr}}\right)C_{Fe} + \cdots$$

The system of linear equations upon which Marti based his calculations is identical to the one mentioned before. Since this system of equations is applied to the analysis of steels, Marti arbitrarily assumes the interaction coefficients of iron on the fluorescent intensity of all other elements, such as chromium, nickel, and molybdenum, to be equal to 1:

$$\left(\frac{\bar{\mu}_{Fe}}{\bar{\mu}_{Cr}}\right) = \left(\frac{\bar{\mu}_{Fe}}{\bar{\mu}_{Ni}}\right) = \left(\frac{\bar{\mu}_{Fe}}{\bar{\mu}_{Mo}}\right) = 1$$

These values, however, are neither equally large nor equal to one and, hence, this method results in somewhat uncertain relationships.

16.5. Formulation According to Traill and Lachance

Traill and Lachance (1965) and Lachance and Traill (1966) formulate the linear system of equations for a three-component system as follows:

$$C_A = R_A(1 + C_B\alpha_{AB} + C_C\alpha_{AC})$$
$$C_B = R_B(1 + C_A\alpha_{BA} + C_C\alpha_{BC})$$
$$C_C = R_C(1 + C_A\alpha_{CA} + C_B\alpha_{CB})$$

where $R_A = I_A/I_{(A)}$, etc.; I_A is the fluorescent intensity of A in the mixture; $I_{(A)}$ is the fluorescent intensity of the pure component A; and $\alpha_{AB}, \alpha_{AC}, \ldots,$ α_{CB} are constants describing the effects of associated components on the respective fluorescent intensities.

After rearranging the equation by using familiar terms, the first line may be written as follows:

$$-C_A\frac{N_{A100}}{N_A} + C_B\alpha_{AB} + C_C\alpha_{AC} = -1$$

We add $+1$ to both sides of the equation (expressed as $C_A + C_B + C_C$ on the left side) and obtain:

$$-C_A\left(\frac{N_{A100}}{N_A} - 1\right) + \underbrace{C_B(\alpha_{AB} + 1)}_{r_{AB}} + \underbrace{C_C(\alpha_{AC} + 1)}_{r_{AC}} = 0$$

A simple relation exists between the constants α_{ij} of Traill and Lachance (1965) and the regression coefficients

$$r_{ij} = \alpha_{ij} + 1$$

16.6. Determination of the Interaction Coefficients

Let us assume a multicomponent mixture to be divided into the individual two-component mixtures (for example, the four-component mixture ABCD may be divided into six two-component mixtures AB, AC, AD, BC, BD, and CD). The effect of an associated component on the fluorescent intensity occurs separately in every two-component mixture: the effect of the associated component B on the fluorescent intensity of A in a two-component mixture AB, for example, may be studied separately. Analogously, the effect of the associated component C on the fluorescent intensity of A in a two-component mixture AC may also be studied independently, etc. The first line in the system of equations for a two-component mixture AB is therefore reduced as follows:

$$-C_A \frac{N_{A100} - N_A}{N_A} + C_B \left(\frac{\bar{\mu}_B(\alpha)}{\bar{\mu}_A(\alpha)} \right) = 0$$

and we obtain for the interaction coefficient of component B for the fluorescent intensity of A

$$\left(\frac{\bar{\mu}_B(\alpha)}{\bar{\mu}_A(\alpha)} \right) = \frac{C_A}{C_B} \frac{N_{A100} - N_A}{N_A} = \frac{C_A}{1 - C_A} \frac{N_{A100} - N_A}{N_A} = r_{AB} \quad (16.8)$$

The interaction coefficient is identical to the regression coefficient (Chapter 12) and, in the following considerations, the symbols of the regression coefficient are used for the interaction coefficient. Thus, r_{AB}, r_{AC}, and r_{AD} are the interaction coefficients of the components B, C, and D for the fluorescent intensity of component A; r_{BA}, r_{BC}, and r_{BD} are the coefficients of the components A, C, and D for the fluorescent intensity of component B; r_{CA}, r_{CB}, and r_{CD} are the coefficients of the components A, B, and D for the fluorescent intensity of component C; etc. The value of the coefficient may be determined experimentally by using samples of known composition. This method was employed by Beattie and Brissey (1954) for the analysis of chromium–iron–nickel–molybdenum alloys. They determined the interaction coefficient experimentally for the individual two-component mixtures (Table 16.1).

Another possibility of determining the interaction coefficients is that of determining the best value via a compensation calculation. For this purpose, several chemically well-analyzed standard samples are required of

Table 16.1. Determination of the Interaction Coefficient for Chromium–Iron–Nickel–Molybdenum Alloys (Beattie and Brissey, 1954).

Chemical composition of the two-component alloys utilized in the experiments

	Associated component			
Cr with	100 % Cr	48.24% Fe	48.07% Ni	74.28% Mo
Fe with	50.83% Cr	100 % Fe	51.53% Ni	65.33% Mo
Ni with	48.19% Cr	46.65% Fe	100 % Ni	53.70% Mo
Mo with	23.53% Cr	34.44% Fe	46.27% Ni	100 % Mo

Intensity ratio N_{100}/N in the samples utilized in the experiments

Element emitting fluorescence	Associated component			
	Cr	Fe	Ni	Mo
Cr	1.000	1.760	1.815	1.841
Fe	3.360	1.000	1.613	1.835
Ni	2.860	3.670	1.000	1.954
Mo	3.770	2.787	2.461	1.000

Influence factors $(\bar{\mu}_i/\bar{\mu}_j)$ or r_{ji}

Element emitting fluorescence	Associated component			
	Cr	Fe	Ni	Mo
Cr	1.000	0.721	0.813	2.660
Fe	2.482	1.000	0.676	1.582
Ni	1.863	2.420	1.000	1.108
Mo	0.877	0.944	1.260	1.000

compositions similar to the multicomponent system whose fluorescent intensities are to be measured. The interaction coefficients in the above-described form are related to concentration and fluorescent intensities by the standard samples. Let us first consider the equation for component A. Since concentration and fluorescent intensities of the calibration curves are experimentally determined numbers, they are subject to small errors so that the regression equation is not strictly fulfilled and the error equation is considered instead

$$-C_A\left(\frac{N_{A100}-N_A}{N_A}\right) + C_B r_{AB} + C_C r_{AC} + C_D r_{AD} = \Delta_A \qquad (16.9)$$

where Δ_A is the deviation from the desired value 0 and results from experimental error. Best values for individual interaction coefficients may be found when the following requirements are made:

$$\Sigma \Delta_A^2 = \text{minimum}$$

i.e.,

$$\frac{\partial \Sigma \Delta_A^2}{\partial r_{AB}} = 0 \qquad \frac{\partial \Sigma \Delta_A^2}{\partial r_{AC}} = 0 \qquad \frac{\partial \Sigma \Delta_A^2}{\partial r_{AD}} = 0 \qquad (16.10)$$

For the value Δ_A^2 we obtain

$$
\begin{aligned}
\Delta_A^2 =& \left[-C_A\left(\frac{N_{A100} - N_A}{N_A}\right) + C_B r_{AB} + C_C r_{AC} + C_D r_{AD} \right]^2 = \\
=& \; C_A^2 \left(\frac{N_{A100} - N_A}{N_A}\right)^2 + C_B^2 r_{AB}^2 + C_C^2 r_{AC}^2 + C_D^2 r_{AD}^2 + \\
& + C_B r_{AB} \left[-2\,C_A\left(\frac{N_{A100} - N_A}{N_A}\right) + C_C r_{AC} + C_D r_{AD} \right] + \\
& + C_C r_{AC} \left[-2\,C_A\left(\frac{N_{A100} - N_A}{N_A}\right) + C_B r_{AB} + C_D r_{AD} \right] + \\
& + C_D r_{AD} \left[-2\,C_A\left(\frac{N_{A100} - N_A}{N_A}\right) + C_B r_{AB} + C_C r_{AC} \right] \qquad (16.11)
\end{aligned}
$$

Partial differentiation of r_{AB} results in all terms not containing the factor r_{AB} going to 0, and the following equation remains:

$$\frac{\partial \Delta_A^2}{\partial r_{AB}} = 2 r_{AB} C_B^2 + 2 C_B \left[-C_A\left(\frac{N_{A100} - N_A}{N_A}\right) + C_C r_{AC} + C_D r_{AD} \right] \qquad (16.12)$$

Analogously, partial differentiation for δr_{AB} and δr_{AD} results in all factors not containing r_{AC} and r_{AD}, respectively, going to 0, and we eventually obtain three equations for the determination of the three interaction coefficients r_{AB}, r_{AC}, and r_{AD}:

$$r_{AB} \Sigma C_B^2 + r_{AC} \Sigma C_B C_C + r_{AD} \Sigma C_B C_D = \Sigma C_A C_B \left(\frac{N_{A100} - N_A}{N_A}\right)$$

$$\text{from} \; \frac{\partial \Sigma \Delta_A^2}{\partial r_{AB}}$$

$$r_{AB} \Sigma C_B C_C + r_{AC} \Sigma C_C^2 + r_{AD} \Sigma C_C C_D = \Sigma C_A C_C \left(\frac{N_{A100} - N_A}{N_A}\right)$$

$$\text{from} \; \frac{\partial \Sigma \Delta_A^2}{\partial r_{AC}}$$

$$r_{AB} \, \Sigma \; C_B C_D + r_{AC} \, \Sigma \; C_C C_D + r_{AD} \, \Sigma \; C_D^2 = \Sigma \; C_A C_D \left(\frac{N_{A100} - N_A}{N_A} \right)$$

$$\text{from} \; \frac{\partial \Sigma \Delta_A^2}{\partial r_{AD}} \quad (16.13)$$

The coefficients r_{AB}, r_{AC}, and r_{AD} may be calculated from these three equations by use of determinants. Analogously, determination of the coefficients for the fluorescent intensity of component B may be achieved by minimizing the sum $\Sigma \Delta_B^2$ in the error equations for B. We again obtain three equations which allow determination of the three interaction coefficients r_{BA}, r_{BC}, and r_{BD} (see Appendix).

When limiting the determination of concentration in a multicomponent system to major components only, the chemically-determined compositions of the major components in the compensation calculation should not be normalized to unity (or 100%). As indicated by difficulties experienced by Burnham, Hower, and Jones (1957) in the calculation of chromium–iron–nickel alloys, the compositions and fluorescent intensities of pure components will have to be changed as well.

The best value for the regression coefficient for the special case of a two-component mixture may be found according to the following formula, provided several calibration standards are at hand for its determination:

$$\left(\frac{\bar{\mu}_B(\alpha)}{\bar{\mu}_A(\alpha)} \right) = r_{AB} = \frac{\sum C_A C_B \dfrac{N_{A100} - N_A}{N_A}}{\Sigma \, C_B^2} \quad (16.14)$$

where C_A, C_B, and N_A are the particular values for a standard sample. In the case of very low concentrations of C_A or C_B, experimental errors influence strongly the calculation of r_{AB}. Acceptable limits for the values of C_A and C_B are approximately 0.1 to 0.9 (10–90%, respectively).

Analysis of Multicomponent Mixtures and Solutions for the Linear System of Equations

Determination of concentration in a multicomponent mixture is mathematically equivalent to the solution of a linear system of equations. The n components of a mixture require n equations with n unknowns. Once the concentrations are determined it is possible to solve the system of equations. Since the mathematical solution of systems of equations with four and more unknowns is timeconsuming without a computer, chemical procedures are used instead of the direct mathematical solution. Chemical procedures involve separation, dilution, admixture, etc. of the samples for the purpose of arriving at a product in which the concentration can easily be determined. We consider here only a few of the possibilities used for the determination of the concentration. On the basis of the respective systems of equations, we also demonstrate the nature of the particular chemical reaction. It is apparent that the particular system of equations also changes and becomes simpler along with the chemical reactions so that its solution is easily possible. If, for example, a mixture is diluted by a solvent it is possible to unequivocally explain the mathematical analog of the dilution process.

17.1. Separation into Subsystems

One of the simplest means of determining the concentration of a component in a multicomponent mixture involves chemical separation of the mixture into subsystems. The number of interacting components in a subsystem is smaller and, hence, the determination of the concentration is simplified. Interfering and less-interesting components frequently can be removed from the mixture. Choice of the particular separation method depends upon the chemical composition of the particular mixture and is strictly an analytical problem. Let us consider the change of the linear system of equations in a subsystem as the result of chemical separation, and let us also assume the original system of equations to contain n unknowns and n equations. Every line corresponds to one component and every line contains an expression of the interaction of the desired component with its associated components. Separation of a multicomponent mixture into subsystems results in the concentration of the separated components in every line becoming 0:

$$0 = -C_A\left(\frac{N_{A100}-N_A}{N_A}\right) + C_B r_{AB} \quad + 0 r_{AC} \cdots \qquad\qquad 0 r_{AN}$$

$$0 = C_A r_{BA} - C_B\left(\frac{N_{B100}--N_B}{N_B}\right) \quad + 0 r_{BC} \cdots \qquad\qquad 0 r_{BN}$$

$$0 r_{CA} + 0 r_{CB} \qquad\qquad -C_C\left(\frac{N_{C100}-N_C}{N_C}\right) + \cdots C_N r_{CN} = 0$$

$$\cdots\cdots\cdots \qquad\qquad \cdots\cdots\cdots$$

$$0 r_{NA} + 0 r_{NB} \qquad\qquad C_N r_{NC} \cdots - C_N\left(\frac{N_{N100}-N_N}{N_N}\right) = 0$$

$$(17.1)$$

The original system of equations is then split into two or more independent subsystems of lower degree making determination of the concentration much simpler. One eventually obtains mixtures consisting only of two or three components, allowing determination of the concentration with calibration curves, correction factors, or simple systems of equations.

In the analysis of niobium- and tantalum-containing ore, such as columbite and tantalite, and of niobium–tantalum compounds and titanium alloys, Mitchell (1958–1960) proceeded by first separating the interfering elements, such as manganese, calcium, chromium, and silicon, from the system. The composition of the remaining mixture, which consists of niobium, tantalum, iron, and titanium, is then determined by calibration curves and with the aid of correction factors. For analysis of heat- and corrosion-resistant ferro-alloys and steels, Luke (1963a) separated the desired elements molybdenum, tungsten, niobium, and tantalum from the other elements and then determined the contents with the borax method.

17.2. Decomposition and Dilution

Another possibility of determining the concentration in a multi-component mixture involves decomposition and dilution of the sample. In the dilution method, the interaction between the individual components is reduced by the solvent. The combined mass absorption coefficient $\bar{\mu}_i$ is replaced by a combined mass absorption coefficient consisting of the contribution by component i and the contribution by the diluent

$$\bar{\mu}_i \rightarrow [\bar{\mu}_i + (w-1)\,\bar{\mu}_L] \quad \text{for all } i$$

In this equation w is the factor of dilution of the sample. The term $\bar{\mu}_i/\bar{\mu}_j$ is replaced by the expression

$$\frac{\bar{\mu}_i}{\bar{\mu}_j} \rightarrow \frac{\bar{\mu}_i + (w-1)\,\bar{\mu}_L}{\bar{\mu}_j + (w-1)\,\bar{\mu}_L} \quad \text{and} \quad \lim_{w=\infty} \frac{\bar{\mu}_i + (w-1)\,\bar{\mu}_L}{\bar{\mu}_j + (w-1)\,\bar{\mu}_L} = 1$$

For strong dilutions, i.e., large values of w, the regression coefficients approach 1. The linear system of equations for the determination of the concentrations of n components in the case of diluted mixtures is as follows:

$$-C_A\left(\frac{N_{A100}-N_A}{N_A}\right) + C_B \qquad\qquad + C_C \qquad\qquad \cdots + C_N \qquad \cong 0$$

$$C_A \qquad\qquad - C_B\left(\frac{N_{B100}-N_B}{N_B}\right) + C_C \qquad\qquad \cdots + C_N \qquad \cong 0$$

$$C_A \qquad\qquad + C_B \qquad\qquad - C_C\left(\frac{N_{C100}-N_C}{N_C}\right) \cdots + C_N \qquad \cong 0$$

$$\cdots\cdots\cdots \qquad\qquad\qquad \cdots\cdots\cdots$$

$$C_A \qquad\qquad + C_B \qquad\qquad + C_C \qquad\qquad \cdots - C_N\left(\frac{N_{N100}-N_N}{N_N}\right) \cong 0$$

$$(17.2)$$

Hence, the determination of the individual concentrations is very simple. When normalizing the sums of all concentrations,

$$C_A + C_B + C_C + \cdots + C_N = 1$$

we find, for example, for the concentration of component A

$$-C_A\left(\frac{N_{A100}-N_A}{N_A}\right) + (1-C_A) \cong 0 \quad \text{or} \quad C_A \cong \frac{N_A}{N_{A100}} \quad (17.3)$$

In diluted mixtures, the concentrations are nearly proportional to the fluorescent intensity of the components and the resulting calibration curves are straight lines. Solvents and fluxes, in which the sample is decomposed, are particularly well-suited for preparation of diluted mixtures. Gunn (1957), however, has diluted powdered samples with a lithium carbonate–starch mixture by a factor of 20 (by weight) and then successfully analyzed for elements with atomic numbers between $Z = 20$ (calcium) and $Z = 42$ (molybdenum). Preparation of the solutions depends upon the nature of the substance that is to be investigated, and decomposition is achieved, for example, by strong acids (HF, H_2SO_4, HCl, and HNO_3) or other solvents. Only clear solutions, however, which during radiation do not cloud, flocculate, or form schlieren are suited for analysis. Measurement in solution is described, for example, by the following authors: Campbell and Carl (1954; 1956) and Hakkila and Waterbury (1960) for the analysis of niobium and tantalum oxide and niobium–tantalum ores; Friedlander and Goldblatt (1959) for steel analyses; and Hakkila, Hurley, and Waterbury (1964) for the analysis of zirconium and molybdenum in uranium carbides.

Borax Method

Preparation of borax melts for analysis of multicomponent mixtures is described extensively by the following authors: Kemp, Hasler, and Jones (1954) and Heinrich and McKinley (no date) for analysis of niobium–tantalum alloys; Claisse (1957) for the analysis of ores (oxides and sulfides); Blavier *et al.* (1960) for the analysis of alnico alloys; Bruch (1962) for the analysis of ferro alloys; Robinson and Gertiser (1964) for the analysis of raw materials used in cement production; and Wang (1962), Hooper (1964), and Welday *et al.* (1964) for rock analysis. Samples are decomposed with borax ($Na_2B_4O_7$) or lithium borate ($Li_2B_4O_7$), and an oxidizing medium (BaO_2, $KClO_3$), an oxidation catalyst (MnO_2), or a strongly absorbing substance is occasionally added to the melt. The homogeneous melt is then poured into a mold, cooled slowly, and the resulting pellet is used in the analysis. Borax pellets are stable for several months, when stored in desiccators (Bruch 1962), and do not show any noticeable change even after 20 h of irradiation. Luke (1963b) describes a procedure to decompose microgram amounts of substance in borax. In order to avoid too high a dilution of the substance, Luke (1963b) uses only 2 g of borax for decomposition, and the melt is flattened with a pistil immediately after it is poured. Hooper and others propose to prepare a powder of the solidified melt after cooling, thus making the pouring of pellets a less critical operation.

Preparation of Pellets

Kemp, Hasler, and Jones (1954). One part of the powdered substance of a grain diameter less than 0.06 mm is mixed with 5 parts borax and 5 parts lithium carbonate, melted in a graphite crucible, and annealed for 5 min at 1000°C.

Heinrich and McKinley (no date). The substance (oxide) is ground and homogenized; 0.77 grams of it are mixed well with 6.93 grams of waterfree borax in a platinum crucible, melted, and annealed for 10 min. The melt is poured onto an aluminum plate which is held at a temperature of

410 $\pm$ 15°C, so that the borax pellet neither cracks nor sticks to the plate. The pellet is slowly cooled until the hot plate reaches approximately 300°C; it is then further cooled on a warm substrate. After cooling, the bottom side of the pellet is polished with SiC paper.

Claisse (*1957*). 100 milligrams of sample are mixed in a platinum crucible with 10 g of molten borax, heated on a Meker burner while turning the crucible occasionally, and then annealed for a few minutes. In the meantime, an aluminum plate is heated on a hot plate to 450°C. Pouring of the pellet is as follows: A ring at room temperature is placed on the aluminum plate, and the melt is immediately poured into the ring. The melt is then cooled for one to two min; the aluminum plate is removed from the hot plate and the pellet is further cooled. The ring is then removed from the pellet. The pellet has to be clear and of uniform color. If the ring has been on the hot plate too long before the pellet is poured, then it will stick to the ring. If the base plate is too hot during pouring, the pellet may stick to it; if it is too cold, the bottom surface of the pellet may become irregular. The hot pellet may crack when making contact with a cold object or if cooled too quickly.

Bruch (*1962*). In a platinum crucible 2 g of bariumperoxide are weighed and 0.2 g of the ferro alloy in question is added and carefully mixed. The amount of the added bariumperoxide depends upon the amount of oxygen required for oxidation. Its amount may be increased to 2.5 or 3.0 g. A layer of 10 g of water-free borax, whose quality in regard to water content has been carefully tested, is added to the mixture. The crucible is positioned on the front rim of a muffle furnace whose temperature is 1000°C and whose door is not completely closed. The reaction begins after a few minutes and may be recognized by the occurrence of bubbles on the surface of the now viscous borax melt. After formation of bubbles has ceased, the melt is translucent. At this point incomplete oxidation may be recognized by luminescent phenomena in the melt. Ten minutes after starting the muffle furnace, the crucible is moved into the hot zone of the furnace, where it remains for 15 min. Thereafter the melt is turned in the crucible for 1 to 3 min over a Meker burner in order to obtain sufficient mixing. Certain substances require an additional 15 to 30 min of decomposition time in the muffle furnace. It is often necessary to place the crucible for an additional 3 min over a Meker burner and to turn it frequently. Before pouring, the melt is again placed into the muffle furnace for a period of 10 min and held at 950°C in order to obtain the required temperature for pouring. A hot plate with a 5-mm-thick aluminum sheet and a 5-mm-thick asbestos plate is held at 300°C and placed next to the muffle furnace. A ring, made of steel containing approximately 13 % chromium, is placed onto the aluminum sheet; the ring is 31 mm in diameter, 10 mm high, and 10 mm thick. The melt, having a temperature of 950°C, is poured into the ring. After 5 min the ring with the solidified melt is pushed onto the asbestos plate, and the ring is removed. After another 3 min the asbestos plate is removed from the hot plate and is left to cool to room temperature in a draft-free environment. Borax pellets which are not clear and transparent indicate incomplete decomposition. The flat side of the borax pellet is polished with corundum paper of grain size 320 using water cooling, then rinsed with ethanol and dried on filter paper.

Wang (1962). A procedure to decompose rock and soil samples with lithium fluoride is described and directions are given for the preparation and removal of the melt from the crucible; the pellet is pulverized before analysis.

Welday, Baird, McIntyre, and Madlem (1964). Various procedures for sample preparation and decomposition of rocks are described and their relative merits are compared; an extensive and detailed description is given of decomposition of 2.8 g of rock sample with 5.2 g of lithium boride ($Li_2B_4O_7$).

17.3. Admixture of a Strongly Absorbing Substance

Interaction between the individual components may also be reduced by adding to the sample a strongly absorbing substance, such as BaO, $BaSO_4$, La_2O_3, or $K_2S_2O_7$. In this procedure, the absorption of the incident primary tube radiation and of the emerging fluorescent radiation is largely determined by the absorption coefficient of the admixed substance. Admixing of a strongly absorbing substance is frequently applied simultaneously with the dilution method. Analogous to the dilution method, the quotients $\bar{\mu}_i/\bar{\mu}_j$ are nearly 1 and the system of equations is identical to that for diluted mixtures [system of equations (17.2)]. Concentration is proportional to fluorescent intensity, and calibration curves are straight lines

$$C_A \cong \frac{N_A}{N_{A100}} , \quad C_B \cong \frac{N_B}{N_{B100}} , \quad C_C \cong \frac{N_C}{N_{C100}} , \quad \text{etc.} \qquad (17.4)$$

For the analysis of sulfide ore with the borax method, Claisse (1957) recommends the addition of BaO_2 and $BaSO_4$ to the melt in order to reduce the "matrix effect." For the same reason, Luke (1963a) adds BaO to the borax melt, while Rose, Adler, and Flanagan (1963) use La_2O_3.

Addink *et al.* (1962) describe procedures to determine the lead content of glasses. As this involves analysis of a heavy element in a light matrix, the calibration curve is so strongly curved that for more than 25 wt% Pb no significant increase in the fluorescent intensity occurs. In order to be able to analyze higher contents and derive a straight-line calibration curve, one part by weight of the lead-containing glass is mixed well with five parts of copper powder. The fluorescent intensity of the lead now changes nearly linearly with the lead content.

17.4. The Internal Standard Method

The effects of associated components on the fluorescent intensity may also be determined experimentally with the aid of an internal standard. The internal standard method is particularly well-suited when determination of only a few elements in a mixture is required, and when the remaining elements can be neglected. An element, which is similarly affected by the associated components as is the desired element, is chosen as a standard, and a known amount of it is mixed with the sample. The standard is designated S and the desired component A. All equations in the linear system of equations which do not concern the determination of A are not considered, and the equation for the standard element is added instead. The system of equations which has to be solved now reads as follows:

$$-C_A\left(\frac{N_{A100}-N_A}{N_A}\right) + C_B r_{AB} + C_C r_{AC} \cdots + C_N r_{AN} + C_S r_{AS} = 0$$

$$C_A r_{SA} \quad + \underbrace{C_B r_{SB} + C_C r_{SC} \cdots + C_N r_{SN}}_{\text{Associated components}} - C_S\left(\frac{N_{S100}-N_S}{N_S}\right) = 0$$

$$(17.5)$$

This is a system of equations consisting of 2 equations with n unknowns. Certain prerequisites have to be fulfilled in order to make the system of equations solvable. Mathematically, all pairs of coefficients have to fulfill the following condition:

$$\frac{r_{Ai}}{r_{Si}} = \frac{r_{Aj}}{r_{Sj}} = \text{constant}$$

or in matrix formulation

$$(r_{AB} r_{AC} \cdots r_{AN}) = k (r_{SB} r_{SC} \cdots r_{SN})$$

This relation is the critical condition for the internal standard method. In order to fulfill this condition, attention has to be given to proper selection of the standard; the two fluorescent lines which have to be measured must always be located on the same side of the absorption edge of the various associated components (Figure 17.1). In this case, the effects of the associated components may be determined experimentally from the equations of the standard [second line of equation (17.5)]; using the critical condition we obtain

$$(C_B r_{AB} + C_C r_{AC} + \cdots C_N r_{AN}) = k\left(C_S \frac{N_{S100} - N_S}{N_S} - C_A r_{SA}\right)$$

and, hence, obtain for line 1 of equation (17.5)

$$C_A\left(\frac{N_{A100} - N_A}{N_A} + k r_{SA}\right) = C_S\left(k\frac{N_{S100} - N_S}{N_S} + r_{AS}\right) \qquad (17.6)$$

The expression $k r_{SA}$ on the left side of the equation can formally be written r_{AA}. This value corresponds to the effect of component A on its own fluorescent intensity $\bar{\mu}_A/\bar{\mu}_A$. Analogously to earlier considerations $\bar{\mu}_A/\bar{\mu}_A$ may be set equal to 1. The term r_{AS} on the right side of the expression is formally equal to k ($r_{AS} = k r_{SS}$ and $r_{SS} = 1$). With the two values $k r_{SA} = 1$, and $r_{AS} = k$, expression (17.6) is simplified and we obtain for the desired concentration C_A

$$C_A = C_S \frac{N_A}{N_S}\underbrace{\left(\frac{k N_{S100}}{N_{A100}}\right)}_{\text{constant}} \qquad (17.7)$$

The concentration C_A is proportional to the intensity ratio of the two fluorescent lines, and the resulting calibration curve (intensity ratio *vs* concentration) is a straight line. The values of the constants and the slopes of the calibration curves can be determined experimentally with the aid of standard samples. Experiments show that in some instances elements may

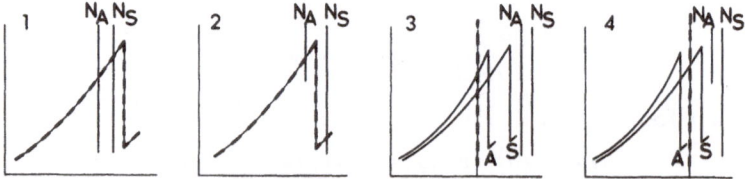

Fig. 17.1. Choice of internal standard. N_A and N_S are the fluorescent lines of the desired element and of the standard, respectively. Dashed lines illustrate the absorption coefficient and a strong emission line of the associated component, respectively. Undesirable effects of the associated component are well-compensated for by the standard [(1) and (3)]; examples given in (2) and (4) illustrate improper choice of standard. For further discussion see text.

be used as standards which do not completely fulfill the critical condition but still yield useful analytical results. When working with highly diluted mixtures, the values for the individual regression coefficients r approach 1 and, hence, the critical condition for the internal standard method is more easily fulfilled.

Jones (1959), Burke *et al.* (1964), and Gunn (1965) modified the method by dipping a solid standard into the liquid which is to be analyzed. The solid body is dipped into the liquid (benzene and liquid hydrocarbons) in such a way that only a very thin layer of liquid remains between the solid body and the window of the sample container. If the distance between the sample window and the solid body is small, then the ratio of the two fluorescent intensities is nearly independent of the absorption efficiency of the matrix. For choice of standard see Figure 17.1.

Case (1). The fluorescent lines of the standards and of the desired element are both located on the short-wavelength side and near the absorption edge of an associated component; both lines are then strongly absorbed by the associated component. The absolute fluorescent intensities vary strongly with varying amounts of associated components, while the intensity ratio of the two elements changes little.

Case (2). If the absorption edge of an associated component is located between the two fluorescent lines, then only one of the two lines is strongly absorbed by the associated component. The intensity of the short-wavelength line is affected strongly by varying amounts of the associated component, while the intensity of the long-wavelength line changes little. Correspondingly, the intensity ratio of the two lines changes with the composition of the sample.

Case (3). When the absorption edges of the standard and the desired component are of longer wavelength than a strong emission line of an associated component, then both are excited to fluorescence by the associated component. Although varying amounts of the associated component cause change in the absolute intensities, the intensity ratio changes little.

Case (4). When a strong emission line of an associated component is located between the absorption edges of the two particular elements, only the element with the long-wavelength absorption edge, in addition, is excited by the associated component and its fluorescent intensity changes with varying amounts of the associated component. The intensity of the radiation emitted by the other element, however, changes little. The intensity ratio of the two lines varies with the amount of the associated component. Application of internal standards was discussed by Hevesy and Alexander (1933) and later by Adler and Axelrod (1955*a*).

Pish and Huffman (1955) determine thorium and germanium contents in aqueous and organic solutions with strontium and bromine as internal standards, thus compensating for varying contents of rare earths, phosphates, and heavy metals. Silverman, Houk, and Moudy (1960) determine the uranium content of stainless steels using the method of internal standards, while Hakkila, Hurley, and Waterbury (1964) determine zirconium and molybdenum contents of uranium carbide with this method. Campbell and Carl (1956) and Mortimore, Romans, and Tews (1954) carried out analyses of niobium- and tantalum-bearing ores with the aid of internal standards. Selenium is used as an internal standard for the bromine determination in hydrocarbons by Kokotailo and Damon (1953). Dwiggins and Dunning (1960) use internal standards in the analysis of petroleum in order to compensate for the effects of sulfur and salt water on the fluorescent intensities of vanadium, iron, and nickel. Louis (1964b) analyzes lubricants for heavy-metal contents: internal standards are added to compensate for the varying absorption efficiency of the matrix and relative errors are held to within 1 to 3%.

17.5. The External Standard Method

In the external standard method, a sample is compared to several well-analyzed standards of similar composition. Unless the intensity emitted by the sample is by chance equal to that of a standard (and, hence, both samples are of the same composition), deviations in the composition are calculated from small deviations in the recorded intensities. For these calculations, it is assumed that small differences in fluorescent intensities, for example, of various samples containing component A, are related to small differences in the content of A and are not caused by associated elements. This assumption is largely correct, since the matrix compositions of the samples are approximately the same. The same assumption is made for components B, C, D, etc. It is further desirable to have as simple a relationship as possible between the measured intensities and the concentrations. Relative intensities are therefore used instead of the directly measured absolute fluorescent intensities

$$\frac{N_A}{N_{A100}} \quad \frac{N_B}{N_{B100}} \quad \frac{N_C}{N_{C100}} \cdots$$

In analogy to the concentrations C_A, C_B, C_C, etc., the relative fluorescent intensities are also normalized to the sum of 1 or 100

$$1 = \frac{N_A}{N_{A100}} k + \frac{N_B}{N_{B100}} k + \frac{N_C}{N_{C100}} k + \cdots$$

Table 17.1. Examples for the Choice of Internal Standards

Fluorescent line	Standard line	Literature[a]
94-PuL_α	ThL_α	Menis; A.C. **35**, 1049
	YK_α	Turnley; Talanta **6**, 189
92-UL_α	ThL_α	Menis; A.C. **35**, 1049
	SrK_α, BrK_α	Pish; A.C. **27**, 1875
	YK_α	Kehl; A.C. **28**, 1350
90-ThL_α	TlL_α	Adler; A.C. **27**, 1002
	SrK_α, BrK_α	Pish; A.C. **27**, 1875. Niekerk; A.S. **15**, 121
	SeK_α	King; Science **122**, 72
	YK_α	Hakkila; Talanta **6**, 46
82-PbL_α	BiL_α	Burke; A.C. **36**, 2404
74-WL_α	BrK_α	Fagel; A.C. **30**, 1918
73-TaL_α	HfL_α, WL_α	Campbell; A.C. **26**; 800. Mortimore; A.S. 8, 24
56-BaL_α	LaL_α	Lewis; A.C. **28**, 1282. Burke; A.C. **36**, 2404
	MnK_α	Louis; Z.a.C. **201**, 336
42-MoK_β	NbK_β	Hakkila; A.C. **36**, 2094
41-NbK_α	ZrK_α	Mortimore; A.S. **8**, 24
	MoK_α	Campbell; A.C. **26**, 800
40-ZrK_α	NbK_α	Hakkila; A.C. **36**, 2094
39-YK_α	ThL_α	Hakkila; Talanta **6**, 46
38-SrK_α	RbK_α	Collin; A.C. **33**, 605. Stone; Analyst. **88**, 56
35-BrK_α	SeK_α	Kokotailo; A.C. **25**, 1185.
		Herrmann; Z.a.C. **181**, 122
30-ZnK_α	BiL_α	Burke; A.C. **36**, 2404
	AsK_α	Lewis; A.C. **28**, 1282
	CuK_α	Louis; Z.a.C. **201**, 336
		Bartkiewicz; A.C. **36**, 833
29-CuK_α	PbL_α	Wood; A.C. **33**, 1344
28-NiK_α	CoK_α	Dwiggins; A.C. **32**, 1137
27-CoK_α	CuK_α	Bartkiewicz; A.C. **36**, 833
26-FeK_α	CoK_α	Dwiggins; A.C. **32**, 1137
	CuK_α	Bartkiewicz; A.C. **36**, 833
23-VK_α	CrK_α	Dwiggins; A.C. **32**, 1137
22-TiK_α	LaL_α	Lewis; A.C. **28**, 1282
20-CaK_α	SnL_α	Louis; Z.a.C. **201**, 336
	SbL_α	Burke; A.C. **36**, 2404
17-ClK_α	SnL_α	Louis; Z.a.C. **201**, 336
16-SK_α	MoL_β	Louis; Z.a.C. **201**, 336
15-PK_α	ZnL_α	Louis; Z.a.C. **201**, 336

[a]Key: A.C., *Anal. Chem.*; A.S., *Appl. Spectrosc.*; Z.a.C., *Z. analyt. Chemie.*

Table 17.2. Usefulness of Internal Standards in the Analysis of Lubricants (Louis, 1964b).

Analysis	Amount of substance, in wt. %	Internal standard	Concentration of associated elements, in wt. %	Recovery (% of amount of substance) without internal standard	with internal standard
Zinc	0.1	Cu	0	100	100
			+ 0.5% Ba	70	101
			+ 0.5% Ca	83	101
			+ 0.5% Cl	87	100
			+ 0.5% S	89	101
			+ 0.5% P	90	101

In the method of external standards, one appropriately uses these relative normalized values of N^Δ

$$N_A^\Delta = \frac{\dfrac{N_A}{N_{A100}}}{\dfrac{N_A}{N_{A100}} + \dfrac{N_B}{N_{B100}} + \dfrac{N_C}{N_{C100}} + \cdots}$$

$$N_B^\Delta = \frac{\dfrac{N_B}{N_{B100}}}{\dfrac{N_A}{N_{A100}} + \dfrac{N_B}{N_{B100}} + \dfrac{N_C}{N_{C100}} + \cdots} \tag{17.8}$$

When using the normalized relative intensities, the following first approximation is obtained

$$\Delta N_A^\Delta = \Delta C_A \qquad \Delta N_B^\Delta = \Delta C_B \qquad \Delta N_C^\Delta = \Delta C_C \quad \cdots \tag{17.9}$$

This equation indicates that the value of N^Δ changes analogously to the concentration; a change of 0.05 in N^Δ, for example, corresponds to a change of 0.05 in the concentration. Ultimately, only one standard sample is required for quantitative analysis, and deviations in the relative normalized intensities N^Δ are equal to deviations in concentrations.

Standards of compositions close to those of the samples are used for more accurate quantitative determinations in order to calculate the magnitude of change in content or in the differential quotient $\Delta C/\Delta N^\Delta$ corresponding to a change in intensity. The content of the sample is again determined by adjusting the concentrations in accordance to intensity differences between standards and sample. A numerical example from the four-component mixture Nb_2O_5–Ta_2O_5–WO_3–Fe_2O_3 is listed in Table 17.3 (borax pellet with 200 mg of substance per 9.8 g of borax; as a rule,

Table 17.3. The External Standard Method. Intensities (counts/sec) and content (wt. %) of standards and test sample No. 9.

Sample	C_{Nb}	NbK_α	C_{Ta}	TaL_α	C_W	WL_α	C_{Fe}	FeK_α
1	100	57 220						
2			100	11 550				
3					100	12 880		
4							100	10 500
5	40	17 650	20	2560	20	2920	20	2050
6	20	8300	40	4850	20	2850	20	1990
7	20	8300	20	2480	40	5530	20	2020
8	20	8900	20	2420	20	2790	40	4110
9	?	10 670	?	3080	?	3490	?	2530

Reduced and normalized intensities (counts/sec) and contents (wt %)

Sample	C_{Nb}	Nb^Δ	C_{Ta}	Ta^Δ	C_W	W^Δ	C_{Fe}	Fe^Δ
5	40	32.4	20	23.3	20	23.9	20	20.4
6	20	15.0	40	43.3	20	22.4	20	19.4
7	20	14.9	20	21.8	40	43.8	20	19.5
8	20	15.7	20	21.6	20	22.3	40	40.5
9	?	19.2	?	27.7	?	28.2	?	25,0

Differential quotient, $\Delta C / \Delta N^\Delta$, of the individual standard samples

Samples	$\Delta C_{Nb}/\Delta Nb^\Delta$	$\Delta C_{Ta}/\Delta Ta^\Delta$	$\Delta C_W/\Delta W^\Delta$	$\Delta C_{Fe}/\Delta Fe^\Delta$
5—6	1.15	1.00	—	—
5—7	1.14	—	1.00	—
5—8	1.20	—	—	1.00
6—7	—	0.93	0.93	—
6—8	—	0.93	—	0.95
7—8	—	—	0.93	0.95
Averages	1.16	0.95	0.95	0.97

Intensity difference and calculated difference in contents between samples 5 and 9, and determination of composition of the test sample 9

Samples	ΔNb^Δ	ΔC_{Nb}	ΔTa^Δ	ΔC_{Ta}	ΔW^Δ	ΔC_W	ΔFe^Δ	ΔC_{Fe}
5—9	— 13.2		+ 4.4		+ 4.3		+ 4.6	
	→	— 15.3	→	+ 4.2	→	+ 4.1	→	+ 4.5
Content		40		20		20		20
Content		24.7		24.2		24.1		24.5
Σ 100		25.3		24.8		24.8		25.1
True composition		25.0		25.0		25.0		25.0

the standards should be closer to the composition of the sample). The method of the external standard is particularly suited for analysis of samples of very similar chemical composition, where at least one well-analyzed standard is required.

17.6. The Double Dilution Method

The double dilution method, which permits reliable and simple analysis of complex mixtures, was proposed by Tertian (1968). The substance which is to be analyzed is separated into two fractions having a $1 : 2$ concentration ratio, using either fluxes or solvents to achieve the dilution. The relative fluorescent intensity of component A in the original mixture before dilution is given as follows [equation (12.2)]:

$$\frac{N_A}{N_{A100}} = \frac{C_A}{C_A + (1 - C_A)r} = \frac{1}{r} \times \frac{C_A}{1 + C_A(1 - r)/r}$$

Let us designate the fluorescent intensity of A in the original mixture by N_A; the concentrations in the two fractions by x and $2x$, respectively; and the corresponding intensities by N_{AX} and N_{A2X}. The relative fluorescent intensities of A for the concentrations x and $2x$, respectively, in the two fractions may then be expressed analogously to the above relation (with $\varphi = (1 - r)/r$):

$$\frac{N_{AX}}{N_A} = \frac{1}{r} \times \frac{X}{1 + \varphi x}$$

$$\frac{N_{A2X}}{N_A} = \frac{1}{r} \times \frac{2x}{1 + 2\varphi x}$$

From these intensities, the so-called "corrected intensities" are calculated; thus, self-absorption of the diluted substance and, hence, curvature in the curve formed by a plot of intensity vs concentration, is eliminated. The same relation between intensity and concentration exists for the corrected intensities as was previously discussed for trace amounts: the corrected intensity is proportional to the concentration, and the proportionality factors are the same (corrected and measured intensities are identical for trace amounts). The corrected intensities N_{AXK} and N_{A2XK} are then [equation (12.5)]:

$$\frac{N_{AXK}}{N_A} = \left(\frac{N_{AX}}{N_A}\right)_{\text{trace}} = \frac{1}{r} \times x$$

$$\frac{N_{A2XK}}{N_A} = \left(\frac{N_{A2X}}{N_A}\right)_{\text{trace}} = \frac{1}{r} \times 2x$$

The following equations relate corrected to measured intensities:

$$N_{AXK} = N_{AX}(1 + \varphi x)$$
$$N_{A2XK} = N_{A2X}(1 + 2\varphi x)$$

The value of φ may be taken from the measured fluorescent intensities

$$\frac{N_{A2X}}{N_{AX}} = \frac{2x}{x} \times \frac{1 + \varphi x}{1 + 2\varphi x}$$

$$\varphi = \frac{1}{2x} \times \frac{2 - N_{A2x}/N_{AX}}{(N_{A2x}/N_{AX}) - 1}$$

Thus, the corrected intensity N_{A2XK} is given by the following simple relation:

$$N_{A2XK} = \frac{N_{A2X}}{(N_{A2x}/N_{AX}) - 1}$$

The corrected intensity N_{A2XK} may easily be calculated from the ratio of the measured fluorescent intensities N_{A2x}/N_{AX} and, hence, we obtain the intensity which is proportional to the concentration $2xC_A$

$$N_{A2XK} = \text{const} \times 2xC_A$$

The desired concentration C_A of A in the original mixture is then given as follows:

$$C_A = N_{A2XK}/(\text{const} \times 2x)$$

The value of the proportionality factor may be obtained with the aid of standards, while the term $2x$ is known from the preparation of the solutions. Known amounts of A (either pure A or A-bearing components) are decomposed with the same flux or solvent as is the sample, and two fractions of concentration ratios $1:2$ are prepared for which the corresponding

Table 17.4. **Double Dilution Method.** Yttrium oxide, yttrium oxide–molybdenum oxide, and yttrium oxide–titanium oxide are decomposed in borax fluxes (Tertian, 1968b). (Intensities in 10^4 counts/min.)

Concentration	Measured intensities	N_{2x}/N_x	N_{2xK}	Const/Y_2O_3
Y_2O_3				Constant
0.005	13.29	—	—	—
0.010	23.46	1.765	30.66	30.66
0.015	31.48	—	—	—
0.020	38.11	1.624	61.07	30.54
0.030	48.10	1.528	91.10	30.37
0.040	55.57	1.458	121.33	30.33
			Average constant	30.48
10% Y_2O_3 + 90% MoO_3				Y_2O_3
0.02	5.10	—	—	—
0.04	8.62	1.690	12.49	10.24%
0.08	13.15	1.526	25.00	10.25%
10% Y_2O_3 + 90% TiO_2				Y_2O_3
0.02	4.93	—	—	—
0.04	8.15	1.653	12.48	10.24%
0.08	12.09	1.483	25.03	10.26%

corrected fluorescent intensity N_{A2XK} is calculated. Thus, the proportionality factor, which is a constant for a particular flux or solvent, may be determined.

Any ratio may be chosen for the concentrations x instead of the $1:2$ value in the example discussed here, and equally simple equations are obtained for ratios such as $1:3$, $1:4$, etc.

$$N_{A3XK} = \frac{2N_{A3X}}{(N_{A3X}/N_{AX}) - 1} = \text{const} \times 3xC_A$$

$$N_{A4XK} = \frac{3N_{A4X}}{(N_{A4X}/N_{AX}) - 1} = \text{const} \times 4xC_A$$

The determination of the proportionality factor is illustrated in the first part of Table 17.4, and the determination of yttrium in mixture with MoO_3 and TiO_2, respectively, in borax fluxes is explained in the second and third parts of that table.

17.7. Effects of Certain Minor Components

Occasionally, multicomponent mixtures consist of two or three major and several minor components which occur in low concentrations. The fluorescent intensities of major components are largely determined by interactions with other major components. In order to simplify the determination of composition, the effect of minor components, in a first approximation, may be neglected. Let us assume a multicomponent mixture consisting of n components, where A, B, and C are the major components. The concentrations of the minor components D, E, ..., N in the linear system of equations are said to be zero. Thus, the system of equations may be divided into a system of equations for the major components and in $(n - 3)$ individual equations for the minor components, and one obtains

$$-C_A\left(\frac{N_{A100} - N_A}{N_A}\right) + C_B r_{AB} \qquad + C_C r_{AC} \qquad + 0 \cdots = 0$$

$$C_A r_{BA} \qquad - C_B\left(\frac{N_{B100} - N_B}{N_B}\right) + C_C r_{BC} \qquad + 0 \cdots = 0$$

$$C_A r_{CA} \qquad + C_B r_{CB} \qquad - C_C\left(\frac{N_{C100} - N_C}{N_C}\right) + 0 \cdots = 0 \tag{17.10}$$

and $(n - 3)$ equations of the type

$$C_A r_{jA} + C_B r_{jB} + C_C r_{jC} + 0 \cdots - C_j\left(\frac{N_{j100} - N_j}{N_j}\right) + 0 \cdots = 0 \tag{17.11}$$

The determination of the concentration of major components is simplified and may be carried out algebraically or with the aid of calibration curves. The fluorescent intensity of minor components is largely determined by interaction with the major components. For known concentrations of major components, the concentration of minor components may be determined, for example, by constructing groups of calibration curves for varying mixing ratios [ABC]. This technique has successfully been applied to the analysis of ores and refining products.

17.8. Semiquantitative Determination of Concentration

For semiquantitative determination of concentration, the bent calibration curves are replaced by straight lines (Figure 17.2). The accuracy of analysis depends upon the actual deviation of the calibration curve from a straight line. The individual interaction coefficients in the linear system of equations are arbitrarily set equal to one and the corresponding system of equations is as follows:

$$-C_A\left(\frac{N_{A100}-N_A}{N_A}\right) + C_B \qquad\qquad + C_C\cdots + C_N \qquad = 0$$

$$C_A \qquad\qquad - C_B\left(\frac{N_{B100}-N_B}{N_B}\right) + C_C\cdots + C_N \qquad = 0$$

$$\cdots\cdots\cdots\cdots\cdots\cdots\cdots\cdots\cdots\cdots\cdots\cdots\cdots\cdots\cdots\cdots$$

$$C_A \qquad\qquad + C_B \qquad\qquad + C_C\cdots - C_N\left(\frac{N_{N100}-N_N}{N_N}\right) = 0$$

$$(17.12)$$

The concentrations of the individual components are

$$C_A = \frac{N_A}{N_{A100}} \qquad C_B = \frac{N_B}{N_{B100}} \qquad C_C = \frac{N_C}{N_{C100}}\cdots \qquad (17.13)$$

In addition, one can estimate whether the true value is larger, equal, or smaller than the value determined semiquantitatively. In this procedure,

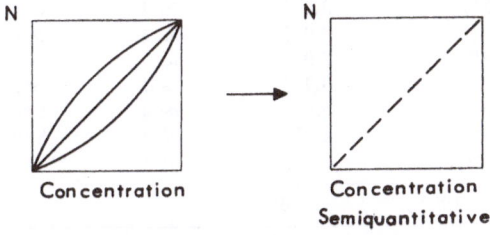

Fig. 17.2. Replacement of calibration curves by straight lines in semiquantitative analysis.

the magnitude of the interaction coefficients $\bar{\mu}_i/\bar{\mu}_j$ or r_{ji}, which are arbitrarily set to be equal to one, is estimated. If the interaction coefficients $\bar{\mu}_i/\bar{\mu}_j$ in a line are, on the average, larger than one, then the sum of the interaction coefficients for the j-line may be written as follows:

$$\sum_{i \neq j} C_i \left(\frac{\bar{\mu}_i}{\bar{\mu}_j} \right) > \sum_{i \neq j} C_i, \quad \text{where} \quad \sum_{i \neq j} C_i = 1 - C_j$$

For the j-line we obtain

$$C_j \left(\frac{N_{j100} - N_j}{N_j} \right) > 1 - C_j \tag{17.14}$$

The true concentration C_j of component j is then

$$C_j > \frac{N_j}{N_{j100}} \tag{17.15}$$

Table 17.5. Semiquantative Determination of Concentration of Rare Earths, Yttrium, and Certain Other Elements in a Synthetic Powdered Mixture

Component	True composition	Semiquantitatively determined composition	Estimates of the value $\Sigma C_i \mu_i/\mu_j$	Relation of semiquantitative value to effective value
CaO	3%	—	> 1	Semiquant < eff
Sc$_2$O$_3$	3%	—	> 1	Semiquant < eff
SrO	3%	1.6%	> 1	Semiquant < eff
Y$_2$O$_3$	42%	27.8%	> 1	Semiquant < eff
Nb$_2$O$_5$	3%	1.6%	> 1	Semiquant < eff
BaO	3%	2.7%	≈ 1	Semiquant ≈ eff
CeO$_2$	3%	3.7%	≈ 1	Semiquant ≈ eff
Nd$_2$O$_3$	3%	2.6%	≈ 1	Semiquant ≈ eff
Sm$_2$O$_3$	3%	3.3%	≈ 1	Semiquant ≈ eff
Eu$_2$O$_3$	3%	3.1%	≈ 1	Semiquant ≈ eff
Gd$_2$O$_3$	3%	3.5%	≈ 1	Semiquant ≈ eff
Tb$_4$O$_7$	3%	3.2%	≈ 1	Semiquant ≈ eff
Dy$_2$O$_3$	3%	3.1%	≈ 1	Semiquant ≈ eff
Ho$_2$O$_3$	3%	3.0%	≈ 1	Semiquant ≈ eff
Er$_2$O$_3$	3%	2.9%	≈ 1	Semiquant ≈ eff
Tm$_2$O$_3$	3%	3.7%	≈ 1	Semiquant ≈ eff
Yb$_2$O$_3$	5%	5.5%	≈ 1	Semiquant ≈ eff
Lu$_2$O$_3$	3%	4.3%	≤ 1	Semiquant ≥ eff
ThO$_2$	5%	7.0%	< 1	Semiquant > eff
Sum	100%	82.6%		

[a]Sample preparation may cause relative errors of $\pm 5\%$.

The true concentration of this component is larger than the semiquantitatively determined value. In the reverse case, where the interaction coefficients $\bar{\mu}_i/\bar{\mu}_j$ in a line, on the average, are smaller than one, the true concentration of the component is smaller than the semiquantitatively determined value. For interaction coefficients in a line which on the average are equal to one, the semiquantitatively determined content is approximately equal to the true content. The interaction coefficients $\bar{\mu}_i/\bar{\mu}_j$ also determine the curvature of the calibration curve of fluorescent intensity plotted vs concentration in a two-component mixture ji. By estimating its value, it is possible to determine whether the calibration curve is located above, below or on a 45° line (Chapter 12).

Such estimates were made for the analysis of rare earths in yttrium-containing concentrates and the semiquantitatively determined contents are compared to the true contents (Table 17.5). In this particular case, analysis of a mixture containing 15 to 20 components was required, and the ratio of semiquantitatively determined values to true values varied between 0.5 to 1.4 for concentrations between 1 and 5 wt. %. Semiquantitative analyses were made by Eichhoff, Beck, and Kiefer (1965) who decomposed samples in acids and diluted the solutions to 0.5 %. As a result of the dilution nearly linear calibration curves that are independent of the sample composition, are obtained. For quick determination of concentration, the proportionality factors k between the concentration C and the intensity N in the equation $C = kN$ are determined experimentally so that, for the same apparatus and method of sample preparation, the content of a sample can be calculated directly from the intensities. In order to avoid chemical decomposition the same authors in another experiment ground one milligram of substance with nine times as much sodium chloride, suspended the powder in water, and filtered over filter paper. Interelement effects in the relatively thin sample layer on a filter of 1000 to 2000 Å thickness are not very pronounced so that the semiquantitative approximation is correct to a factor of 1/2 or 2. Witmer and Addink (1965) sprinkled the substance onto a mylar foil and analyzed the substance which stuck to the foil. As these are very thin layers, interelement effects are minimal and calibration curves are straight lines. Thus, relative concentrations of the individual elements can easily be calculated with the aid of the proportionality factors k.

17.9. Direct Mathematical Solution

Mathematical solution of the linear system of equations leads directly to the desired concentrations. The solution can be carried out manually as long as the mixtures consist of only three or four components. For mixtures

of five or more components and, correspondingly, five and more equations
and unknowns, computers have to be employed. In this chapter, the matrix
formulation is used for the linear system of equations because the coefficient
matrix, which is of importance to the solution of the linear system of equations, is clearly separated from the unknowns. For the purpose of brevity we
assume further

$$N^* \equiv \frac{N_{100} - N}{N}$$

The linear system of equations in matrix formulation for a mixture consisting of four components is as follows:

$$\begin{pmatrix} -N_A^* & r_{AB} & r_{AC} & r_{AD} \\ r_{BA} & -N_B^* & r_{BC} & r_{BD} \\ r_{CA} & r_{CB} & -N_C^* & r_{CD} \\ r_{DA} & r_{DB} & r_{DC} & -N_D^* \end{pmatrix} \begin{pmatrix} C_A \\ C_B \\ C_C \\ C_D \end{pmatrix} = \begin{pmatrix} 0 \\ 0 \\ 0 \\ 0 \end{pmatrix} \qquad (17.16)$$

This is a linear homogeneous system of equations. In general, the solution
for this is trivial, namely $C_A = C_B = C_C = C_D = 0$. We know, however,
that there exists another solution for which the determinant in the coefficient
matrix must be equal to zero (this condition can easily be checked provided
the four numbers N_A^*, N_B^*, N_C^*, and N_D^* are known from the measurement of
their respective fluorescent intensities; if this condition is poorly fulfilled,
i.e., if the determinant deviates greatly from zero, then the calculated
contents contain errors because there are no values which fulfill the system
of equations. Similar difficulties are encountered when using calibration
curves and correction factors; however, these difficulties are then often
ignored). The four unknowns C_A, C_B, C_C, and C_D in this system of equations
are determined only to a constant factor. The value for this factor and,
hence, the individual concentrations may be found by using the normalizing
condition:

$$C_A + C_B + C_C + C_D = 1$$

Besides the four equations for the four unknowns, a fifth equation for the
normalizing condition has to be considered, and we are therefore dealing
with an inhomogeneous linear system of equations consisting of five
equations of fourth order with four unknowns. This system of equations
consists of four linearly independent equations, namely the normalizing
equation and three component equations; the fourth component equation is
simply the linear combination of the other equations.

In order to solve this system of equations, we subtract the normalizing
condition from every one of the four component equations of the original
system of equations and obtain the reduced system of equations (the familiar

term N_{100}/N is again used instead of $N^* - 1$)

$$\begin{pmatrix} -N_{A100}/N_A & r_{AB} - 1 & r_{AC} - 1 & r_{AD} - 1 \\ r_{BA} - 1 & -N_{B100}/N_B & r_{BC} - 1 & r_{BD} - 1 \\ r_{CA} - 1 & r_{CB} - 1 & -N_{C100}/N_C & r_{CD} - 1 \\ r_{DA} - 1 & r_{DB} - 1 & r_{DC} - 1 & -N_{D100}/N_D \end{pmatrix} \begin{pmatrix} C_A \\ C_B \\ C_C \\ C_D \end{pmatrix} = \begin{pmatrix} -1 \\ -1 \\ -1 \\ -1 \end{pmatrix}$$

$$(17.17)$$

The individual concentrations are determined by stepwise substituting according to Cramer's rule, a column in the coefficient matrix by the column vector of the right side of the equation and by calculating the value of this determinant. The following determinant applies to the determination of the first concentration C_A:

$$D_1 = \begin{vmatrix} -1 & r_{AB} - 1 & r_{AC} - 1 & r_{AD} - 1 \\ -1 & -N_{B100}/N_B & r_{BC} - 1 & r_{BD} - 1 \\ -1 & r_{CB} - 1 & -N_{C100}/N_C & r_{CD} - 1 \\ -1 & r_{DB} - 1 & r_{DC} - 1 & -N_{D100}/N_D \end{vmatrix} \quad (17.18)$$

The calculation is best done by developing the 4×4 determinant for the first column or the column which originally was on the right side of the system of equations (for the determinants, see Appendix). The concentration C_A is equal to the quotient

$$C_A = D_1/D \qquad (17.19)$$

where D is the value of the determinant for the unsubstituted coefficient matrix.

The determinant D_2 is calculated for the determination of the concentration C_B

$$D_2 = \begin{vmatrix} -N_{A100}/N_A & -1 & r_{AC} - 1 & r_{AD} - 1 \\ r_{BA} - 1 & -1 & r_{BC} - 1 & r_{BD} - 1 \\ r_{CA} - 1 & -1 & -N_{C100}/N_C & r_{CD} - 1 \\ r_{DA} - 1 & -1 & r_{DC} - 1 & -N_{D100}/N_D \end{vmatrix} \quad (17.20)$$

The concentration C_B may then be calculated according to the equation:

$$C_B = D_2/D \qquad (17.21)$$

The third and fourth concentrations are determined by analogous procedures.

Mathematical determination of the concentration in a four-component mixture is possible with a slide rule or desk calculator by simply solving the determinants of the $|3 \times 3|$ formate.

Beattie and Brissey (1954) have numerically written the system of equations for the four-component mixture chromium–iron–nickel–molybdenum and determined the concentration of the individual elements by direct mathematical solution of the equations. The interaction coefficients were first determined experimentally for the respective two-component mixtures (Table 16.1). For the particular x-ray fluorescence unit used by these authors, the system of equations, with Cr, Fe, Ni, and Mo as symbols for the concentrations of these elements, is written as follows:

$$-\left(\frac{N_{\text{Cr100}}}{N_{\text{Cr}}} - 1\right)\text{Cr} + \quad 0.721\ \text{Fe} + \quad 0.813\ \text{Ni} + \quad 2.660\ \text{Mo} = 0$$

$$2.482\ \text{Cr} - \left(\frac{N_{\text{Fe100}}}{N_{\text{Fe}}} - 1\right)\text{Fe} + \quad 0.676\ \text{Ni} + \quad 1.582\ \text{Mo} = 0$$

$$1.863\ \text{Cr} + \quad 2.420\ \text{Fe} - \left(\frac{N_{\text{Ni100}}}{N_{\text{Ni}}} - 1\right)\text{Ni} + \quad 1.108\ \text{Mo} = 0$$

$$0.877\ \text{Cr} + \quad 0.944\ \text{Fe} + \quad 1.260\ \text{Ni} - \left(\frac{N_{\text{Mo100}}}{N_{\text{Mo}}} - 1\right)\text{Mo} = 0$$

$$(17.22)$$

Table 17.6 compares calculated and chemically determined contents. The validity of the system of equations has been tested in numerous examples from the literature (Traill and Lachance, 1965; 1966).

The linear system of equations may also be solved with the aid of a programmed computer or an analog calculator. Marti (1962) has designed an analog calculator to calculate chromium, iron, nickel, and molybdenum (cobalt–manganese) contents of steels. Measured fluorescent intensities are read on potentiometers and the voltage corresponding to a certain concentration is read on calibrated voltmeters.

17.10. Mathematical Determination of Concentration of a Single Component in a Multicomponent Mixture

Occasionally, determination of the concentration of only one component in a multicomponent system is required, while the concentrations of the other elements in the mixture are of no interest. However, the concentrations of the associated components have to be known in order to determine the concentration of a single component from its fluorescent intensity. Although these concentrations are not normally known, the intensities of the associated components may be used to solve this problem. Let us consider the fluorescent intensity of component A in a mixture with associated components B, C, D, etc. and compare it to the fluorescent intensity of pure component A:

$$\frac{N_A}{N_{A100}} \cong \frac{C_A \bar{\mu}_A(\alpha)}{C_A \bar{\mu}_A(\alpha) + C_B \bar{\mu}_B(\alpha) + C_C \bar{\mu}_C(\alpha) \cdots} \qquad (17.23)$$

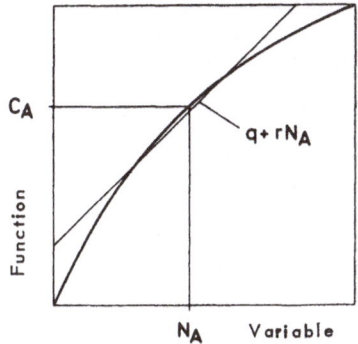

C_A

Function

N_A Variable

$q + r N_A$

Fig. 17.3. Concentration as a function of fluorescent intensity, and replacement of the calibration curve by a straight line so that in the area of interest the deviation between curve and straight line is at a minimum.

We obtain for the concentration of component A in a mixture

$$C_A \cong \frac{N_A}{N_{A100}} \left[C_A \left(\frac{\bar{\mu}_A(\alpha)}{\bar{\mu}_A(\alpha)} \right) + C_B \left(\frac{\bar{\mu}_B(\alpha)}{\bar{\mu}_A(\alpha)} \right) + C_C \left(\frac{\bar{\mu}_C(\alpha)}{\bar{\mu}_A(\alpha)} \right) + \cdots \right] \quad (17.24)$$

The fluorescent intensities of the particular components are introduced on the right side of the equation in place of the concentrations C_A, C_B, C_C, etc. The concentration is thought to be presented as a function of the intensity, where the fluorescent intensity and concentration are chosen as the variable and the function, respectively (Fig. 17.3). Every value of the concentration may be presented as a function of the corresponding intensity N. The values of C_A, C_B, C_C, etc. are expressed as

$$C_A = q_A + r_A N_A + s_A N_A^2 + t_A N_A^3 + \cdots$$
$$C_B = q_B + r_B N_B + s_B N_B^2 + t_B N_B^3 + \cdots$$
$$C_C = q_C + r_C N_C + s_C N_C^2 + t_C N_B^3 + \cdots, \text{ etc.} \quad (17.25)$$

where the factors q, r, s, and t have different values for every component. All higher powers of N are neglected, resulting in replacement of the calibration curve by a straight line (the latter, however, does not necessarily pass through the origin)

$$C_i \approx q_i + r_i N_i \quad \text{(substituted calibration curve)} \quad \text{with} \quad i = A, B, C \cdots$$

The concentration C_A is approximately expressed as follows (substitution of the concentrations on the right side of the equation by the intensities):

$$C_A \approx \frac{N_A}{N_{A100}} \left[\sum_{i=A}^{i=N} \left(\frac{\bar{\mu}_i(\alpha)}{\bar{\mu}_A(\alpha)} \right) q_i + \sum_{i=A}^{i=N} \left(\frac{\bar{\mu}_i(\alpha)}{\bar{\mu}_A(\alpha)} \right) r_i N_i \right] \quad (17.26)$$

Table 17.6. Elemental Concentration in Steel Calculated by Direct Mathematical Solution of the Linear System of Equations as Compared to Chemically Determined Values (Beattie and Brissey, 1954).

Sample number		Content in wt. % Calculated	Chemically determined	Relative deviation %
1	Cr	20.29	19.85	+ 2.2
	Fe	65.95	65.94	± 0
	Ni	13.76	13.98	— 1.6
2	Cr	20.39	21.23	— 4.0
	Fe	15.04	13.85	+ 8.6
	Ni	64.57	64.84	— 0.4
3	Cr	32.44	32.13	+ 1.0
	Fe	35.86	35.43	+ 1.2
	Ni	31.66	31.86	— 0.6
4	Cr	16.96	16.45	+ 3.1
	Fe	51.35	50.20	+ 2.3
	Ni	25.54	26.80	— 4.7
	Mo	6.15	6.55	— 6.1
5	Cr	19.63	18.43	+ 6.5
	Fe	64.93	66.26	— 2.0
	Ni	13.23	12.99	+ 1.8
	Mo	2.22	2.32	— 4.3

The first sum is a constant and, hence

$$\frac{1}{N_{A100}} \sum_{i=A}^{i=N} \left(\frac{\bar{\mu}_i(\alpha)}{\bar{\mu}_A(\alpha)} \right) q_i = \text{constant} = k_0$$

The second sum is arranged in terms of the intensities of the individual elements

$$\frac{1}{N_{A100}} \sum_{i=A}^{i=N} \left(\frac{\bar{\mu}_i(\alpha)}{\bar{\mu}_A(\alpha)} \right) r_i N_i = k_A N_A + k_B N_B + k_C N_C + \cdots$$

The concentration C_A of component A is therefore obtained as follows:

$$C_A \approx N_A \, (k_0 + k_A N_A + k_B N_B + k_C N_C + \cdots) \tag{17.27}$$

Thus, in order to determine the concentration of component A it is only necessary to know the fluorescent intensities of the associated components and not their concentrations. Using this procedure, Lucas-Tooth and Price (1961) determined the copper content of copper–zinc–tin alloys, where the values for k_0, k_{Cu}, k_{Zn}, and k_{Sn} are determined with the aid of calibration curves (Table 17.7).

Determination of k. For this purpose standards of known contents of the desired component A are required for which the fluorescent intensities of all components have been measured. Best values for k may be found by equalization calculation. We consider the relative deviation between the chemically determined content C_A and the content C'_A, calculated according to the following formula:

$$\Delta = \frac{C_A - C'_A}{N_A} = \frac{C_A}{N_A} - k_0 - k_A N_A - k_B N_B - k_C N_C - \cdots \quad (17.28)$$

It is required that the sum of all Δ^2 of the standards is a minimum or

$$\frac{\partial \Sigma \Delta^2}{\partial k_0} = 0, \qquad \frac{\partial \Sigma \Delta^2}{\partial k_A} = 0, \qquad \frac{\partial \Sigma \Delta^2}{\partial k_B} = 0, \qquad \text{etc.}$$

Thus, a linear system of equations is obtained from which the values for k may be calculated. When designating the intensities for the components by N_A, N_B, N_C, etc. and the chemically determined content by C_A, the system

Table 17.7. Chemical Composition and Determination of Copper in Copper–Zinc–Tin Alloys (Lucas-Tooth and Price, 1961).

Chemical composition			Fluorescent intensity			Calculated copper content	
% Cu	% Zn	% Sn	N_{Cu}	N_{Zn}	N_{Sn}	% Cu	deviation
96.62	0	3.34	23.320	334	14.320	96.78	+ 0.16
94.54	0	5.42	22.430	334	22.470	94.53	— 0.01
92.62	0	7.35	21.670	313	30.490	92.85	+ 0.23
90.61	0	9.36	20.870	295	38.330	90.82	+ 0.21
65.27	34.73	0	17.740	50.800	790	65.27	0
62.95	37.05	0	17.250	53.620	775	63.03	+ 0.08
67.61	32.39	0	18.240	47.800	785	67.58	— 0.03
69.98	30.02	0	18.750	44.680	840	69.94	— 0.04
100.0	0	0	24.720	349	800	99.82	— 0.18
87.88	5.01	7.11	20.770	7.778	29.030	87.64	— 0.24
87.93	2.96	9.11	20.380	4.552	37.180	87.83	— 0.10
87.93	0.83	11.24	20.030	1.659	45.020	87.94	+ 0.01
89.85	3.06	7.09	21.220	4.845	28.520	89.88	+ 0.03
86.73	1.99	11.28	19.740	3.028	45.220	86.60	— 0.13
84.00	4.99	11.01	19.260	7.151	44.960	83.97	— 0.03

of equations for the calculation of k_0, k_A, k_B, k_C, etc. is as follows (summation over the values of all standards):

$$\sum \frac{C_A}{N_A} = k_0 \qquad + k_A \sum N_A \quad + k_B \sum N_B \quad + k_C \sum N_C \quad + \cdots$$

$$\sum C_A = k_0 \sum N_A + k_A \sum N_A^2 \quad + k_B \sum N_A N_B + k_C \sum N_A N_C + \cdots$$

$$\sum \frac{C_A}{N_A} N_B = k_0 \sum N_B + k_A \sum N_A N_B + k_B \sum N_B^2 \quad + k_C \sum N_B N_C + \cdots$$

$$\sum \frac{C_A}{N_A} N_C = k_0 \sum N_C + k_A \sum N_A N_C + k_B \sum N_B N_C + k_C \sum N_C^2 \quad + \cdots$$

$$(17.29)$$

The values for the unknowns may be found according to Cramer's rule.

17.11. Appendix

Complete Formulas for the Determination of Regression Coefficients by Equalization Calculation

Let us assume a system of equations consisting of four homogeneous equations with four unknowns just as we did for the mathematical treatment of quantitative x-ray fluorescence analysis

$$- C_A N_A^* + C_B r_{AB} + C_C r_{AC} + C_D r_{AD} = 0$$
$$C_A r_{BA} - C_B N_B^* + C_C r_{BC} + C_D r_{BD} = 0$$
$$C_A r_{CA} + C_B r_{CB} - C_C N_C^* + C_D r_{CD} = 0$$
$$C_A r_{DA} + C_B r_{DB} + C_C N_{DC} - C_D N_D^* = 0$$

N_i^* are values which are obtained from the experimentally measured fluorescent intensities, r_{ij} are the regression coefficients, and the concentrations C_i are the variables. We set

$$N_i^* = \frac{N_{i100} - N_i}{N_i}$$

where N_{i100} and N_i are the fluorescent intensities of the 100% pure component i and the component i in a mixture, respectively. The best values for the regression coefficients r_{ji} in this system of equations are to be determined first (this system of equations is later used to determine the concentration C_i). Three coefficients have to be determined per line and, hence, at least three standards are required. Better values for the coefficients, however, may be obtained by using approximately two to three times as many standards as are required as a minimum, and the best values for the regression coefficients r_{ji} are found from all of these standards by a compensation calculation. The following systems of equations are derived for the determination of r_{ji} by equalization calculation (Σ = sum over all standards):

r_{AB}, r_{AC}, r_{AD}:

$$r_{AB} \Sigma C_B^2 \quad + r_{AC} \Sigma C_B C_C + r_{AD} \Sigma C_B C_D = \Sigma C_A C_B N_A^*$$
$$r_{AB} \Sigma C_B C_C + r_{AC} \Sigma C_C^2 \quad + r_{AD} \Sigma C_C C_D = \Sigma C_A C_C N_A^*$$
$$r_{AB} \Sigma C_B C_D + r_{AC} \Sigma C_C C_D + r_{AD} \Sigma C_D^2 \quad = \Sigma C_A C_D N_A^*$$

r_{BA}, r_{BC}, r_{BD}:

$$r_{BA} \Sigma C_A^2 \quad + r_{BC} \Sigma C_A C_C + r_{BD} \Sigma C_A C_D = \Sigma C_A C_B N_B^*$$
$$r_{BA} \Sigma C_A C_C + r_{BC} \Sigma C_C^2 \quad + r_{BD} \Sigma C_C C_D = \Sigma C_B C_C N_B^*$$
$$r_{BA} \Sigma C_A C_D + r_{BC} \Sigma C_C C_D + r_{BD} \Sigma C_D^2 \quad = \Sigma C_B C_D N_B^*$$

r_{CA}, r_{CB}, r_{CD}:

$$r_{CA} \Sigma C_A^2 \quad + r_{CB} \Sigma C_A C_B + r_{CD} \Sigma C_A C_D = \Sigma C_A C_C N_C^*$$
$$r_{CA} \Sigma C_A C_B + r_{CB} \Sigma C_B^2 \quad + r_{CD} \Sigma C_B C_D = \Sigma C_B C_C N_C^*$$
$$r_{CA} \Sigma C_A C_D + r_{CB} \Sigma C_B C_D + r_{CD} \Sigma C_D^2 \quad = \Sigma C_C C_D N_C^*$$

r_{DA}, r_{DB}, r_{DC}

$$r_{DA} \Sigma C_A^2 \quad + r_{DB} \Sigma C_A C_B + r_{DC} \Sigma C_A C_C = \Sigma C_A C_D N_D^*$$
$$r_{DA} \Sigma C_A C_B + r_{DB} \Sigma C_B^2 \quad + r_{DC} \Sigma C_B C_C = \Sigma C_B C_D N_D^*$$
$$r_{DA} \Sigma C_A C_C + r_{DB} \Sigma C_B C_C + r_{DC} \Sigma C_C^2 \quad = \Sigma C_C C_D N_D^*$$

These are systems of equations, each of which consists of three equations with three unknowns; the unknowns may be found, for example, according to the rules of Cramer and Sarruss.

Determinants and Linear Systems of Equations

A determinant of second order is defined as follows:

$$\begin{vmatrix} a_{11} & a_{12} \\ a_{21} & a_{22} \end{vmatrix} \equiv a_{11} a_{22} - a_{12} a_{21}$$

The horizontal sequence is called a line, and the vertical sequence is referred to as a column. The symbols a_{11}, a_{12}, a_{21}, and a_{22} are the elements of the determinant.

A determinant of third order is defined as follows:

$$\begin{vmatrix} a_{11} & a_{12} & a_{13} \\ a_{21} & a_{22} & a_{23} \\ a_{31} & a_{32} & a_{33} \end{vmatrix} \equiv \left\{ \begin{array}{l} + a_{11} a_{22} a_{33} + a_{12} a_{23} a_{31} + a_{13} a_{21} a_{32} \\ - a_{11} a_{23} a_{32} - a_{12} a_{21} a_{33} - a_{13} a_{22} a_{31} \end{array} \right\}$$

This may also be written

$$\begin{vmatrix} a_{11} & a_{12} & a_{13} \\ a_{21} & a_{22} & a_{23} \\ a_{31} & a_{32} & a_{33} \end{vmatrix} \equiv a_{11} \begin{vmatrix} a_{22} & a_{23} \\ a_{32} & a_{33} \end{vmatrix} - a_{21} \begin{vmatrix} a_{12} & a_{13} \\ a_{32} & a_{33} \end{vmatrix} + a_{31} \begin{vmatrix} a_{12} & a_{13} \\ a_{22} & a_{23} \end{vmatrix}$$

In the latter formulation the determinant after the first column is developed into subdeterminants. The determinant may also be developed into sub-determinants after the first line

$$\begin{vmatrix} a_{11} & a_{12} & a_{13} \\ a_{21} & a_{22} & a_{23} \\ a_{31} & a_{32} & a_{33} \end{vmatrix} \equiv a_{11} \begin{vmatrix} a_{22} & a_{23} \\ a_{32} & a_{33} \end{vmatrix} - a_{12} \begin{vmatrix} a_{21} & a_{23} \\ a_{31} & a_{33} \end{vmatrix} + a_{13} \begin{vmatrix} a_{21} & a_{22} \\ a_{31} & a_{32} \end{vmatrix}$$

All three formulations are identical, as can easily be tested by completing multiplication in the equations. In order to calculate a determinant of third order numerically, Sarruss' rule is applied as follows:

$$\begin{vmatrix} a_{11} & a_{12} & a_{13} \\ a_{21} & a_{22} & a_{23} \\ a_{31} & a_{32} & a_{33} \end{vmatrix} \begin{matrix} a_{11} & a_{12} \\ a_{21} & a_{22} \\ a_{31} & a_{32} \end{matrix} \equiv \begin{matrix} + a_{11} a_{22} a_{33} - a_{13} a_{22} a_{31} \\ + a_{12} a_{23} a_{31} - a_{11} a_{23} a_{32} \\ + a_{13} a_{21} a_{32} - a_{12} a_{21} a_{33} \end{matrix}$$

The positive products are parallel to the main diagonal, $a_{11} a_{22} a_{33}$, and the negative products are parallel to the minor diagonal, $a_{13} a_{22} a_{31}$. Thus, six products of three terms have to be formed.

Every determinant may be developed into subdeterminants after any line or column. The subdeterminants are formed by assuming that the respective line and column of the element for which the particular sub-determinant is to be formed is eliminated, and the remaining elements are grouped to form the subdeterminant. Thus, a determinant of fourth order may be developed after the first column

$$D \equiv a_{11} D_{11} - a_{21} D_{21} + a_{31} D_{31} - a_{41} D_{41}$$

Four subdeterminants are obtained corresponding to the number of elements per column. All that has to be carefully considered is that the signs during development for subdeterminants alternate. The sign of an element for which the subdeterminant is formed is, in general,

$$(-1)^{i+j}$$

where i and j are the two indices of the element. The sign distribution for the determinant of fourth order is as follows:

$$\begin{vmatrix} + & - & + & - \\ - & + & - & + \\ + & - & + & - \\ - & + & - & + \end{vmatrix}$$

If, for example, a determinant of fourth order is developed for the second line, we obtain

$$D \equiv \begin{vmatrix} \cdot & \cdot & \cdot & \cdot \\ a_{21} & a_{22} & a_{23} & a_{24} \\ \cdot & \cdot & \cdot & \cdot \\ \cdot & \cdot & \cdot & \cdot \end{vmatrix} \quad \begin{vmatrix} \cdot & \cdot & \cdot \\ a_{21} & & & \\ \cdot & \cdot & \cdot \\ \cdot & \cdot & \cdot \end{vmatrix} \quad \begin{vmatrix} \cdot & & \cdot & \cdot \\ & a_{22} & & \\ \cdot & & \cdot & \cdot \\ \cdot & & \cdot & \cdot \end{vmatrix} \quad \begin{vmatrix} \cdot & \cdot & & \cdot \\ & & a_{23} & \\ \cdot & \cdot & & \cdot \\ \cdot & \cdot & & \cdot \end{vmatrix} \quad \begin{vmatrix} \cdot & \cdot & \cdot & \\ & & & a_{24} \\ \cdot & \cdot & \cdot & \\ \cdot & \cdot & \cdot & \end{vmatrix}$$

$$D \equiv \qquad - a_{21} D_{21} \qquad + a_{22} D_{22} \qquad - a_{23} D_{23} \qquad + a_{24} D_{24}$$

Let us consider a linear system of equations of the form

$$x_1 a_{11} + x_2 a_{12} + x_3 a_{13} + x_4 a_{14} = m_1$$
$$x_1 a_{21} + x_2 a_{22} + x_3 a_{23} + x_4 a_{24} = m_2$$
$$x_1 a_{31} + x_2 a_{32} + x_3 a_{33} + x_4 a_{34} = m_3$$
$$x_1 a_{41} + x_2 a_{42} + x_3 a_{43} + x_4 a_{44} = m_4$$

x_1, x_2, x_3, and x_4 are the unknowns or independent variables, $a_{11}, \ldots, a_{44}$ are the coefficients or constants, and m_1, m_2, m_3, and m_4 are the dependent variables or the free terms. The system of equations is said to be homogeneous when all of the right sides of the equations are equal to zero, and it is referred to as inhomogeneous when at least one of the right sides is different from zero. An identical formulation for the linear system of equations is the matrix formulation, where the coefficients appear separated from the unknowns in every matrix:

$$\begin{pmatrix} a_{11} a_{12} a_{13} a_{14} \\ a_{21} a_{22} a_{23} a_{24} \\ a_{31} a_{32} a_{33} a_{34} \\ a_{41} a_{42} a_{43} a_{44} \end{pmatrix} \begin{pmatrix} x_1 \\ x_2 \\ x_3 \\ x_4 \end{pmatrix} = \begin{pmatrix} m_1 \\ m_2 \\ m_3 \\ m_4 \end{pmatrix}$$

Coefficient matrix   Variable matrix   Matrix or column vector of the right side

In order to find the first unknown x_1 we substitute the first column of the coefficient matrix by the column vector of the right side and calculate determinant D_1 of the substituted matrix (whose first line is $|m_1 \, a_{12} \, a_{13} \, a_{14}|$). Furthermore, we calculate determinant D of the unsubstituted original coefficient matrix (whose first line is $|a_{11} \, a_{12} \, a_{13} \, a_{14}|$). The value of x_1 is then found by formation of the quotients

$$x_1 = \frac{D_1}{D} \equiv \frac{\begin{vmatrix} m_1\,a_{12}\,a_{13}\,a_{14} \end{vmatrix}}{\begin{vmatrix} a_{11}\,a_{12}\,a_{13}\,a_{14} \end{vmatrix}}$$

The second unknown x_2 is found by replacing the second column of the coefficient matrix by the column vector of the right side and calculating the determinant D_2 of the substituted matrix (whose first column is $|a_{11}\,m_1\,a_{13}\,a_{14}|$). The value for x_2 is again found by forming the quotient

$$x_2 = \frac{D_2}{D} \equiv \frac{\begin{vmatrix} a_{11}\,m_1\,a_{13}\,a_{14} \end{vmatrix}}{\begin{vmatrix} a_{11}\,a_{12}\,a_{13}\,a_{14} \end{vmatrix}}$$

For determination of the third unknown x_3 we replace the third column of the coefficient matrix and form

$$x_3 = \frac{D_3}{D} \equiv \frac{\begin{vmatrix} a_{11}\,a_{12}\,m_1\,a_{14} \end{vmatrix}}{\begin{vmatrix} a_{11}\,a_{12}\,a_{13}\,a_{14} \end{vmatrix}}$$

and x_4 is determined analogously

$$x_4 = \frac{D_4}{D} \equiv \frac{\begin{vmatrix} a_{11}\,a_{12}\,a_{13}\,m_1 \end{vmatrix}}{\begin{vmatrix} a_{11}\,a_{12}\,a_{13}\,a_{14} \end{vmatrix}}$$

For the special case of a homogeneous linear system of equations, where all right sides are equal to zero, the trivial solution $x_1 = x_2 = x_3 = x_4 = 0$ is obtained. In general, there exist no other values of x_i which would bring the system of equations to equality other than the case where the determinant D itself is equal to zero. Only in this case is it possible to find values for the unknown which are different from zero. The unknowns, however, can only be determined to a constant factor. In homogeneous linear systems of equations consisting of n equations with n unknowns, only $(n - 1)$ equations are independent of each other while one equation is the linear combination of one or of all other equations, and the coefficient matrix of the $(n - 1)$ independent equations is used for the determination of the unknown [in that case the matrix consists of $(n - 1)$ lines and n columns]. In order to determine the first unknown, one assumes the first column of the remaining coefficient matrix to be deleted and calculates the determinant from the remaining $(n - 1) \times (n - 1)$ elements (whose first line is $|a_{12}\,a_{13} \ldots a_{1n}|$). The unknown x_1 is proportional to this determinant

$$x_1 = k\,\begin{vmatrix} a_{12}\,a_{13} \cdots a_{1n} \end{vmatrix}$$

The second unknown is determined by assuming the second column of the remaining coefficient matrix to be deleted and by calculating the determinant from the remaining elements (whose first line is $|a_{11}\,a_{13} \ldots a_{1n}|$).

The unknown x_2 is proportional to this determinant

$$x_2 = k \, |a_{11} a_{13} \cdots a_{1n}|$$

Analogously, unknowns x_3 to x_n are determined. The proportionality factor may have any value and the unknowns can only be determined to this factor. Other special cases are not of importance for the problem of the mathematical determination of the concentration.

Part III
EXAMPLES OF APPLICATIONS AND ABSTRACTS

Part III

Chapter 18

Analysis of Mixtures Which Are Difficult to Separate Chemically

One of the major areas of application of the x-ray fluorescence method is the analysis of mixtures which are difficult to separate chemically, particularly when determinations of concentrations in the range of 0.1 to 100% are required. Chemical separation is often unnecessary because spectra of the individual elements can clearly be differentiated. Coster and Hevesy (1923) were the first to detect the element hafnium in zircon on the basis of its emission spectrum. Today, it is easily possible to determine by x-ray fluorescence analysis the presence and the ratios of zirconium to hafnium in minerals, ores, refining products, and industrial materials. Powders and solid materials may be analyzed directly, or solutions and borax decompositions may be used. Internal standards are often applied to compensate for the effects of varying amounts of associated components. Owing to improved spectrometers with high-resolution crystals, proportional or scintillation counters, and pulse-height discriminators, it is now possible to resolve the undesirable overlap of the hafnium *L* lines by second-order zirconium *K* lines. Likewise, the elements niobium and tantalum can easily be determined in ores, minerals, refining products, industrial materials, and ultrapure metals. These samples are either analyzed directly, where detection limits may be as low as 10 to 20 ppm, or the samples are decomposed in solvents with or without internal standards. The same is true for the elements molybdenum and tungsten. Rare earths are often associated with scandium and yttrium: x-ray fluorescence analysis allows quick and simple identification of the individual elements and quantitative analysis without previous chemical separation. Other than analysis of lanthanides, the measurement of actinides is a preferred area of application of the technique: thorium, uranium, neptunium and plutonium can be determined without difficulty. Reference to selected papers from the literature is made in the following sections.

18.1. Zirconium–Hafnium

On the Missing Element of Atomic Number 72

D. Coster and G. Hevesy (University Institute for Theoretical Physics, Copenhagen)

Nature **111**, 79 (1923).

In this paper the discovery of element $Z = 72$ (hafnium) is described. The x-ray spectrum is used by Coster and Hevesy for the detection of the unknown element, where the wavelengths of the expected lines are predicted accurately on the basis of Moseley's work. They suspected the unknown element of $Z = 72$ to be related to the element zirconium and not to belong to the group of the rare earths. Detailed study of the x-ray spectra of zirconium minerals indeed resulted in the discovery of the expected lines of the L spectrum of hafnium. The wavelengths agree within 0.001 Å with the calculated values. The old name for Copenhagen, Hafnia, is suggested for the new element.

Hafnium–Zirconium and Tantalum–Columbium Systems: Quantitative Analysis by X-ray Fluorescence

L. S. Birks and E. J. Brooks (U.S. Naval Research Laboratory, Washington, D.C.)

Anal. Chem. **22**, 1017 (1950).

In this paper, x-ray fluorescence analysis of minor amounts of hafnium in zircon and of minor amounts of tantalum in niobium is discussed. The intensity ratio of the second-order lines $TaL_\beta + NbK_\beta$ to first-order NbK_α are measured for the determination of tantalum, because at the time the equipment was not sufficient to resolve the overlapping lines of first-order TaL_β and second-order NbK_β. A similar procedure is used for the determination of hafnium and zirconium. Synthetic samples serve as standards. The intensity ratio of the lines, $HfL_{\beta2}$ to second-order ZrK_β, may also be used for the determination of hafnium. Another possibility for analysis is to lower the excitation potential of the tube to 30 kV in order to limit as much as possible excitation of the K series of zirconium. At that time (1950) an accuracy of approximately 0.02 at.% tantalum or 4% relative was achieved for a content of 0.5 at.% Ta.

X-ray Spectroscopy as a Control Method in the Production of Zirconium and Hafnium

D. M. Mortimore and P. A. Romans (Northwest Electrodevelopment Laboratory, Bureau of Mines, Albany, Oregon)

J. Opt. Soc. Am. **42**, 673 (1952).

X-ray fluorescence techniques are used to analyze mixtures of hafnium and zirconium. The oxides are ground to a fine powder, mixed with an equal amount of starch, and then pressed into pellets. Known amounts of hafnium and zirconium are dissolved in hydrofluoric acid, precipitated simultaneously, and converted into the oxides to serve as standards. The intensity ratio of the $HfL_{\beta 1}$ line to the second-order ZrK_{β} line is measured for the determination of hafnium, because the two lines partially overlap. For higher hafnium contents the intensity ratio of the lines ZrK_{α} to $HfL_{\beta 1}$ is measured and the zirconium content is determined. The method is used to analyze amounts of about 0.5% hafnium in zircon, and 0.5% zirconium in hafnium. The relative standard deviation for a hafnium content of 1 wt.% is 3.8%, and for a hafnium content of 40 wt.% it is 0.8% relative.

The Separation and Fluorescent X-ray Spectrometric Determination of Zirconium, Molybdenum, Ruthenium, Rhodium, and Palladium in Solution in Uranium-Base Fission Alloys

J. O. Karttunen (Argonne National Laboratory, Argonne, Illinois)

Anal. Chem. **35**, 1045 (1963).

An x-ray fluorescence method is developed to determine quantitatively the elements Zr, Mo, Ru, Rh, and Pd in uranium-containing reactor fuels. In order to largely eliminate interferring line overlap and absorption-excitation effects, the samples are first dissolved in HF-containing aqua regia, and palladium is precipitated with dimethyl glyoxim. Molybdenum and ruthenium are then concentrated with ion-exchange columns, separated from zirconium and rhodium, and fluorescent intensities are measured in the respective fractions. Synthetic standard samples are used for calibration. Accuracy of the method is determined by the reliability of the chemical separation.

X-ray Fluorescence Spectrometric Determination of Zirconium and Molybdenum in the Presence of Uranium

E. A. Hakkila, R. G. Hurley, and G. R. Waterbury (University of California, Los Alamos Scientific Laboratory, Los Alamos, New Mexico)

Anal. Chem. **36**, 2094 (1964).

An x-ray fluorescence method is developed to determine, without previous separation, zirconium and molybdenum in uranium–zirconium–molybdenum carbides. The carbides are dissolved in nitric and hydrofluoric acids and niobium is added as internal standard. The intensity ratios of the lines ZrK_{α} to NbK_{α}, and MoK_{β} to NbK_{β} are measured. Since the lines $UL_{\beta 6}$ and $UL_{\gamma 5}$ overlap with the ZrK_{α} and MoK_{β} lines, corrections have to be made in the intensity ratios, the magnitude of which is a linear function of the uranium content. This method of analysis is suitable for the determination of 5 to 50% zirconium or molybdenum, and with a relative standard deviation of 1 to 2%. The effects of other associated elements on the analytical results are investigated.

18.2. Niobium–Tantalum

X-ray Spectroscopic Determination of Columbium and Tantalum in Rare-Earth Ores

D. M. Mortimore, P. A. Romans, and J. L. Tews (Northwest Electro-development Laboratory, Albany, Oregon)

Appl. Spectroscopy **8**, 24 (1954).

An x-ray fluorescence method for the determination of niobium and tantalum in monazite-rich ores is described and use is made of the internal standard principle. The ores are first decomposed, known amounts of zirconium and hafnium are added as internal standards, and the elements Zr, Nb, Ta, and Hf are precipitated simultaneously as hydroxides. The dried precipitate is then ground to a fine powder, mixed with 1 g of starch, and pressed into pellets. Synthetic mixtures of Nb_2O_5 and Ta_2O_5, prepared by the same procedure, are used for calibration. The effects of iron, titanium, and molybdenum content are compensated for by internal standards. Detection limits are 0.05% Nb and 0.2% Ta, and the relative standard deviation is 5% or less, depending upon the concentration.

Quantitative Analysis of Niobium and Tantalum in Ores by Fluorescent X-ray Spectroscopy

W. J. Campbell and H. F. Carl (Eastern Experiment Station, Bureau of Mines, College Park, Maryland)

Anal. Chem. **26**, 800 (1954).

An x-ray fluorescence method is developed to determine quickly and accurately the concentration of niobium and tantalum in ores. Three procedures are described. In the first method, the two oxides are chemically separated from the other constituents of the ores; the concentrates are analyzed for niobium and tantalum and the intensities of the NbK_α and TaL_α lines are measured. In order to avoid the undesirable overlap of the TaL_α line by the second-order NbK_α line, an excitation potential of only 19 kV is used and, hence, the NbK series is not excited. For amounts of sample of less than 200 mg, the sample is mixed with starch in order to obtain a larger amount of material. Synthetic reference samples serve for calibration. In the second procedure, molybdenum or zirconium, and hafnium or tungsten are added to the finely ground samples as internal standards. Synthetic samples of approximately the same bulk composition serve as standards. The third technique, which is particularly suited for the analysis of niobium- and tantalum-poor ores, involves mixing a known amount of Nb_2O_5 to part of the sample, and the original Nb_2O_5 and Ta_2O_5 contents are extrapolated from the change in the intensities. Many different niobium and tantalum ores may be analyzed quantitatively with either of these three procedures, and other elements such as Fe, Mb, Sn, and Ti in columbite–tantalite or Nb in pyrochlore may be determined as well.

Fluorescent X-ray Spectrographic Determination of Tantalum in Commercial Niobium Oxides

W. J. Campbell and H. F. Carl (Eastern Experiment Station, Bureau of Mines, College Park, Maryland)

Anal. Chem. **28**, 960 (1956).

An improved x-ray fluorescence method for the determination of tantalum in niobium oxide is described. Mechanically mixed standard samples are compared to chemically mixed samples; both methods of sample preparation yield the same results with $\pm 2\%$ relative. In order to determine the tantalum content in so-called pure niobium oxides, zirconium oxide samples containing known amounts of tantalum oxide are used to obtain a calibration curve for the trace element region. In the case of chemical mixtures of standard samples, particular care is taken to prevent separation during simultaneous precipitation of niobium and tantalum. This is achieved by, for example, soaking the niobium and tantalum containing solutions in a cellulose powder, burning the cellulose, and then annealing the residue. The effects of contaminations such as TiO_2, SnO_2, FeO, Na_2SO_4, and SiO_2 are considered numerically by using correction factors. Relative accuracy of analysis is $\pm 5\%$ in the range 0.5 to 10% Ta_2O_5, and the lower limit of detectability is around 0.03% Ta_2O_5.

X-ray Spectrophotographic Determination of Tantalum, Niobium, Iron, and Titanium Oxide Mixtures

B. J. Mitchell (Research and Development Analytical Laboratory, Electro-Metallurgical Co., Division of Union Carbide Corp., Niagara Falls, New York)

Anal. Chem. **30**, 1894 (1958).

An x-ray fluorescence method for quantitative determination of tantalum, niobium, iron, and titanium in columbite–tantalite ores, tantalum–niobium metals, and titanium alloys is developed. Before analysis, these elements are separated chemically by precipitation with cupferron from associated components such as manganese, calcium, chromium, and silicon. The composition in terms of wt.% of the tantalum–niobium–iron–titanium mixture is determined using calibration curves, where interaction of elements is mathematically compensated for by using correction factors. Content of

Ta_2O_5, Nb_2O_5, Fe_2O_3, and TiO_2 in the range of 10 to 100% may be determined to a relative accuracy of 1%, while in the range of 1 to 10% and 0.05 to 1%, relative accuracies are 5 and 10%, respectively. The developed correction procedure, which involves mathematical compensation for effects of a third component in a three-component mixture on the fluorescent intensities of the two other components, may in principle be applied to any three-component system.

X-ray Spectrographic Determination of Zirconium, Tungsten, Vanadium, Iron, Titanium, Tantalum, and Niobium Oxides: Application of the Correction Factor Method

B. J. Mitchell (Research and Development Analytical Laboratory, Union Carbide Metals Co., Division of Union Carbide Corp., Niagara Falls, New York)

Anal. Chem. **32**, 1652 (1960).

An x-ray fluorescence method is developed for simultaneous determination of the elements Zr, W, V, Fe, Ti, Ta, and Nb in oxide mixtures. Solutions, ores, and metals are analyzed, and the elements are precipitated simultaneously and transferred into oxide form. Metals and ores are first decomposed, the oxide powders are ground to a fine mesh, and pressed into a metal ring to form a pellet. Chemical as well as mechanical mixtures of powders serve as standards. Determination of concentration is carried out using calibration curves, and correction factors are used to compensate for the effects of third and fourth components on fluorescent intensities of the desired component. In principle, every mixture is treated as a two-component system with interfering admixtures of third and fourth components. In 1959, approximately 10,000 element determinations were made with this method.

X-ray Fluorescence Spectrographic Determination of Impurities and Alloying Elements in Tantalum Container Materials

E. A. Hakkila and G. R. Waterbury (University of California, Los Alamos Scientific Laboratory, Los Alamos, New Mexico)

Talanta **6**, 46 (1960).

An x-ray fluorescence method is developed to determine the elements Mo, Nb, Th, W, Y, and Zr which occur as contaminants in tantalum. The samples are decomposed with hydrofluoric acid and the solutions are then analyzed. The line to background ratio is measured for trace element determinations of the elements Mo, Nb, and Zr and the standard deviations are around 25 ppm. For the determination of tungsten in the range 0.5 to 10 wt. %, the intensity ratio of the WL_α line to the diffusely and incoherently scattered WL_α line of the tube is measured and the standard deviation for the determination of tungsten is found to be $\pm$ 0.07–0.09 wt. % W. The elements Th and Y are not soluble in hydrofluoric acid and are therefore separated as a residue fróm the tantalum-containing solution, put in solution with sulfuric acid, and then measured. Yttrium is added as an internal standard for the thorium determination, and thorium serves as an internal standard for the yttrium analysis. The relative standard deviations are between 2 to 13 %.

Quantitative Determination of Tantalum, Tungsten, Niobium, and Zirconium in High-Temperature Alloys by X-ray Fluorescent Solution Method

M. L. Tomkins, G. A. Borun, and W. A. Fahlbusch (Sierra Metals Corp., Division of Martin Marietta Corp., Wheeling, Illinois)

Anal. Chem. **34**, 1260 (1962).

An x-ray fluorescence method for the determination of the elements Ta, W, Nb, and Zr in alloys is described. The samples are first dissolved and the four elements are separated from associated elements by precipitation with cupferron. The precipitate is then decomposed with potassium pyrosulfate, the melt dissolved in diluted hydrofluoric acid, an organic complex-forming compound added to the solution for the purpose of stabilization, and the fluorescent intensities measured. Synthetic samples which are brought into solution in the same way as the sample, serve as standards to compensate for drift in the apparatus and inhomogeneities in the sample composition. Relative fluorescent intensities are determined in relation to the intensity of the diffusely scattered background. Relative standard deviations of 2 % may easily be achieved. The method is applicable to all alloys in which these elements can be separated from the other associated elements.

Röntgenspektrometrische Bestimmung von Tantal und Niob und einiger Begleitelemente in tantalniobhaltigen Stoffen (X-ray Spectrometric Determination of Tantalum, Niobium, and Several Associated Elements in Tantalum–Niobium Containing Substances)

H. Rothmann, H. Schneider, J. Niebuhr, and C. Pothmann (Laboratorien der Gesellschaft für Elektrometallurgie mbH., Zentrale Forschung, Nürnberg, Germany)

Archiv Eisenhüttenwesen **33**, 17 (1962).

Three procedures are described for x-ray spectrometric analysis of various tantalum–niobium containing substances: (1) Analysis of tantalum and niobium in hydrofluoric acid solutions, after separation from associated elements; applicable to ores, slags, and alloys. (2) Analysis of niobium, tantalum, and their associated elements in acid-soluble ferro- and special alloys, after the samples have been dissolved in hydrofluoric acid. And (3) analysis of niobium, tantalum, and their associated elements in ores after borax decomposition and addition of internal standards. In all three cases, synthetic standards are used for calibration. Results obtained with the three procedures are compared to chemically determined values. Time required for analysis is considerably less than for chemical methods of analysis, and reproducibility is good.

18.3. Molybdenum–Tungsten

Determination of Tungsten and Molybdenum by X-ray Emission Spectrography

J. E. Fagel, H. A. Liebhafsky, and P. D. Zemany (Research Laboratory, General Electric Co., Schenectady, New York)

Anal. Chem. **30**, 1918 (1958).

An x-ray fluorescence method for quantitative determination of tungsten and molybdenum in alkali solutions and in powders is developed. Bromine serves as internal standard for the tungsten determination and is added to the solution. The intensity ratio $MoK_\alpha/WL_{\gamma 1}$ is measured in the case of samples of known tungsten contents for the determination of molybdenum; thus, the degree of contamination of pulverized tungsten ores and residues by molybdenum is determined. More accurate molybdenum values, however, are obtained when measurements are carried out in solution.

18.4. Scandium–Yttrium–Rare Earths

X-ray Fluorescent Spectrometric Determinations of Yttrium in Rare Earth Mixtures

R. H. Heidel and V. A. Fassel (Institute for Atomic Research and Department of Chemistry, Iowa State College, Ames, Iowa)

Anal. Chem. **30**, 176 (1958).

Yttrium and rare earths have very similar chemical properties which make impossible application of classical chemical methods of separation. An x-ray fluorescence method is therefore developed to quantitatively determine yttrium in the presence of rare earths. The oxides are dissolved in hydrochloric acid and strontium is added to the solution as an internal standard. The solution is then dried and the residue dissolved in water. For the analysis of ores the rare earths, yttrium, and thorium are first separated from the other elements. Synthetic standards are used for calibration. Effects of varying amounts of rare earths and thorium are compensated for by the use of an internal standard. The relative standard deviation for a content of 50% Y is 0.74%. The values are within + 2.8% relative of data determined by optical emission spectroscopy.

X-ray Emission Spectrographic Analysis of High-Purity Rare Earth Oxides
F. W. Lytle and H. H. Heady (Rare and Precious Metals Experiment Station, U.S. Bureau of Mines, Reno, Nevada)
Anal. Chem. **31**, 809 (1959).

An x-ray fluorescence method is developed to determine quantitatively minor amounts of rare earths in pure rare earth oxides. The oxides are ground to a fine mesh and the fluorescent intensities of the desired elements are measured. Pure oxides are used to determine accurately the intensity of the background radiation, and the latter values are then subtracted from the measured intensities. Chemically mixed samples serve as standards, where the content of rare earth contaminants varies between 0.005 to 1%. Relative standard deviations range from 10 to 15%.

Determination of Lanthanum, Cerium, Praseodymium, and Neodymium as Major Components by X-ray Emission Spectroscopy
D. R. Maneval and H. L. Lovell (Department of Mineral Preparation, College of Mineral Industries, Pennsylvania State University, University Park, Pennsylvania)
Anal. Chem. **32**, 1289 (1960).

X-ray fluorescence methods are applied to quantitative analysis of the elements $Z = 57$ (La), 58 (Ce), 59 (Pr), and 60 (Nd), which occur in one and the same sample. The samples are ground, 0.1 g are decomposed with 10 g of borax, and the melt is poured into a pellet. If necessary, interferring elements such as Th, Hf, and Zr are chemically separated before borax decomposition. Samples of known contents of the particular oxides, which in addition contain silicon dioxide to bring the total weight to 0.1 g, serve as standards and linear calibration curves result. Interaction with other rare earths are, for the most part, negligible. Analysis of standard samples of known composition indicate an average relative standard deviation of less than 5%.

Fluorescent X-ray Spectrometric Determination of Scandium in Ores and Related Materials
R. H. Heidel and V. A. Fassel (Institute for Atomic Research and Department of Chemistry, Iowa State University, Ames, Iowa)
Anal. Chem. **33**, 913 (1961).

An x-ray fluorescence method for direct quantitative determination of scandium in minerals, ores, refining products, and mixtures of rare earths is developed. The samples are ground to a fine mesh and five parts are mixed well with three parts V_2O_5 as internal standard and three parts of SiC; the intensity ratio of the lines ScK_α/VK_α is measured. Experiments to decompose the samples in borax do not indicate an improvement in accuracy. The weak overlap in the case of tantalum- and erbium-containing samples of the second order of these lines with the ScK_α line is corrected for mathematically by simultaneous measurement of the intensity of the $ErL_{\beta2}$ and TaL_α lines. The relative standard deviation of the intensity ratio in the range 50% Sc_2O_3 and 1.5% Sc_2O_3 is 1.3% and 3.7%, respectively. The x-ray fluorescence method is experimentally shown to be equal in accuracy to the optical emission and the photometric titration methods.

Die Röntgenfluoreszenzanalyse seltener Erdmetall-Mangan-Legierungen (X-ray Fluorescence Analysis of Rare Earth–Manganese Alloys)

H. R. Kirchmayr and D. Mach (Institut für Angewandte Physik, Technische Hochschule, Vienna, Austria)

Z. Metallkunde **55**, 247 (1964).

X-ray fluorescence methods for quantitative analysis of rare earth–manganese alloys are developed. The method allows accurate analysis of even smallest amounts of substance (1–20 mg) by decomposing the alloy in nitric acid and direct measurement of the intensity of the fluorescent lines. Calibration curves are straight lines. In the analysis of solid pressed powders of alloys, contents of 10 ppm Mn in gadolinium and 60 ppm Gd in manganese (3 σ values) can still be detected. Accurate analyses in the concentration range of 1 to 100% are carried out in nitric acid.

18.5. Thorium–Uranium–Plutonium

Analysis of Uranium Solutions by X-ray Fluorescence

L. S. Birks and E. J. Brooks (U.S. Naval Research Laboratory, Washington, D.C.)

Anal. Chem. **23**, 707 (1951).

This publication is part of a series of papers which deal with the application of x-ray fluorescence to practical analysis. In the present case, it is shown that with x-ray fluorescence techniques, contents of 0.05 g uranium per liter may be detected in aqueous solution, and that the relative accuracy for contents of 1 gU/liter is approximately 5%. Other elements have no effect on the determination of uranium, except for the heavy elements such as lead, provided their content exceeds 10% of the uranium content. The study indicates that determination of minor amounts of uranium is possible without knowledge of contaminants or necessity of their separation from uranium. The speed of the method makes it suitable for use in continuous control of separation processes.

Determination of Thorium by Fluorescent X-ray Spectrometry

I. Adler and J. M. Axelrod (U.S. Geological Survey, Washington, D.C.)

Anal. Chem. **27**, 1002 (1955).

An x-ray fluorescence method for quantitative determination of thorium contents of rocks, minerals, and ores is described. The samples are ground to fine mesh and 0.9 g of it are mixed well with 0.1 g thallium chloride as internal standard as well as with 0.67 g of aluminum powder and 0.33 g of silicon carbide. Pellets are pressed after mixing and a well-analyzed monazite standard is used for calibration. The thorium content is then determined mathematically by comparison of the intensity ratios in the sample to the intensity ratio in the standard. Comparison with chemically determined values indicates agreement within 10 % relative for contents between 0.2 to 10 wt. % ThO_2.

Quantitative Determination of Thorium and Uranium in Solutions by Fluorescent X-ray Spectrometry

G. Pish and A. A. Huffman (Mound Laboratory, Monsanto Chemical Co., Miamisburg, Ohio)

Anal. Chem. **27**, 1875 (1955).

A quick and accurate x-ray fluorescence method for quantitative determination of thorium and uranium in aqueous and organic solutions is developed. No sample preparation other than admixture of an internal standard is required. Strontium is used as an internal standard in the case of aqueous solutions and bromine is added when organic solutions are analyzed. Calibration curves are straight lines, and the effects of contaminants are negligible in the concentration ranges studied. The accuracy of the method is equal to that of chemical analysis, and an analysis may be carried out in as little as one half hour.

Combined Radiometric and Fluorescent X-ray Spectrographic Method of Analyzing for Uranium and Thorium

W. J. Campbell and H. F. Carl (Eastern Experiment Station, U.S. Bureau of Mines, College Park, Maryland)

Anal. Chem. **27**, 1884 (1955).

Radioactivity and x-ray fluorescence measurements are combined to determine the thorium and uranium contents of ores. The samples are ground to a fine mesh and their radioactivities are determined. The intensity ratio of ThL_α/UL_α lines are then measured and from it the weight ratio of thorium to uranium is calculated. As the relative radioactivity of the individual elements is known, the content of thorium and uranium in the ore may be calculated (2 equations with 2 unknowns). Analysis takes approximately 20 min, and the accuracy is about 10% relative for contents larger than 0.5 wt. %, and the detection limit is near 0.01 to 0.03%.

Quantitative Analysis for Thorium by X-ray Fluorescence

A. G. King and P. Dunton (U.S. Geological Survey, Denver Federal Center, Colorado)

Science **122**, 72 (1955).

An x-ray fluorescence method for quantitative determination of thorium in rocks, minerals, and ores is proposed. The samples are ground to a fine mesh and 40 mg of a mixture consisting of 90% Si and 10% Se are added as an internal standard to 2 g of the sample. Furthermore, 0.4 g of silicon carbide are added to increase the grinding effectiveness. The powder is ground for one half hour and then analyzed. For this purpose, the intensity ratio of the ThL_α and SeK_β lines are determined. Synthetic samples of known thorium contents serve as standards. The standard deviation for ore analyses in the range of 0.4 to 0.8% Th is $\pm$ 0.0026% Th.

Fluorescent X-ray Spectrographic Determination of Uranium in Waters and Brines

W. L. Kehl and R. G. Russel (Gulf Research & Development Co., Pittsburgh, Pennsylvania)

Anal. Chem. **28**, 1350 (1956).

An x-ray fluorescence method is developed to determine uranium contents of water and brine. For this purpose the uranium of the original solution is first precipitated as a phosphate and the precipitate is then ashed. Yttrium nitrate may be added to the solution before precipitation to serve as an internal standard. Since small amounts of material are involved, the ash must be analyzed in a specially designed sample container. As little as 0.01 mg of uranium may be detected corresponding to an original content of 0.01 ppm U provided, however, the sample does not contain large amounts of strongly absorbing elements.

The Determination of Uranium in Solution by X-ray Spectrometry

H. M. Wilson and G. V. Wheeler (Phillips Petroleum Company, Atomic Energy Division, Idaho Falls, Idaho)

Appl. Spectroscopy **11**, 128 (1957).

A quick and accurate method for quantitative determination of uranium in a nitric acid solution is presented. All solutions are diluted to the same concentrations and the fluorescent intensity of the UL_α line is directly measured. Synthetic standards serve for calibration and a uranium-free sample is used to measure the background scattering. A molybdenum tube, scintillation counter, and pulse-height discriminator are used. Comparison of mass spectrometrically and x-ray fluorimetrically determined values indicates 10% higher values for the x-ray technique. The effects of other elements on the results are discussed.

Determination of Uranium Dioxide in Stainless Steel: X-ray Fluorescent Spectrographic Solution Technique

L. Silverman, W. W. Houk, and L. Moudy (Atomics International, Division of North American Aviation, Inc., Canoga Park, California)

Anal. Chem. **29**, 1762 (1957).

An x-ray fluorescence method for quick quantitative determination of the uranium content of uranium–steel bars used in nuclear reactors is described. The samples are dissolved in aqua regia and a known amount of strontium is added to the solution to serve as an internal standard, thus compensating for the effects of varying amounts of iron, chromium, and nickel. The intensity ratio of UL_α to SrK_α is measured, and synthetic mixtures prepared from uranium dioxide and steel serve as standards. The standard deviation, in the range of 15 to 25% UO_2, is 0.5 to 1%, and the agreement with chemical analyses is good.

X-ray Fluorescence Analysis of Plutonium

W. S. Turnley (The Dow Chemical Company, Rocky Flats Plant, Denver, Colorado)

Talanta **6**, 189 (1960).

An x-ray fluorescence method is developed to determine plutonium quantitatively in solids, liquids, and plutonium-poor muds. Solid samples are ground to a fine mesh and diluted by a mixture consisting of $Al(OH)_3$–MgO. Yttrium is added as an internal standard, and in the case of liquid samples, the internal standard is added without previous sample preparation. The intensity ratio of the PuL_α line to the YK_α line is measured, and synthetic samples serve as standards. The determination of concentration ranges from 10 ppm to 100% Pu in the case of solids, and from 5 to 250 g Pu/liter for liquids. The relative standard deviation in the analysis of solids is 0.63%; for liquids it is 1.24%, and for plutonium-poor muds with contents of 10 to 1800 ppm it is approximately $\pm$ 10%. The method is nearly unaffected by the presence of other elements and independent of isotope ratios.

X-ray Spectrographic Determination of Thorium in Low Concentration

J. N. Van Niekerk, F. W. E. Sterlow, and F. T. Wybenga (National Physical Research Laboratory, Council for Scientific and Industrial Research, Pretoria, South Africa)

Appl. Spectroscopy **15**, 121 (1961).

An x-ray fluorescence method is described for quick quantitative determination of low thorium contents. In the case of ore analyses, thorium is first brought into solution, concentrated in a suitable ion exchange column, and then the thorium content of the column is determined. Bromine is used as an internal standard and added to the ion-exchange column in order to compensate for the effects of varying amounts of iron on the fluorescent intensity of thorium. Synthetic samples serve as standards. The procedure is apparently particularly suitable for the analysis of thorium-poor ores and for production control. The results agree well with chemically determined values.

X-ray Spectrographic Determination of Thorium in Uranium Ore Concentrates

W. C. Stoecker and C. H. McBride (Uranium Division, Mallinckrodt, Chemical Works, Saint Charles, Missouri)

Anal. Chem. **33**, 1709 (1961).

An x-ray fluorescence method for quick and accurate determination of thorium contents in uranium ore concentrates is described. The samples are ground to a fine mesh, the intensity ratio of the ThL_α to the UL_λ line is measured, and the relative weight ratio of thorium to uranium in the ore is determined. The absolute thorium content may also be determined, provided the uranium content is known. An internal standard of uranium may be used to compensate for the varying composition of the ore concentrate. The strong UL_α line overlaps in part the weak ThL_α line and, hence, mathematical correction of the intensity of the thorium line is necessary. The results agree well with chemically determined values.

Dosage précis et rapide de l'uranium dans ses composés par fluorescence X-Application aux carbures d'uranium (*Accurate and Rapid Uranium Determination in Uranium Fuels by X-ray Fluorescence Techniques*)

R. Tertian, F. Gallin, and R. Geninasca (Cie Péchiney, Centre de Recherches d'Aubervilliers, France)

Rev. univers. Mines [9] **17**, 298 (1961).

An x-ray fluorescence method for the determination of uranium in uranium fuels is described. The samples are ground to a fine mesh and 100 mg of substance are mixed and decomposed with 9 g of borax and 1 g of BaO_2, the latter of which serves as an oxidation medium. The borax melt is poured to form a pellet. Synthetic samples serve as standards, and nearly linear calibration curves are obtained as a result of the strong dilution of the sample. Uranium determination is possible with a relative error of $+$ 0.2 % U. The method appears to be suitable as well for the uranium determination in other substances.

X-ray Emission Analysis of Plutonium and Uranium Compound Mixtures
O. Menis, E. K. Halteman, and E. E. Garcia (Advanced Materials Center, Nuclear Materials and Equipment Corp., Apollo, Pennsylvania)
Anal. Chem. **35**, 1049 (1963).
A time saving and reliable x-ray fluorescence method is developed for quantitative determination of plutonium and uranium in reactor fuels. The samples are ground to a fine mesh and decomposed with 10 g of potassium pyrosulfate. The melt is poured into a pellet, cooled slowly, or ground and again pressed into a pellet. In case of samples of simple compositions, the intensity ratio of lines PuL_α to UL_α is measured. For samples which contain a third component, thorium is added before decomposition to serve as an internal standard, and the intensity ratios of lines PuL_α/ThL_α and UL_α/ThL_α are then measured. Synthetic samples are used as standards. The analytical range is from 0.5 to 25 % PuO_2, and 75 to 99.5 % UO_2. The relative standard deviation for samples of simple composition is less than 0.4 %. Secondary excitation and self-excitation effects for transuranium elements are discussed.

Chapter 19

Steel and Iron Industry

X-ray fluorescence methods have successfully been applied to the analysis of numerous materials in the iron and steel industry, as well as for industrial of iron and noniron based materials such as zirconium alloys, alnico, ferro alloys, etc. Major components can be determined with an accuracy equal to that of chemical methods, and x-ray fluorescence techniques have the added advantage of being more economical and less time consuming. Depending upon the problem, the samples are either pulverized, ground, and pressed into pellets, where a binding medium is occasionally added, or polished. Care has to be taken in this procedure so as not to plug up softer substances or to smear them over the surface of the sample. Furthermore, attention has to be given to possible segregation and local enrichment of components during solidification of melts and slags. Interaction between the individual components is most pronounced in the case of solids. A multitude of procedures have therefore been developed to correct for the effects of associated components on the fluorescent intensity of the desired elements. This often results in a linear system of equations consisting of several equations and unknowns, the solutions of which yield the desired concentration. Mathematical correction procedures may be avoided when dissolving the samples in acids or fluxes, where the sample is strongly diluted and effects of interaction are reduced in magnitude. Strongly absorbing substances such as BaO_2 may be added to the melt in order to completely suppress interelement effects (borax technique). As a result, linear calibration curves are obtained, and the composition of the sample in terms of associated components may be neglected. Although sensitivity decreases when applying dilution procedures, accuracy of the results is usually improved. Reference to the literature is made in the following sections.

19.1. Steel and Iron

Some Problems in the Analysis of Steels by X-ray Fluorescence

E. Gillam and H. T. Heal (British Iron and Steel Research Association, Battersea, London, England)

Brit. J. Appl. Physics **3**, 353 (1952).

Application of x-ray fluorescence techniques to analysis of alloyed steels is investigated with particular reference to the determination of iron, chromium, and nickel. It is shown that the fluorescent intensity of the elements not only depends on their concentration, but also on the nature and amounts of associated elements. Therefore, calibration curves are required for the determination of composition. The mathematical relationship between fluorescent intensity, concentration, and composition of the sample is derived and interelement effects are discussed.

Analysis of High-Temperature Alloys by X-ray Fluorescence

R. M. Brissey (Thomson Laboratory, General Electric Co., Lynn, Massachusetts)

Anal. Chem. **25**, 190 (1953).

Principles and possibilities of application of the x-ray fluorescence method are demonstrated in this paper with particular reference to production control of high-alloy steels.

Calibration Method for X-ray Fluorescence Spectrometry

H. J. Beattie and R. M. Brissey (Thomson Laboratory, General Electric Co., Lynn, Massachusetts)

Anal. Chem. **26**, 980 (1954).

Fluorescent intensity of elements in the analysis of multicomponent systems not only depends on elemental concentration but also on overall composition of the sample. Complicated interactions between individual elements of a mixture consisting of n components may be expressed mathematically as a linear system of equations consisting of $n + 1$ equations with n unknowns of rank n. The unknowns correspond to the desired concentrations, the diagonal coefficients are determined by the fluorescent intensities of sample and pure element, and the nondiagonal coefficients are calculated from intensity measurements of two-component mixtures. The system of equations is solvable unequivocally. The procedure which, for calibration, requires only pure components and two-component mixtures, is tested for molybdenum, nickel, iron, and chromium analyses of high-alloy steels. The calculated values agree in nearly all cases within $\pm 5\%$ relative with chemically determined values.

Generalized X-ray Emission Spectrographic Calibration Applicable to Varying Composition and Sample Forms

H. G. Burnham, J. Hower, and L. C. Jones (Wood River Research Laboratory, Shell Oil Co., Wood River, Illinois)

Anal. Chem. **29**, 1827 (1957).

Determination of concentration leads mathematically to a linear system of equations. Additional transformations and extensions of the system of equations according to Sherman is necessary. The system of equations for a three-component mixture consists of three equations which are linear in regard to concentration and reciprocal intensities, and which are of third order in regard to the geometrical factor. The system of equations is solvable unequivocally and, hence, yields the desired concentrations. The value of this method increases when graphical methods are used for the solution of the three equations. Samples of different forms such as plates, wires, chips, and drill and file shavings may be studied by the x-ray method with seemingly the same success. Samples of known composition serve as standards. Factors accounting for interelement effects are determined by equalization calculation. The method is applied to the analysis of chromium, iron, and nickel in steels.

Quantitative Ermittlung von Legierungselementen im Stahl mit der Röntgen-Fluoreszenz-Spektralanalyse (*Quantitative Analysis of Alloy Elements in Steel by X-ray Fluorescence Spectral Analysis*)

H. Krächter and W. Jäger (Mannesmann Forschungsinstitut, Duisburg, Germany)

Archiv Eisenhüttenwesen **28**, 633 (1957).

The principles of the x-ray fluorescence method are presented in the introduction. The effects of sample thickness, surface properties, and x-ray counting time on the accuracy of the measurements are investigated. Experiments with low- and high-alloy steels indicate strong interelement effects. It is therefore desirable in quantitative analysis of alloyed steels to refer intensity measurements to standards of well-known composition similar in elemental content to the steels. Analysis of commerical steels with this procedure indicates fair agreement with wet-chemically determined values. In the case of standards not too different in composition than the samples and for contents of more than 1%, an accuracy of at least 2% relative may be expected. Advantages and disadvantages of the x-ray method in comparison to optical emission spectral analysis are discussed.

Determination of Iron, Chromium, and Nickel by Fluorescent X-ray Analysis
W. W. Houk and L. Silverman (Atomics International, Division of North American Aviation, Inc., Canoga Park, California)
Anal. Chem. **31**, 1069 (1959).
Iron, chromium, and nickel contents of various stainless steels and nickel–chromium alloys are determined using x-ray fluorescent techniques. Samples are dissolved in aqua regia and hydrofluoric acid, vaporized with perchloric acid, and diluted with water. Fluorescent intensity and background radiation are measured and the intensity ratios of (line + background)/background are used. A well-analyzed NBS steel serves as standard. Analytical results are within ± 1.5% of the true composition when using a scintillation counter. Pulse-height discrimination improves the detection limit so that 5 to 10 μg Fe, Ni, and Cr can be determined in 1 ml of solution. Calibration curves for stainless-steel analysis are also applied to the analysis of chromium–nickel alloys.

X-ray Fluorescence Analysis of Stainless Steel in Aqueous Solutions
R. W. Jones and R. W. Ashley (Chemistry and Metallurgy Division, Chalk River Project, Atomic Energy of Canada Ltd., Chalk River, Ontario, Canada)
Anal. Chem. **31**, 1629 (1959).
Nickel, chromium, molybdenum, and niobium are determined quantitatively in stainless steel using x-ray fluorescence. The samples (0.5 g) are dissolved in aqua regia and dried after adding sulfuric acid. Sulfates are dissolved in an aqueous solution, and niobium is precipitated by hydrolysis and separated chemically. Nickel, chromium, and molybdenum contents are determined directly in the solution. The separated niobium is mixed with 1 g of cellulose powder and pressed into a pellet. Synthetically mixed samples serve as standards. This method yields accurate results and the relative standard deviation, as indicated by comparison to chemically analyzed standards, is less than 1% for the four elements studied. The method is somewhat more involved than direct analysis of solid samples; however, strong interelement effects are reduced by the solvent so that calibration is simpler and independent of the sample form.

Line Interference Corrections for X-ray Spectrographic Determination of Vanadium, Chromium, and Manganese in Low-Alloy Steels
P. D. Zemany (Research Laboratory, General Electric Co., Schenectady, New York)
Spectrochim. Acta **16**, 736 (1960).
Line overlap such as between CrK_α–VK_β or MnK_α–CrK_β occasionally occurs when analyzing low-alloy steels. In order to determine in such

cases the intensity of the individual lines, the integrated intensity of the two overlapping lines is measured and the intensity of one line is then subtracted from the integrated intensity. The intensity value that is to be subtracted is found by measuring a second line of the same spectrum which is free of overlap, and the true intensity of the overlapped line may be calculated from the relative line intensities within the spectrum. The method is explained using the example of manganese–chromium–vanadium containing steels.

Röntgenspektroskopie im Eisenhüttenlaboratorium (*X-ray Spectroscopy on Laboratories of Steel Mills*)

H. De Laffolie (Deutsche Edelstahlwerke AG, Krefeld, Germany)

DEW-Techn. Ber. **1**, 161 (1961).

Principles of x-ray fluorescence analysis are presented and certain requirements in the apparatus such as dispersion and resolution of the analyzing crystals, detectors, and associated pulse-height distribution of the registered radiation, effects of air or vacuum in the spectrometer chamber, effects of sample distance on the measured values, roughness of sample surfaces, accuracy of measurements, and procedures for quantitative analysis are discussed.

Application de la fluorescence X au contrôle analytique dans une aciérie de moulage (*Application of X-ray Fluorescence to Analytical Control in a Steel Mill*)

R. Berger and P. Deceuleneer (Usines Emile Henriot, Court-Saint-Etienne, France)

Rev. univers. Mines [9] **15**, 207 (1961).

The x-ray fluorescence method is applied on a large scale to production control and monitoring of a steel mill. Solid samples turned to the proper dimensions are analyzed with 400 specially prepared standards which, on the basis of their composition, may be divided into several groups. The two component systems Fe–Cr, Fe–Ni, Fe–W, and Fe–Mn are investigated and the respective calibration curves are constructed. Next follow the three-component systems Fe–Cr–Cu, Fe–Ni–Mo, and Fe–Ni–Cu. Groups of binary calibration curves are plotted for varying amounts of a third component, and interelement effects are discussed. In the four-component systems Fe–Ni–Cr–Mn, Fe–Ni–Cr–Mo, Fe–Ni–Cr–Cu, and Fe–Cr–V–W, calibration curves are constructed, e.g., for the determination of chromium in the system Fe–Ni–Cr–Mo, where concentrations of associated elements vary (8%, 20%, and 36% nickel; and 0 and 5% Mo). Finally, the determination of niobium and titanium in multicomponent systems is investigated. On the basis of all the data, a sequence is established for the effects of one element on another during analysis, and the respective elements are determined in this sequence in the actual analysis. The first element to be determined is molybdenum, whose determination is practically independent of the composition of the sample as a whole. Then follows the determination of the nickel content, with molybdenum serving as a reference. Nickel content, in turn, serves as a reference parameter for the copper determination, and nickel and molybdenum are used as references for the chromium measurement. Chromium is the standard for the manganese determination. Niobium and titanium contents are independent of the overall sample composition. Numerous analyses made according to this procedure are compared to chemical analyses. The method is most suitable for production control of Fe–Cr–Ni–Mo–Cu–Mn steels as well as for Fe–W–Cr–V steels.

X-ray Fluorescence Spectroscopy in Metallurgical Research
T. A. Davies (The Mond Nickel Company Limited, Birmingham, England)
Rev. univers. Mines [9] **15**, 228 (1961).
X-ray fluorescence is not only applied to quantitative analysis but also to routine analysis. The latter concerns nickel–copper alloys, nickel–iron alloys, low-alloy steels, stainless steels, and high-temperature alloys on nickel basis. Analytical procedures are developed for the study of solid, powdered, and liquid samples. Strong interelement effects are occasionally noted when analyzing solid samples, the most noticeable of which are the effects of molybdenum on the chromium determination in nickel-containing alloys and stainless steels, and of chromium on the cobalt determination in high-temperature alloys. The interfering elements are chemically separated in order to eliminate these effects. When working with solutions, certain elements may be separated and concentrated, thus making possible analysis of amounts which ordinarily are below the detection limit. With the aid of synthetic standard solutions, analyses are possible which are most difficult to carry out with other analytical methods.

Determination of the Interelement Effect in the X-ray Fluorescence Analysis of Cr in Steels
W. Marti (Physikalisches Laboratorium Gebrüder Sulzur AG, Winterthur, Switzerland)
Spectrochim. Acta **17**, 379 (1961).
A procedure is described for accurate chromium determination in steels by x-ray fluorescence analysis involving minimum requirements for standards. The fluorescence intensity of chromium not only depends on the Cr concentration but also on the overall composition of the sample, and this inter-

action is expressed numerically in the form of an equation. The interaction coefficients of the individual elements are determined experimentally by systematically varying the content of a given element in a steel. With this procedure, the interaction coefficients for Si, Mn, Ni, Mo, V, Ti, Nb, Cu, W, and Co are determined. When applying, for example, the correction to a chromium analysis (this, however, requires knowledge of the overall sample composition), the corrected fluorescent intensity of chromium is proportional to the Cr concentration.

On the Determination of the Interelement Effect in the X-ray Fluorescence Analysis of Steels

W. Marti (Physikalisches Laboratorium, Gebrüder Sulzer AG, Winterthur, Switzerland)

Spectrochim. Acta **18**, 1499 (1962).

Interelement effects in a high-alloy steel are expressed in the form of a linear system of equations. Interaction coefficients for the determination of the elements Cr, Mn, Fe, Co, Ni, Cu, and Mo in steels are calculated using well-analyzed steel samples. Since the interaction coefficients are known, measured fluorescent intensities are subsequently corrected for interelement effects; solution of the systems of equations yields all desired concentrations simultaneously. The corrected intensities are then strictly proportional to the concentrations. An analog computer may be used to solve the system of equations, and its principle is briefly discussed.

Die Anwendung der Röntgenfluoreszenzanalyse bei der Untersuchung verschiedener Ferrolegierungen (Application of X-ray Fluorescence Analysis to Study of Various Ferro Alloys)

J. Bruch (Metallurgische Abteilung, Gusstahlwerk Witten AG, Witten)

Archiv Eisenhüttenwesen **33**, 5 (1962).

X-ray fluorescence techniques are described to analyze quantitatively and reliably various powdered alloys such as Fe–W, Fe–Mo, Fe–Nb–Ta, and Fe–Cr. Two procedures for sample preparation are described extensively, namely decomposition of the sample in borax with addition of BaO_2, and dissolution in acids. Both procedures are tested extensively for reproducibility, and large sets of measurements are obtained to estimate scatter in the data. Synthetic samples prepared by analogous procedures serve as standards. The following determinations are made: W, Mo, Mn, and Cr in ferro tungsten; Mo, Cu, and Mn in ferro molybdenum; Ta, Nb, Mn, and Ti in ferro niobium–tantalum; and Cr and Ni in ferro chromium. Absolute scatter of $\pm 0.25\%$ may be obtained for major elements with 95% certainty, and accuracy is also sufficient for the determination of minor components. Usefulness of the analytical procedure is indicated by numerous comparative analyses. Time required for analysis including sample preparation varies between 45 m and $3\frac{1}{2}$ h.

Beitrag zur röntgenspektrometrischen Analyse legierter Stähle (Contribution to X-ray Spectrometric Analysis of Alloyed Steels)
H. De Laffolie (Forschungsinstitut, Deutsche Edelstahlwerke AG, Krefeld, Germany)
Archiv Eisenhüttenwesen **33**, 101 (1962).
Feasibility and prerequisites for exact and accurate analysis of solid steel samples are discussed. Experimental calibration curves obtained with various accelerating potentials are compared to theoretically derived curves for the binary system iron–nickel. Use of an external standard for the determination of the composition is described, and enhancement and absorption effects of numerous elements are reported. Finally, results of x-ray spectrometric analyses are compared to wet-chemically determined values.

Beitrag zur röntgenspektrochemischen Analyse niedriglegieter Stähle (Contribution to X-ray Spectrochemical Analysis of Low-Alloy Steels)
H. De Laffolie (Forschungsinstitut, Deutsche Edelstahlwerke AG, Krefeld, Germany)
DEW-Techn. Berichte **2**, 119 (1962).
It is shown that calibration curves for minor elements are straight lines in x-ray fluorescence analysis. The slope of the calibration curves changes with the excitation conditions and the atomic number of the desired element, and simple relations exist between the calibration curves for various elements. A method for background correction is discussed and systematic errors which occur when neglecting the background are presented quantitatively.

Determination of Lead in Leaded Steels by X-ray Spectroscopy
B. A. Kilday and R. E. Michaelis (National Bureau of Standards, Washington, D.C.)
Appl. Spectroscopy **16**, 137 (1962).
Lead content of carbon-poor steels is determined by x-ray fluorescence analysis. Lead is only poorly soluble in steel (0.005%), and in the case of contents of 0.15 to 0.35% Pb small lead particles are present in the steel. The samples are first polished with normal polishing compounds; in order to avoid smearing of the soft lead over the total sample surface the specimens are given a final metallographic polish with diamond powder of 6 μ

and 0.25 μ grain size using oil as a lubricant. With this procedure, suitable surfaces even for varying grain sizes of the lead particles are obtained. A well-analyzed NBS steel is used as a standard, and homogeneity of the sample in terms of the lead content is investigated by selecting small areas for analysis. The studies serve to select a new NBS standard steel.

General X-ray Spectrographic Solution Method of Analysis of Iron, Chromium, and/or Manganese Bearing Materials

B. J. Mitchell and H. J. O'Hear (Technology Department, Union Carbide Metals Co., Division of Union Carbide Corp., Niagara Falls, New York) *Anal. Chem.* **34**, 1620 (1962).

A simple and accurate x-ray fluorescence method is described to determine iron, chromium, and manganese in soluble substances. The procedure is applied to the analysis of the following materials: manganese and chrome alloys, slags and ores, alnico, steels, and special alloys containing iron, chromium, manganese, nickel, copper, and cobalt. Contents in the range of 0.1 to 99.9% may be determined. The samples are either dissolved in acid or decomposed with sodium peroxide and then dissolved in acids. Fluorescent intensities' vary with the amount of solvent, temperature, acid and pH value, and manganese, nickel, and copper are therefore added as internal standards. Hence, only one calibration curve is necessary per element for all substances and solvents. Synthetic or other well-analyzed samples serve as standards. Comparison of x-ray spectrometric and wet-chemically determined contents indicate good agreement.

Determination of Refractory Metals in Ferrous Alloys and High-Alloy Steel by the Borax Disc X-ray Spectrochemical Method

C. L. Luke (Bell Telephone Laboratories, Inc., Murray Hill, New Jersey) *Anal. Chem.* **35**, 56 (1963).

An x-ray fluorescence method is developed for quantitative determination of the elements Mo, W, Nb, and Ta in temperature and corrosion resistant alloys, and for determination of Mo and W in high-alloy steels. The samples are dissolved in acid and the rare metals are chemically separated from associated elements by precipitation. Rare metals are transferred into oxide form and the oxides are decomposed with 10 g of borax and an additional 2 g of BaO; the melt is finally poured into pellet form. Addition of BaO tends to largely compensate for interelement effects. Comparison with chemical analyses shows that the method is well suited for the analyses in question and yields accurate results in a comparatively short time.

The Determination of Microgram Quantities of Zirconium in Iron, Cobalt, and Nickel Alloys by X-ray Fluorescence

O. Kriege and J. S. Rudolph (Westinghouse Research Laboratories, Pittsburgh, Pennsylvania)

Talanta **10**, 215 (1963).

An x-ray fluorescence method is developed to quantitatively determine lowest amounts of zirconium in iron, cobalt, and nickel alloys. The samples are dissolved in aqua regia and zirconium is precipitated and separated with *p*-bromium mandelic acid. The precipitate is analyzed directly on the filter paper. Samples prepared in the same manner serve as standards, and the calibration curve is a straight line. Low-alloy NBS steels of known zirconium contents are analyzed for test purposes, and the method is found suitable for the determination of concentrations in the range of 0.003 to 0.5% Zr.

19.2. Slags

Zur Schnellbestimmung des Eisens in der Schlacke (Rapid Determination of Iron in Slags)

H. J. Kopineck (Versuchsanstalt der Hoesch AG, Westfalenhütte, Dortmund, Germany)

Archiv Eisenhüttenwesen **27**, 753 (1956).

An x-ray fluorescence analysis unit for quick determination of the iron content of Thomas slags is developed and described extensively. After sampling, the slag is hammered during cooling, sieved, and pressed to a pellet. Natural variations in iron content of Thomas slags are $\pm 5\%$ relative and, hence, determination of the iron content to $\pm 2\%$ relative is sufficient. Chemically well-analyzed slag samples are used as standards, and a linear calibration curve is obtained in the range of 5–20% Fe. Time required for analysis from sampling to transmission of the results is approximately 4 mins.

Zur Anwendung der Röntgenfluoreszenz-Spektralanalyse in der Eisenhütten-industrie (Application of X-ray Fluorescence Spectral Analysis in the Steel Industry)

H. J. Kopineck and P. Schmitt (Versuchsanstalt der Hoesch AG, Westfalenhütte, Dortmund, Germany)

Archiv Eisenhüttenwesen **32**, 19 (1961).

Use of x-ray fluorescence analysis in the steel industry for production monitoring, research, and routine as well as special analyses is discussed. Of particular importance in this work are the detection limits that can be achieved. Detection limits for the elements Al to Ti in iron-

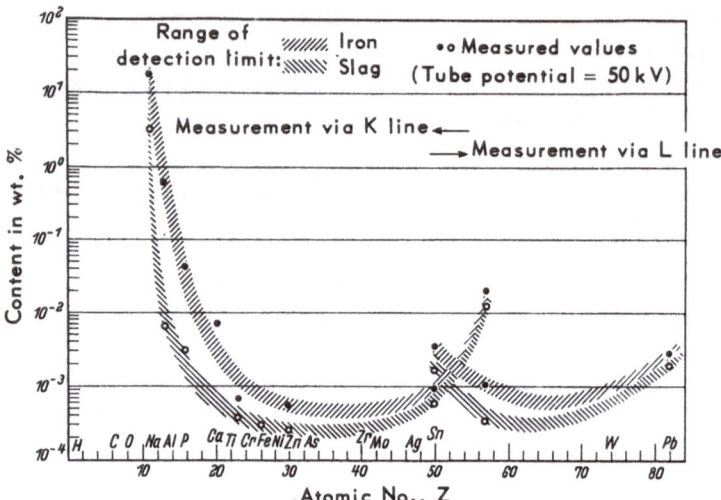

Fig. 19.1. Detection limit for the elements $Z = 11$ (sodium) to $Z = 82$ (lead) in iron slags and fireproof substances (from Kopineck and Schmitt, 1961).

containing substances and in slags are in the range of 0.01 to 0.001 %, and for Cr to Pb they are in the range of 1 to 10 ppm. Study of Siemens–Martin slags and, in particular, quick determination of alkalinity which involves determination of the $CaO-SiO_2-P_2O_5$ contents, is discussed extensively. Furthermore, studies of binary systems with particular reference to the systems iron–molybdenum and iron–manganese are described. Application of x-ray fluorescence techniques in place of radioactive measurements to determination of traces of lanthanum in Siemens–Martin slags is discussed.

Bestimmung der Zusammensetzung von Stahlwerkschlacke während des Schmelzens mit einem selbsttätigen Röntgenfluoreszenzspektrometer (*Determination of the Composition of Steel Slags During Melting with an Automatic X-ray Fluorescence Spectrometer*)

P. Dickens, P. König, and Th. Dippel (Chemisches Laboratorium, Hütten-werk Huckingen der Mannesmann AG, Duisburg, Germany)

Archiv Eisenhüttenwesen **34**, 519 (1963).

In view of the importance of the iron content and the degree of alkalinity of Siemens–Martin slags for the process of steel production, an x-ray fluores-cence procedure is developed to determine quantitatively and accurately the iron, silicon, calcium, phosphorus, magnesium, and aluminum contents in slags. The samples are ground, sieved, and pressed into pellets. The PK_α,

SiK_α, AlK_α, and MgK_α lines, as well as the higher order reflections of the FeK_α (third-order) and CaK_α (second-order) lines are measured in order to keep the running time of the goniometer from one position to the next at a minimum. Chemical analyses made simultaneously with x-ray spectrographic determinations serve as standards and determine the calibration curves for the following groups of samples. The results are, however, somewhat dependent upon the crystalline texture of the slags.

Spektrometrische Untersuchung nichtmetallischer Stoffe in Eisenhütten-laboratorien. III. Die Untersuchung von Eisensinter und Hochofenschlacke mit einem programmgesteuerten Röntgenfluoreszenz-Spektrometer (Spectrometric Study of Nonmetallic Substances in Steel Mill Laboratories. III. Investigation of Iron Sinter and Blast Furnace Slags with a Programmed X-ray Fluorescence Spectrometer

A. Stetter and H. Kern (Laboratorium der Röchlingschen Eisen- und Stahlwerke GmbH, Völklingen, Germany)

Archiv Eisenhüttenwesen **35**, 867 (1964).

Sample preparation for the analysis of blast furnace slags and sinters is studied in detail. Best results for powdered samples are obtained when 20 g of sample (less than 125μ grain size) are ground and mixed in a disc mill with 4 g of Linters powder (short-fiber cotton), and 7 to 8 g are used to prepare a pellet. Elements Fe, Ca, Si, Mn, and P are determined in iron sinters, and the elements Fe, Ca, Si, Mn, and Al in blast furnace slags. Samples of known composition serve as standards. Time required for analysis is between 25 and 35 min.

Chapter 20

Base Metals and Ores

Base metal alloys are those numerous special alloys containing nickel, cobalt, zirconium, copper, tin etc. whose quantitative composition may be determined accurately and quickly by x-ray fluorescence techniques. The samples are either studied directly as solids or are first decomposed in proper fluxes or solvents. Ore samples are often decomposed and chemically treated before analysis to remove undesirable elements. If, however, continuous and quick production control is anticipated, then finely-ground powders with or without internal standards are analyzed. Reference to the literature is made in the following sections.

20.1. Alloys and Industrial Materials

Some Applications of the Fluorescence X-ray Spectrometer in Ceramics
R. F. Patrick (Pemco Corporation, Baltimore, Maryland)
J. Am. Ceramic Soc. **35**, 189 (1952).
Principles and applications of x-ray fluorescence analysis in the ceramics industry are discussed in this early paper. Examples are cited to show that x-ray fluorescence is well-suited for quantitative determination of nickel which is applied to steel before the enamel is deposited. Furthermore, diffusion and adhesion of nickel and cobalt to the substrate during firing of the enamel may be studied quantitatively. The method is also applied to determination of lead in enamels, iron in pyrophyllite, and to testing of other raw materials for trace element contaminants such as niobium and titanium. Advantages and disadvantages of the method are discussed.

Rapid Chemical Analysis with the X-ray Spectrograph
M. Reith and E. D. Weisert (Haynes Stellite Co., Union Carbide and Carbon Corp., Kokomo, Indiana)
Metal Progress **70**, July 1956, p. 83.

The x-ray fluorescence technique is applied to quick quantitative analysis of stainless steels and nickel and cobalt alloys. Samples are polished for analysis and for every group of alloys, specially selected and well-analyzed standards are used for calibration. The x-ray method is fast, cheap, nondestructive, requires little standard preparation, and yields good results for high concentrations.

X-ray Spectrographic Analysis of Nickel-Containing Alloys with Varied Sample Forms

C. M. Davis and G. R. Clark (Research Laboratory, The International Nickel Company, Inc., Bayonne, New Jersey)

Applied Spectroscopy **12**, 123 (1958).

X-ray fluorescence procedures are developed to quantitatively analyze turnings and bore and file shavings. The samples are either pulverized or dissolved in acid and transformed into oxides. Standard samples are prepared by the same procedures. The powders are mounted with lacquer on glass slides and analyzed. The copper content of certain samples, such as Zn–Ni containing bronze, is practically constant so that copper may serve as an internal standard. The intensity ratios of ZnK_α/CuK_α and NiK_α/CuK_α are measured and evaluated. When studying alloys of W, Mo, and Nb, the samples are dissolved, dried, oxidized, and mixed with an equal amount of graphite powder, and then pressed into pellets. Synthetic samples serve as standards. Various analytical procedures are compared to chemical analyses and the method is found to be well-suited for quick identification of industrial materials.

Über ein Auswertungsverfahren bei der quantitativen Röntgenfluoreszenz-Spektralanalyse von Mehrstoffsystemen (An Evaluation Procedure for Quantitative X-ray Fluorescence Spectral Analysis of Multicomponent Systems

H. Preis and A. Esenwein (Eidg. Materialprüfungs- und Versuchsanstalt Zürich, Switzerland)

Schweizer Archiv **25**, 415 (1959).

An x-ray fluorescence method is described to consider interelement effects in the analysis of multicomponent systems. Groups of calibration curves are constructed for varying amounts of admixed third components. The method is discussed using the example of a lead–antimony–tin alloy (type metal). Agreement with chemically determined values is within $\pm 1.4\%$ relative.

Application of X-ray Fluorescence Methods to the Analysis of Zircaloy

R. W. Ashley and R. W. Jones (Chemistry and Metallurgy Division, Chalk

River Project, Atomic Energy of Canada Ltd., Chalk River, Ontario, Canada)

Anal. Chem. **31**, 1632 (1959).

X-ray fluorescence is applied to quantitative determination of the elements Sn, Fe, Cr, and Ni which, together with zirconium, are alloyed in zircaloy. The samples are transformed into oxides, mixed with an equal amount of cellulose powder, and pressed into pellets. Another method of sample preparation involves dissolving the sample in acids, chemical separation of zirconium, and precipitation of the desired elements as hydroxides. After adding cellulose powder, the hydroxides are pressed into pellets. Samples of known composition, which are prepared following the same procedure, are used as standards. Comparison to chemically determined values indicates that both procedures are sufficiently accurate and reliable; the second method is, however, twice as sensitive as the first. Experiments to determine traces of hafnium indicate that because of overlap with strong zirconium lines, the detection limit is near 200 ppm Hf.

X-ray Fluorescence Analysis and Its Application to Copper Alloys

F. R. Bareham and J. G. M. Fox (The British Non-Ferrous Metals Research Association)

J. Inst. Metals **88**, 344 (1959/60).

X-ray fluorescence is applied to the analysis of copper alloys. The samples are finely ground and analyzed with the grooves parallel to the plane formed by the incident and emerging x-ray beam. The two-component systems bronze (Cu–Sn) and brass (Cu–Zn) are studied, and synthetic samples of known composition are used as standards. The iron content of the three-component system Cu–Ni–Fe is determined first because iron fluorescence is practically independent of the mixing ratio Cu–Ni. The CuK_β line is measured when determining copper, because this line is less sensitive to changes in the Ni–Fe ratio, and groups of calibration curves for various amounts of Fe are constructed for the determination of the concentration. In the three-component system Cu–Zn–Pb, lead is determined first because its fluorescence intensity is independent of the Cu–Zn mixing ratio. For copper and zinc determinations, groups of calibration curves for various amounts of admixed Pb are constructed. Analogous procedures are used in the three-component system Cu–Zn–Sn, where tin is determined first. With these procedures, quick and reliable quantitative analyses may be carried out even in systems with strong interelement effects. The standard deviation for the determination of copper in the range of 60 to 98 % is approximately 0.15 % Cu.

Elemental Analysis of Alnico-Type Alloys by X-ray Fluorescence
P. Blavier, A. Hans, P. Tyou, and I. Houbart (Centre National de Recherches
Métallurgiques, Lüttich, France)
Cobalt **7**, 33 (1960).
Principles of x-ray emission and x-ray fluorescence methods are described,
and use of diluents (solvent or flux) for the purpose of reducing interelement
effects is discussed. For the analysis of alnico alloys, 200 mg of sample are
dissolved in hydrochloric acid. The solution is dried and the residue is
annealed. The oxides are then mixed well with 2 g of BaO_2 to which is added
a flux of 10 g of borax. The melt is poured to form a pellet and analyzed.
Synthetic samples, which are prepared in the same way, serve as standards.
Linear calibration curves are obtained for the determination of the elements
Fe, Ni, Co, and Cu when alloyed with aluminum; it is indicated that
interelement effects are largely eliminated due to borax decomposition with
BaO_2.

*Sampling Errors in the X-ray Fluorescent Determination of Titanium in a
High-Temperature Alloy*
R. F. Stoops and K. H. McKee (Metallurgical Products Department,
General Electric Co., Detroit, Michigan)
Anal. Chem. **33**, 589 (1961).
An x-ray fluorescence procedure is developed to determine quickly and
accurately the titanium contents of high-temperature resistant alloys. It is
found that titanium is locally enriched during solidification of the ingot;
further treatment, however, does result in an even distribution of Ti in the
ingot. In order to assure representative sampling a small pear-shaped
sample is poured before the final pouring of the melt into an ingot. The
sample is cut for investigation in such a way that as much as one-half of the
cut surface is irradiated simultaneously; thus, the bulk titanium content
may be determined accurately, independent of the local distribution of
titanium. Using this method of sampling, x-ray fluorimetrically and
chemically determined values are found to be in good agreement. A well-
analyzed sample of known composition serves as standard.

*A Mathematical Method for the Investigation of Interelement Effects in
X-ray Fluorescent Analyses*
H. J. Lucas-Tooth and B. J. Price (The Solarton Electronic Group Limited,
Farnborough, Hampshire, England)
Metallurgia **64**, 149 (1961).
A method is developed to compensate mathematically for interelement
effects during analysis of multicomponent mixtures. The method is illustrated

using the example of copper determination in copper–tin–zinc alloys, and it
is shown to be particularly useful when determination of the content of only
one component is required.

X-ray Fluorescence Analysis of Polyphase Metals. Use of Borax Glass Matrix
S. Sarian and H. W. Weart (Olin Hall, Cornell University, Ithaca, New
  York)
Anal. Chem. **35**, 115 (1963).

An x-ray fluorescence method for quantitative analysis for alloys of predominantly eutectic compositions of Cd–Sn, Cd–Zn, and Sn–Zn is described.
The metals are transformed with sulfur into sulfides, excessive sulfur is
removed, and the sulfides are decomposed with borax and potassium
pyrosulfate. Synthetic samples serve as standards.

*Einsatz des Röntgen-Fluoreszenzverfahrens zur Schnellanalyse von Messing-
schmelzen (Use of X-ray Fluorescence Techniques for Quick Analysis
of Brass Melts)*
K. Wassmann (Metall-Laboratorium der Kabel- and Metallwerke Neumeyer
  AG, Nürnberg, Germany)
Metall **18**, 1178 (1964).

Usefulness of x-ray fluorescence for quick analysis of copper, lead, iron,
certain contaminants, and other alloy elements in brass melts is tested. The
sample surfaces are carefully machined for analysis. For small compositional ranges, conversion of measured counting rates to concentrations is
achieved by calculation when measurements are made with calibration
curves or external standards of known copper, lead, and tin contents.
Accuracy of copper determination is ≤ 0.2 wt.% with a confidence of
99.7%.

20.2. Ores

*Analyse rapide du Cuivre, du Zink et du Cobalt par fluorescence X dans les
poudres de minerais et produits de concentration (Rapid Copper, Zinc, and
Cobalt X-ray Fluorescence Analysis of Mineral Powders and Dressing
Products*
A. Hans (Centre National de Recherches Métallurgiques, Lüttich, France)
Rev. univers. Mines [9] **15**, 545 (1959).

An x-ray fluorescence method is developed for quick, simple, and reliable
determination of copper, zinc, and cobalt contents of ores and dressing
products of the copper–zinc mine Prince Leopold and the copper–cobalt
deposit of Kolwezi in High-Katanga. The samples (2 g) are dissolved in

aqua regia, transformed into chlorides, dried, and dissolved in nitric acid. The solutions are analyzed and the intensities of the CuK_α, ZnK_α and CuK_α, CoK_α lines, respectively, are measured. Well-analyzed samples serve as standards. One calibration curve is obtained for every element indicating that interelement effects may be neglected. Analytical range is from 0 to 60 % for Zn, 0 to 50 % for Cu, and 0 to 5 % for Co.

Sur le dosage du Plomb et du Zink dans les minerais pauvres par spectrométrie de fluorescence X (Determination of Minor Lead and Zinc Contents of Minerals)

M. Maurette and J. Despujols (Laboratoire de Chimie-Physique de la Faculté des Sciences, Paris, France)

J. Chimie physique **57**, 1099 (1060).

Usefulness of x-ray fluorescence methods for determination of minor amounts of lead and zinc in minerals and ores is tested. Time-consuming sample preparations such as dissolving or melting of the samples is purposely avoided. Either absolute or relative intensities in relation to strontium, which is added as internal standard, are measured; in addition, the diffusely-scattered background radiation is used as internal standard. Samples of well-known composition serve as standards. Comparison with data obtained polarographically and wet-chemically indicates that the x-ray fluorescence method applied here is superior in speed but inferior in accuracy.

The Determination of Iron in Sphalerite by X-ray Fluorescence Spectrometry

B. R. Doe, A. A. Chodos, A. W. Rose, and E. Godijn (Division of Geological Sciences, California Institute of Technology, Pasadena, California)

Am. Mineral **46**, 1056 (1961).

An x-ray fluorescence method is developed to determine the iron content of sphalerite. From 0.1 to 20 mg of finely ground sphalerite is used for analysis, and the intensity ratio of ZnK_α/FeK_α is measured. Chemically well-analyzed sphalerites serve as standards. The effects of small contents of manganese and copper on the intensity ratio is investigated and it is shown that natural variations in manganese and copper contents of sphalerites have no effect on the analytical results. Comparison to wet-chemically obtained values is good and reproducibility is better than 5 %.

X-ray Spectrographic Determination of Copper in Low-Grade Porphyry Ores

R. E. Wood and E. R. Bingham (Kennecott Research Center, Kennecott Copper Corp., Salt Lake City, Utah)

Anal. Chem. **33**, 1344 (1961).

X-ray fluorescence techniques are applied to the analysis of copper ores, and ten parameters are systematically varied in order to find the most suitable procedure of sample preparation. Weighing out 1.300 g of finely ground sample ($-$100 mesh), and admixing of 0.400 g PbO_2 as an internal standard and oxidation medium and 0.300 g SiC as grinding medium is found to be the best method of sample preparation. The sample is mixed by hand and then heated for 30 min to 800°C. In this process all copper minerals are oxidized to CuO. The sinter product is finely ground by hand and analyzed. The intensity ratio of the CuK_α to PbL_γ second-order lines is measured, and copper-poor samples to which known amounts of copper have been added serve as standards; the resulting calibration curve is a straight line. Comparative experiments indicate that the x-ray spectrographic method is nearly as accurate as the chemical method. Standard deviation in the range between 0.5 and 1.2% Cu is ± 0.01%.

Bestimmung von Kupfer in Schwefelkies durch Röntgenfluoreszenzanalyse (*Determination of Copper in Pyrite by X-ray Fluorescence Analysis*)

E. Brodkorb (Applikationslaboratorium der Siemens & Halske AG, Munich, Germany)

Siemens Z. **35**, 451 (1961).

For accurate determination of copper, 1 g of pyrite is mixed with 9.6 g potassium pyrosulfate and 2.4 g sodium chlorate, and decomposed at 600°C in a quartz container. Nickel oxide (50 mg) is added to the sample for use as an internal standard. After cooling, the sample is pulverized and pressed into a pellet. The intensities of the CuK_α and NiK_α lines are measured and corrected for the intensity of the background radiation. The corrected intensity ratio is proportional to the copper content. Pyrite samples to which known amounts of copper oxide are added serve as standards, and comparison between data determined by x-ray spectrographic and electrolytic techniques indicate good agreement. The standard deviation in the range between 0 and 5 wt.% Cu is ± 0.02 wt.%. If content of the pyrite is high, so as to cause interference, then the nickel intensity may be measured.

Automatic X-ray Determination of Lead and Zinc in the Tailings of an Ore Dressing Plant

G. J. Sundkvist, F. O. Lundgren, and L. J. Lidström (Research Laboratory Boliden Mining Co., Skelleftehamn and Laisvall Mine, Laisvall, Sweden)

Anal. Chem. **36**, 2091 (1964).

An x-ray fluorescence unit is developed for continuous production control in the dressing of lead–zinc ores. Six to 12 g of ore mud are filtered in vacuum, the remaining water is removed using acetone, and the sample is dried in hot air. It is then homogenized and delivered to the spectrograph for measurement, where the intensities of the PbL_α and ZnK_α lines are measured simultaneously. Homogenization of the sample is necessary because various grain-size fractions have different lead–zinc contents. The standard deviations for 0.23 % Pb and 0.03 % Zn are, respectively, ± 0.01 % and ± 0.002 %.

Chapter 21

Light-Metal Industry

X-ray fluorescence analysis is applied in the light-metal industry for determination of contaminants in pure and ultrapure products in concentrations from 1 to 100 ppm as well as for the determination of major components. A further area of application is the analysis of light-metal alloys such as silicon–iron, aluminum–iron, or silicon–aluminum. X-ray fluorescence intensities are sometimes affected by texture and grain size of the alloy; in these cases correct analyses cannot be obtained without previous chemical decomposition. Reference to the literature is made in the following sections.

Automatische Analysenmethoden, insbesondere in der Leichtmetallindustrie (Automatic Methods of Analysis with Particular Reference to the Light-Metal Industry)

H. Pfundt (Forschungslaboratorium, Vereinigte Aluminiumwerke AG, Bonn, Germany)

Erzbergbau u. Metallhüttenwesen **13**, 110 (1960).

Usefulness of optical emission and x-ray fluorescence analysis for production control in the light-metal industry is discussed in detail. The principles of the method as well as instrumentation and application of the x-ray fluorescence method are presented.

Probleme und Möglichkeiten der Röntgenfluoreszenzspektrokopie bei der Untersuchung von Aluminium, insbesondere Aluminiumgußlegierungen (Problems and Possibilities of X-ray Fluorescence Spectroscopy in the Study of Aluminum and, in particular, Aluminum Casting Alloys)

H. Kessler and H. Rammensee (Laboratorium der Aluminiumwerke Nürnberg, Nürnberg, Germany)

Z. Metallkunde **51**, 548 (1960).

Several basic questions of x-ray fluorescence analysis of aluminum and aluminum–silicon alloys are discussed. Formation of an oxide layer on the surface may change the fluorescent intensities of light elements. As an example, the decrease in the fluorescent intensity of magnesium and iron for increasing duration of anode oxidation is mentioned. When quantitatively analyzing silicon in Al–Si alloys, the fluorescent intensity of silicon changes continuously with increasing amounts until the eutectic composition is approached; then the fluorescent intensity increases much more strongly, which is probably due to the formation of silicon crystals in the alloy.

Die Anwendung der Röntgenfluoreszenzanalyse auf dem Gebiete der Leicht-metalle (Application of X-ray Fluorescence to Analysis of Light Metals)
H. Pfundt (Forschungslaboratorium, Vereinigte Aluminiumwerke AG, Bonn, Germany)
Rev. univers. Mines [9] **17**, 203 (1961).
Application of x-ray fluorescence to the analysis of light metals is described. An important area of application is the determination of trace elements such as Si, Fe, and Ca in pure magnesium; Al, Fe, Ca, and Ti in pure silicon and quartz; and Si, Fe, Cu, Zn, Ti, Mg, Mn, and Cr in pure and

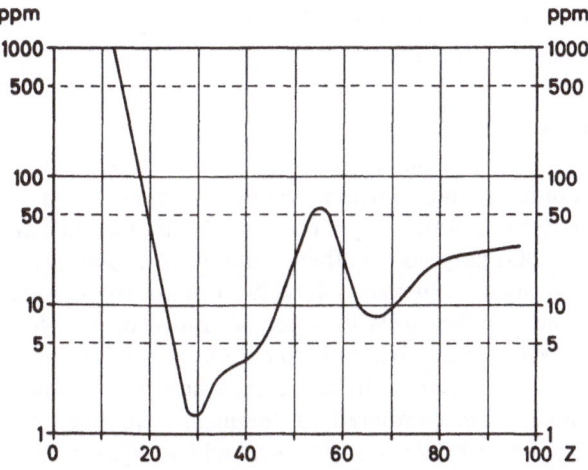

Fig. 21.1. Detection limits of the elements $Z = 12$ (Mg) to $Z = 92$ (U) in aluminum and alloys with light matrices (Pfundt, 1961). [Au-anode tube, 40 kV. Gypsum crystal for $Z = 11$ (Na) to $Z = 19$ (K); and LiF crystal for $Z = 20$ (Ca) to $Z = 92$ (U). Flow proportional counter for $Z = 11$ (Na) to $Z = 28$ (Ni), and scintillation counter for $Z = 29$ (Cu) to $Z = 92$ (U)].

ultrapure aluminum. Solid samples are machined and sometimes etched, while powdered samples are finely ground and, after adding wax, pressed into pellets. Well-analyzed samples serve as standards, and detection limits of 1 to 10 ppm are achieved. Another area of application is the analysis of light-metal alloys; the contents of silicon and iron in Fe–Si alloys, for example, may be determined quickly, quantitatively, and routinely.

Bestimmungsgrenzen für kleine und kleinste Beimengungen in Reináluminium und Raffinal bei Anwendung der Röntgenspektroskopie (Detection Limits for Determination of Low and Lowest Contents of Admixtures in Pure Aluminum and Refined Aluminum Using X-ray Fluorescence)

H. Ginsberg and H. Pfundt (Forschungslaboratorium, Vereinigte Aluminium-Werke AG, Bonn, Germany)

Z. Metallkunde **53**, 695 (1962).

Detection limits and minimal amounts of sample for trace element analysis in pure aluminum and "Raffinal" (99.99% pure aluminum) are determined using well-analyzed standards. Calculated detection limits in pure aluminum are 2 ppm Zn, 10 ppm Cu, 5 ppm Fe, 2 ppm Ti, 4 ppm Mn, and 2 ppm V; detection limits in Raffinal are 1 ppm Zn, 6 ppm Cu, 1 ppm Fe, and 1 ppm Ti. Minimal sample amounts are between 0.4 and 12 mg.

Rapid, Automatic X-ray Analysis of Magnesium Alloys

G. A. Stoner (Nuclear and Basic Research Laboratory, The Dow Chemical Co., Midland Michigan)

Anal. Chem. **34**, 123 (1962).

An x-ray fluorescence method is developed to determine quantitatively low amounts of zirconium, zinc, thorium, cerium, neodymium, praseodymium, and lanthanum in magnesium alloys. The samples are polished and analyzed directly using well-analyzed synthetic standards. Disagreement with chemically determined values are largely due to inhomogeneities and segregations in the samples. Relative intensity ratios of sample to standard are measured, where it is of advantage to choose a standard of composition similar to that of the sample. In some instances correction factors are required. The results are compared to chemically obtained values in the range 0 to 5% Th, 0 to 10% Zn, 0 to 1% Zr, and 0 to 5% Ce, Pr, Nd and La.

Zur Frage der röntgenfluoreszenzanalytischen Bestimmung des Siliziums in Aluminium-Silizium-Legierungen (On the Determination of Silicon in Aluminium–Silicon Alloys by X-ray Fluorescence Analysis)

F. Lihl and J. Fischhuber (Institute für angewandte Physik, Techn. Hochschule, Vienna, Austria)

Z. Metallkunde **53**, 186 (1962).
Silicon determination in aluminum–silicon alloys is discussed extensively. Various grinding and polishing techniques are mentioned which serve to obtain samples with well-polished surfaces. Experiments show that even in the subeutectic area the fluorescent intensity of silicon depends strongly on the conditions of casting of the alloys and on the texture. It appears, therefore, that silicon determination in solid industrial materials by x-ray fluorescence is not possible, although conclusions may be drawn as to the degree of refinement and texture.

Determination of Aluminum in Aluminum–Iron Alloys
J. V. Gilfrich and D. C. Sullivan (U.S. Naval Ordnance Laboratory, White
   Oak, Silver Spring, Maryland)
Norelco Rep. **10**, 127 (1963).
An x-ray fluorescence method is described for the determination of aluminum in aluminum–iron alloys. The samples are ground, polished, and well-analyzed samples serve as standards. The calibration curve is linear in the concentration range studied (0 to 17 % Al). Accuracy of analysis is approximately 0.1 % Al and equal to that of chemical analysis.

Die quantitative Aluminiumoxidanalyse mit Hilfe der Röntgenspektralanalyse
   (*Quantitative Analysis of Aluminum Oxide by X-ray Spectral Analysis*)
H. P. Pfundt (Forschungslaboratorium, Vereinigte Aluminiumwerke AG,
   Bonn, Germany)
Metall **18**, 1067 (1964).
X-ray fluorescence analysis is applied to determine quantitatively traces of zinc, iron, titanium, silicon, calcium, sulfur, vanadium, and chromium in pure aluminum oxide. Finely ground powder (10 g) is mixed repeatedly with 2 g of wax. The mixture is heated to the melting point of the wax, cooled, mixed again, and finally pelleted. Well-analyzed samples serve as standards, and calibration curves are linear. Standard deviations are within 0.3 to 3.5 % relative. Detection limits are 1 ppm ZnO, 2 ppm Fe_2O_3, 2 ppm TiO_2, 30 ppm SiO_2, 6 ppm CaO, 2 ppm V_2O_3, and 120 ppm SO_3. Comparison to wet-chemical analyses indicates that the x-ray fluorescence technique is well-suited for routine production control.

Chapter 22

Determination of Thicknesses of Thin Films

X-ray fluorescence analysis may be used for determination of the thicknesses of thin films. There are two ways to determine the thickness of a thin film, namely direct measurement of the fluorescent intensity of the film, or determination of its absorption. In the latter case the substrate is excited to fluorescence and the absorption of the emerging radiation, after passage through the thin film, is measured. Besides x-ray tubes, radio-active substances may be applied for excitation. Since radioactive sources have considerably lower photon efficiencies, dispersive geometry (i.e., analysis of x-ray fluorescence radiation with crystals) is not applied, and the radiation is collected directly by a counter. Pulse-height discriminators or filters, which allow selective discrimination of the desired radiation, may be used for unequivocal identification of the radiation. The x-ray fluorescence method was applied to investigate very thin nickel–iron films (200–2000 Å) for their compositions (about 80% Ni–20% Fe) and thicknesses. Massive iron and nickel samples are frequently used for calibration, where the fluorescent intensities of thin films are calculated theoretically. The thickness of the film in the production of tinplate can be controlled continuously. In this case, the fluorescent intensity of the tin-plated iron sheet is usually measured. Thicknesses in question range from 3000–30,000 Å. Other areas of application are the determination of thicknesses of titanium, vanadium, chromium, copper, silver, and cadmium films. Reactor fuels are cladded by zirconium, aluminum, or steel which prevent chemical reaction of uranium or plutonium with the surrounding medium. Cladding thick-nesses vary, in general, between 0.1 and 0.5 mm and may be controlled by x-ray fluorescence. Reference to the literature is made in the following sections.

Chemical Analysis of Thin Films by X-ray Emission Spectrography
T. N. Rhodin (Engineering Research Laboratory, E. I. DuPont de Nemours
   & Co. Inc., Wilmington, Delaware)
Anal. Chem. **27**, 1857 (1955).
Methods are described to study thin films of iron, nickel, chromium, and
stainless steel on mylar polyester foils. Particular emphasis is given to thin
films of approximately 100 Å thickness. Furthermore, oxide films which
form on stainless steel are studied. Excellent agreement is found between
the thicknesses determined by x-ray fluorescence and microcolorimetric
analysis, and it is concluded that the accuracy of the x-ray method is within
$\pm 2\%$.

*A New Approach to the Measurement of Coating Thickness by Fluorescent
   X-ray Absorption*
F. A. Achey and E. J. Serfass (Department of Chemistry, Lehigh University, Bethlehem, Pennsylvania)
J. Electrochem. Soc. **105**, 204 (1958).
Fluorescent intensity of the substrate is measured to determine the thickness
of thin films. In cases where the fluorescent radiation of the substrate is of
longer wavelength than the fluorescent radiation of the thin film, the
shorter wavelength radiation of the film may be suppressed using appropriate filters. Use of filter pairs is demonstrated by a "reverse" example
where zirconium and silver serve as substrates and are covered by a thin film
of iron. The fluorescent radiation of the substrate can be separated from that
of the associated iron radiation using properly tuned filter pairs, and the
intensities of the zirconium and silver radiations can be obtained. In the
opposite case, with iron as the substrate and zirconium and silver as the
thin films, the fluorescent intensity of the iron may be separated out.

Use of ^{55}Fe *For Measuring Titanium Coating Thickness*
P. D. Zemany (General Electric Research Laboratory, Schenectady, New
   York)
Rev. Sci. Instr. **30**, 292 (1959).
The radioactive iron isotope ^{55}Fe which largely emits MnK radiation, is
used to excite titanium to fluorescence. Excitation probability is high since
MnK radiation is slightly more energetic than the absorption edge of titanium. The radioactive source is mounted directly on a flow proportional
counter with the active side of the source pointing away from the counter. A
specimen holder is constructed in such a way, that the sample is placed
approximately 1 cm away from the source and the titanium radiation,
which originates in the sample, is directed towards the counter. Samples

which are weighed before and after titanium deposition, serve as standards; weight difference allows one to calculate the amount of material deposited. A calibration curve for the titanium determination in the range of 0 to 1.2 mg Ti/cm^2 is constructed. The same method and radioactive source may also be used for the determination of vanadium.

Determination of Cladding Thicknesses of Nuclear Fuel Elements by X-rays
P. Lublin (Sylvania Research Laboratories, Bayside, New York)
Norelco Rep. **6**, 57 (1959).
Uranium fuels are clad by zirconium, aluminum, and stainless steel to prevent chemical reaction between uranium and the surrounding medium. Cladding thickness is determined by measuring the intensity of the L_α line of the uranium. Thicknesses of up to 0.1 mm may be measured when zirconium is used as a cladding medium; in the case of aluminum larger thicknesses of up to 0.5 mm may be measured.

Measurement of Thin Metal Layers. Fluorescent X-ray Production by Radioisotope Sources
G. B. Cook, C. E. Mellish, and J. A. Payne (Isotope Research Division, Atomic Energy Research Establishment, Harwell, England)
Anal. Chem. **32**, 590 (1960).
Application of x-ray fluorescence to determination of the thicknesses of films is discussed using the examples of chromium, tin, and copper films. A radioactive source is used for excitation and the radiation is registered directly in a counter with associated pulse-height analyzer. Thus, most of the radiation can be separated from interfering radiation without taking recourse to an analyzing crystal which would result in a loss of intensity. Determination of chromium deposited on nickel in the range of 500 to 2500 Å is illustrated; for still thicker chromium films, the intensity of the chromium radiation reaches saturation at about 4000 Å, so that even thicker films will not cause an increase in the chromium intensity. Three procedures are discussed for the determination of tin, silver, and cadmium films (excitation with americium-241); first, measurement of the relatively hard K radiation which is particularly well-suited for relatively thick layers of up to 20 μ (200,000 Å); second, use of the relatively soft L radiation which, particularly for very thin films, results in better sensitivity while the saturation point is reached quickly for thicker films; and third, the indirect method involving measurement of the fluorescent intensity of the substrate (iron) which is applicable to thicknesses of up to 2 μ. Finally, the effects of the nature of the radioactive source on the measurement are illustrated using the example of the copper determination. Copper films of up to 20 μ thickness may be analyzed.

Measurement of Tinplate Thickness Using Fluorescent X-rays Excited by a Radioactive Source

J. F. Cameron and J. R. Rhodes (Isotope Division, Atomic Energy Research Establishment, Harwell, England)

Brit. J. Appl. Phys. **11**, 49 (1960).

A method is developed to measure the thickness of tinplate. The technique involves measurement of the fluorescence excited by a radioactive source (tritium–zirconium) and emitted by the steel substrate; absorption of the iron radiation is used to determine the thickness of the tinplate. Emerging iron radiation is measured directly by a counter without making use of an analyzing crystal, because the SnK_α line is not excited by tritium and the soft SnL_α radiation is largely lost in air. Samples of known mass density serve as standards. Thicknesses in the range from 0 to 15,000 Å are measured. A 1% change in the thickness of the tinplate, in the range of 7,800 Å, corresponds to a change in intensity of 0.4% which can be measured in 70 sec with a confidence of 95%.

Zur zerstörungsfreien Messung von Zusammensetzung und Schichtdicke kleiner Bereiche in dünnen Schichten mit Hilfe der Röntgenfluoreszenz (Nondestructive Measurement of Composition and Thickness of Small Areas in Thin Films by X-ray Fluorescence)

R. Weyl (Forschungslaboratorium der Siemens & Halske AG, Munich, Germany)

Z. angew. Physik **13**, 283 (1961).

Mass density of thin nickel–iron films with approximately 80% nickel is determined. Theoretical principles are discussed in detail and it is shown that the fluorescent intensity of each component in a multicomponent film is proportional to the partial mass density, and that secondary excitation amongst the components may be neglected. The fluorescent intensity of a massive sample is measured for calibration and, on the basis of this measurement, the mass density in the thin film is determined mathematically. Film thicknesses in the range of 500 to 1500 Å with mass densities of 40 to 120 $\mu g/cm^2$ are analyzed.

Composition Studies of Evaporated Nickel–Iron Layers

R. J. Heritage, A. S. Young, and I. B. Bott (Royal Radar Establishment, Malvern, Worcestershire, England)

Brit. J. Appl. Phys. **14**, 439 (1963).

X-ray fluorescence analysis is used to systematically investigate the effects of preparation conditions on nickel–iron layers produced by vacuum deposition. Original composition, speed of deposition, thickness of the layer, and temperature are varied and the respective fluorescent intensities of iron and nickel are measured. Samples of known mass density serve as standards, where the thicknesses of the layers are determined via absorption of the calcium and silicon radiations of the substrate (glass). Thin films between 200 and 2,000 Å and of an average composition of 80% Ni–20% Fe are studied.

Recherche d'une methode absolue de dosage par fluorescence X applicable aux lames minces de fer-nickel obtenues par evaporation sous vide (Development of an X-Ray Fluorescence Method to Determine the Composition of Thin Nickel–Iron Films Produced by Vacuum Deposition)

M. M. Pluchery (Centre d'Etudes Nucléaires de Grenoble, Grenoble, France)

Spectrochim. Acta **19**, 533 (1963).

A method is developed to analyze thin nickel–iron films containing approximately 80% nickel and to determine accurately their thicknesses and compositions. Theoretical principles for calculation of the resulting fluorescent intensities are presented in detail and secondary excitation of iron by nickel is discussed. Massive samples are used as standards.

Thin Films Analysis by X-ray Fluorescence

R. R. Stone and K. T. Potts (Development Lab., General Products Division, IBM Corp., Endicott, New York)

Norelco Rep. **10**, 94 (1963).

Thicknesses of thin iron–nickel films ranging in thickness from 500 to 6,000 Å and from 82 to 88% in nickel content are determined. The fluorescent intensities of both elements are measured and corrected for contributions by background radiation. Samples for which the mass densities were determined colorimetrically are used as standards, and the calibration curves are straight lines. Determination of mass density and thin film thickness is possible to better than 5% relative.

Some Advances in the Application of X-ray Fluorescence to the Continuous Measurement of Tin Coating Weight

D. E. Cass and J. H. Kelly (The Steel Company of Canada Ltd., Hamilton, Ontario, Canada)

Norelco Rep. **10**, 49 (1963).

A method for continuous control in the production of tinplate is described. In this technique, not only the fluorescence of the tin layer but also that of the substrate steel sheet is measured. The effects of surface roughness of the steel sheet and of incident and emergence angles of the x-ray radiation on the determination of the thickness of the thin film are studied in detail. Variations in roughness of the steel sheet affect the fluorescent intensity; however, this interference is smaller the steeper the incident and emergence angles. A specially constructed aperture compensates largely for differences in distance between steel sheet and x-ray tube, where the unavoidable fluttering of the passing steel sheet has no influence on the results. Layers of mass densities from 20 to 200 μg Sn/cm^2 and thicknesses of 3000 to 30,000 Å are determined.

Chapter 23

Cement Industry and Silicate and Rock Chemistry

X-ray fluorescence analysis is applied in the cement industry to quantitative analysis of major and minor components of raw materials and of cement. The elements Mg, Al, Si, Ca, and Fe may be determined directly in finely ground powders. Better accuracies are obtained, however, when the samples are first decomposed with sodium borate or lithium borate. Linear calibration curves universally applicable to all raw materials are desired, and occasionally certain correction factors are applied in order to obtain a higher accuracy. With recent advances in the efficiencies of commercially available apparatus for the analysis of light elements such as Na, Mg, and Al, it is now possible to determine these elements with the same high accuracy. Minor components such as Sr, Mn, Ti, P, and S can also be determined with comparative ease. In silicate and rock analyses, x-ray fluorescence is applied to the determination of the major components Na_2O, MgO, Al_2O_3, SiO_2, P_2O_5, K_2O, CaO, TiO_2, MnO, and Fe_2O_3. Depending upon compositional variations of the rocks, the finely ground powders may either be analyzed directly or they are first decomposed with sodium borate or lithium borate, thus eliminating mineralogical and textural variations. Occasionally, addition of an absorbing component such as La_2O_3 makes possible analysis of rocks of considerable variation in composition with uniformly linear calibration curves. Chemically well analyzed rocks serve as standards. The x-ray fluorescence method is particularly well suited for routine analysis, where accuracies equal or better than those of normal wet-chemical methods are obtained. A particularly successful area of application concerns the trace element analysis of rocks: fluorescent intensities in the trace-element range are directly proportional to the concentration, and variations in absorption by different matrices of the particular rocks may be considered mathematically or experimentally. The elements Ba, Ti, and Zn, for example, were determined quantitatively in rock powders of sediments, and regional variations in the elements K, Rb, Ti, Zr, P, Sr, and Ba were determined in a large granite pluton. Detection limits for direct analysis of finely ground powders is in the range of 1 to 20 ppm. The x-ray fluorescence method is also well-suited for the analysis of catalysts on an aluminum–silicon basis. X-ray fluorescence techniques are also applied to determine the coordination of aluminum in the structures of various alumino silicates: the wavelength of the aluminum fluorescent radiation is somewhat dependent upon the bonding conditions, and accurate measurement of the wavelength and comparison to known structures allows determination of the coordination number.

23.1. Cement and Clay

Fluorescent X-ray Spectrometric Estimation of Aluminum, Silicon, and Iron in the Flotation Products of Clays and Bauxites

S-Ch. Sun (College of Mineral Industries, Pennsylvania State University, University Park, Pennsylvania)

Anal. Chem. **31**, 1322 (1959).

An x-ray fluorescence method for quick determination of aluminum, silicon, and iron in flotation products of clays and bauxites is developed (W-anode tube, EDDT and LiF crystals, flow proportional counter). The samples are ground to 200 mesh (74 μ), pressed into the sample container, and measured. Synthetic samples prepared from aluminum oxide, silicon oxide, and iron oxide serve as standards. Comparison to wet-chemically analyzed samples indicates an average relative error of the method of approximately 5%.

Über den Einsatz der Röntgenfluoreszenz-Analyse in der Zementchemie (On the Application of X-ray Fluorescence Analysis in Cement Chemistry)

H. Schloemer (Institut für Werkstofftechnologie u. Allgemeine Hüttenkunde der Universitat des Saarlandes, Saarbrücken, Germany)

Zement-Kalk-Gips **13**, 522 (1960).

The method of x-ray fluorescence analysis is described in an introduction. Powdered samples are investigated. It is shown that the x-ray fluorescence method is well suited for production control in cement chemistry, and a number of applications are discussed.

Determination of Iron, Calcium and Silicon in Calcium Silicates by X-ray Fluorescence

J. W. Meyer (Johns–Manville Research Center, Manville, New Jersey)

Anal. Chem. **33**, 692 (1961).

X-ray fluorescence is applied to the determination of calcium, silicon, and iron in cement and calcium silicates. The samples are finely ground and analyzed as powders. Chemically analyzed samples as well as synthetic samples prepared from the oxides of silicon, calcium, and iron serve as standards. It is shown that the accuracy depends strongly on the matrix composition, and a simple mathematical relation is derived which allows one to calculate correction factors; thus, the required number of standards is held to a minimum. The correction method is illustrated by the example of an iron measurement made when various amounts of calcium were mixed with the iron. The method may also be applied to correct for the effects of aluminum in silicon determination. As indicated by comparison to chemically analyzed samples, the average error is approximately 7% relative for the Fe_2O_3 determination in the range of 0.4 to 4 wt. % Fe_2O_3; 2.5% relative for the SiO_2 determination in the range 0 to 100 wt. % SiO_2; and 2.3% relative for the CaO determination in the range 0 to 65 wt. % CaO.

Improvements in the X-ray Emission Analysis of Cement Raw Mix

G. Andermann (Applied Research Laboratories, Inc., Glendale, California)

Anal. Chem. **33**, 1689 (1961).

An x-ray fluorescence procedure for accurate determination of silicon, calcium, iron, aluminum, and magnesium in raw mix for cement production is described. A decomposition procedure with lithium borate is developed in order to overcome inaccuracies encountered in the analysis of powdered samples. One gram of finely ground sample is mixed with 1 g of $Li_2B_4O_7$, decomposed in a graphite crucible at approximately 1350°C in an electric furnace, quenched, finely ground, and pressed into pellets. Chemically well-analyzed samples of various origin are investigated. In order to obtain optimal results, the heat loss during decomposition is considered in the analysis and, simultaneously, a correction procedure is applied to correct for the effects of varying matrix compositions of the samples on silicon and iron fluorescent radiation. Furthermore, synthetically mixed samples may also be analyzed. The decomposition procedure described here increases considerably the accuracy of the method in comparison to direct analysis of powdered samples.

X-ray Emission Analysis of Finished Cements

G. Andermann and J. D. Allen (Applied Research Laboratories, Inc., Glendale, California)

Anal. Chem. **33**, 1695 (1961).

An x-ray fluorescence procedure for quantitative determination of the elements Sr, Fe, Mn, Ti, Ca, S, P, Si, Al, and Mg in finished cement is developed. The samples are decomposed with lithium borate at approximately 950°C, ground, and pressed into pellets. Chemically well-analyzed samples are analyzed in order to test the method. Determination of the content of every individual element is carried out with only one calibration curve each. In order to obtain optimal results, the heat loss during decomposition is considered. Correction factors are used to compensate for the varying interelement effects on the fluorescent intensities of silicon, iron, and phosphorus. The accuracy of this method is much improved in comparison to the direct analysis of powders and approaches closely the accuracy of chemical methods.

X-ray Spectrographic Analysis of Raw Materials for the Cement Industry

J. M. M. Robinson and E. P. Gertiser (Technische Forschungs-und Beratungsstelle der Schweiz. Zementindustrie, Holderbank-Wildegg, Switzerland),

ASTM Materials Research & Standards 4, 228 (1964).

A method is developed to determine quantitatively the elements Si, Al, Fe, Ca, and Mg in various raw materials for cement production using x-ray fluorescence. In order to obtain an analytical procedure suitable for all raw materials, sacrifices in the accuracy have to be made. Two pellets are prepared from every substance: first, 1 g of substance is mixed with 10 g of water-free borax ($Na_2B_4O_7$) and decomposed at 1050°C; this pellet serves for the determination of CaO, Fe_2O_3, SiO_2, and Al_2O_3; second, 3 g of substance are decomposed with 10 g $Li_2B_4O_7$; this pellet serves for the determination of MgO. An external standard of well-known composition, which is decomposed in the same fashion as the samples, serves as standard. In order to test the accuracy of the method, chemically determined values are compared to the ones obtained by x-ray fluorescence: with the exception of Fe_2O_3, most values are within the 3σ confidence limit. The accuracy may further be improved by proper treatment of the particular raw materials.

23.2. Silicates and Rocks

X-ray Spectrochemical Analysis: An Application to Certain Light Elements in Clay Minerals and Volcanic Glass

M. W. Molloy and P. F. Kerr (Columbia University, New York, New York),

Am. Mineral. 45, 911 (1960).

The x-ray fluorescence method is applied to study the quantitative composition of tuffs, volcanic glasses, and rhyolites. Major elements Al, Si, K, Ca, Ti, Mn, and Fe are determined quantitatively, and the elements P, S, Sc, V, Cr, Zn, As, Rb, Sr, Zr, Pb, and U are measured semiquantitatively. Synthetically prepared mixtures serve as standards, and samples of known composition are analyzed to check the accuracy.

Fluorescent X-ray Spectrographic Analysis of Amphibolite Rocks

A. A. Chodos and C. G. Engel (California Institute of Technology, Pasadena, and U.S. Geological Survey, University of California, La Jolla, California),

Am. Mineral. **46**, 120 (1961).

X-ray fluorescence analyses of 27 amphibolites are made to determine quantitatively the oxides Fe_2O_3, CaO, MgO, K_2O, TiO_2, and MnO. The amphibolites vary strongly in their mineral content but have approximately the same basaltic bulk chemistry. Variation in composition ranges from hornblende–andesine- amphibolites to pyroxene–labradorite–hornblende gneisses with several % of opaque minerals. The rock samples are finely ground until the complete sample appears as a homogeneous powder at 20 × magnification; measurements indicate that the average grain diameter is approximately 15 μ and all grains are less than 45 μ in diameter. Biotites require the longest grinding time. Well analyzed rocks are used as standards and the respective calibration curves are plotted (calibration is simplified due to similarity in major chemistry of all rocks and, hence, small matrix variation). The method allows quick and accurate routine investigation of amphibolites with an accuracy equal to or better than wet chemical methods.

X-ray Fluorescence Analysis of the Light Elements in Rocks and Minerals

H. J. Rose, Jr., I. Adler, and F. J. Flanagan (U.S. Geological Survey, Washington, D.C.)

Appl. Spectroscopy **17**, 81 (1963).

Advances in x-ray fluorescence analysis procedures allow quick analysis of light elements with accuracies equal to those of wet-chemical analysis. Besides the elements Fe, Mn, Ti, Ca, and K, it is now possible to quantitatively determine the elements P, Si, Al, and Mg. Decomposition of the samples with proper fluxes makes x-ray fluorescence analysis independent of grain size and mineralogical inhomogeneities. Simple decomposition does not, however, completely eliminate differences in absorption which result from variations in the compositions of the rocks. If, however, a strongly absorbing substance such as La_2O_3 is added to the melt, the original differences in the absorption are masked. Therefore, uniformly linear calibration curves can be used, and the method appears well-suited for nearly all types of rocks and silicates.

Determination of Major and Minor Constituents in Ceramic Materials by X-ray Spectrometry

R. Longobucco (Electron Tube Division, Radio Corp. of America, Harison, New Jersey)

Anal. Chem. **34**, 1263 (1962).

An x-ray spectrochemical method is developed to determine magnesium, aluminum, silicon, potassium, calcium, titanium, iron, and barium in forsterite and in materials, such as talcum, magnesium oxide, feldspar, and clay, which are used in the production of forsterite-type ceramics. Major components are determined in a borax flux; 5 g of substance are decomposed at 1350°C together with 7.5 g $Na_2B_4O_7$ and small amounts of Li_2CO_3, and the melt is then poured into a pellet. Minor components are determined directly in the finely ground powders whereby a known amount of the desired element is mixed to part of the powder; the original content may then be calculated from changes in the fluorescent intensities. The accuracy appears to be excellent in most cases and compares well with that of wet-chemical methods.

The Analysis of Aluminosilicates by X-ray Fluorescence

E. W. Orrell and P. J. Gidley (Technical Branch, The Carborundum Co. Ltd., Manchester, England)

Trans. Brit. Ceramic Soc. **63**, 19 (1964).

An x-ray fluorescence method for quantitative analysis of SiO_2, Al_2O_3, TiO_2, Fe_2O_3, and CaO in raw materials for the ceramic production and for production control is described. One gram of the finely ground sample is decomposed at 1000°C with 5 g lithium borate, the clear melt is quenched, ground, and pelleted. Well-analyzed samples serve as standards. In order to compensate for the strong effect of Al on the fluorescent intensity of Si, correction factors are used which are determined with the aid of synthetically prepared samples. Linear calibration curves are obtained for all elements. The x-ray method is equal in accuracy to chemical analysis, but much simpler and quicker.

Rapid Analysis of Rocks by X-ray Fluorescence

P. R. Hooper (Department of Geology, University College of Swansea, Swansea, Wales)

Anal. Chem. **36**, 1271 (1964).

A suite of rocks and residues of different chemical composition are analyzed by x-ray fluorescence spectroscopy for the elements Mg, Al, Si, P, K, Ca, Ti, Mn, Fe, and Sr. In all cases, accuracies are found to be better than those which are reported for the standard rocks W1 and G1. Error estimates indicate that matrix effects can be overcome by employing the borax technique. For low contents of Mg and P and for trace element analyses, the powders are ground to below 74 μ and analyzed directly. For borax decomposition, 1 g of substance is melted with 9 g of borax, ground, and then pelleted. Well-analyzed rocks serve as standards and linear calibration curves are obtained in most cases.

Silicate Sample Preparation for Light-Element Analysis by X-ray Spectrography

E. E. Welday, A. K. Baird, D. B. McIntyre, and K. W. Madlem (Department of Geology, Pomona College, Claremont, California)

Am. Mineral. **49**, 889 (1964).

Various methods of sample preparation for determination of light elements from $Z = 11$ (sodium) to $Z = 26$ (iron) are compared in terms of accuracy, range of applicability, and simplicity. More than 20 different rocks from leucogranite to gabbro serve as test materials. It was found that for the elements Na, Mg, Al, Si, K, Ca, and Fe, decomposition with 65% $Li_2B_4O_7$ and 35% substance yields best results. The borax technique is particularly advantageous in the analysis of mineralogically heterogeneous samples. Addition of a strongly absorbing component is unnecessary for the examples studied; it only results in a poor signal to background ratio. The particular method of decomposition is described in detail and has already been employed in the analysis of several hundred rocks.

Dual Grinding and X-ray Analysis of All Major Oxides in Rocks to Obtain True Composition

A. Volborth (University of Nevada, Reno, Nevada)

Appl. Spectroscopy **19**, 1 (1965).

The magnitude of contamination of rocks during pulverization is studied with the aid of x-ray fluorescence analysis. For this purpose, rock samples are finely ground with two different ball mill materials and the powders are then pelleted. Well-analyzed samples are used as standards and linear calibration curves are constructed. Analyses for Si, Al, Ti, Fe, Mn, Mg, Ca, K, and Na of 6 extrusive rocks are presented. Analytical accuracies are high; an uncertainty exists, however, for conversion of measured pulse rates into contents unless the particular bulk compositions are very similar to those of the standards.

X-ray Fluorescence Determination of Barium, Titanium, and Zinc in Sediments

G. I. Lewis and E. D. Goldberg (Scripps Institution of Oceanography, La Jolla, California)

Anal. Chem. **28**, 1282 (1956).

The x-ray fluorescence method is applied for quick and quantitative determination of barium, titanium, and zinc in marine sediments which range from nearly pure calcium carbonate through glauconite to deep sea mud. Lanthanum and arsenic are mixed to the samples to serve as internal standards and to compensate for matrix effects. The samples are ground to 300 mesh and mixed with 0.6 wt.% arsenic trioxide and 2.5 wt.% lanthanum trioxide. Synthetically prepared samples of known trace element contents serve as standards. The BaL_α and TiK_α lines partly overlap and could not be separated sufficiently; this interference is considered during analysis. Comparison to zinc contents determined by emission spectroscopy indicates good agreement. Detection limits are 0.01% Ti and Ba, and 0.004% Zn.

X-ray Fluorescence as Applied to Cyrtolite

D. A. Norton (Bryn Mawr College, Bryn Mawr, Pennsylvania)

Am. Mineral. **42**, 492 (1957).

The x-ray fluorescence method is applied to determine the content of U, Th, rare earths, Y, Zr, and Hf in zircons. The samples are finely ground and 5 wt.% each of Bi, Ta, and Nb are added as internal standards. For analysis 0.45 g are used where the fluorescent lines are registered in strip chart recordings. The intensity ratio of the fluorescent lines is set directly equal to the weight ratios. Bismuth is the standard for U and Th, tantalum is used as standard for Hf and the rare earths; and niobium serves as standard for Zr, Y, and Fe. The closer the two lines for standard and sample, the higher is the accuracy. The method is well-suited for comparative investigations of zircons; the hafnium–zirconium ratios are calculated and found to agree with those reported in the literature.

Matrix Corrections in the X-ray Spectrographic Trace Element Analysis of Rocks and Minerals

J. Hower (Montana State University, Missoula, Montana)

Am. Mineral **44**, 19 (1959).

This theoretical paper deals with the matrix effects on trace element determinations by x-ray fluorescence. Absorption of the incident primary tube radiation and of the emerging fluorescent radiation can be characterized by the mass absorption coefficients of the rocks and minerals. Graphical presentation of mass absorption coefficients as a function of wavelength is given for the most important rock-forming minerals. It is shown that relative absorption coefficients are generally constant for all wavelengths. It is therefore possible to present the absorption efficiency of the various rocks relative to an aluminum oxide standard. A matrix correction may be carried out by calculation of the relative absorption efficiency of the rock in relation to aluminum oxide, provided the bulk composition of the rock is known. If this is not the case, the relative absorption efficiency has to be determined experimentally by use of an internal standard. Matrix effects may be calculated numerically for analysis of elements with an atomic number Z larger or equal to 28 (nickel). Only the major portion of the matrix effect would, however, be corrected for the elements Co, Mn, Cr, V, and Sc by this procedure, and correction for secondary excitation of these elements by iron would not be achieved. Both procedures are illustrated by examples.

Age Determination by X-ray Fluorescence Rubidium–Strontium Ratio Measurement in Lepidolite

L. F. Herzog (Department of Geophysics and Geochemistry, Pennsylvania State University, University Park, Pennsylvania)

Science **132**, 293 (1960).

An x-ray fluorescence method is developed to determine the rubidium–strontium ratios and, hence, the ages of lepidolites (lithium micas). The samples are finely ground and the intensities of the lines RbK_α and SrK_α are measured and corrected for the contribution of background radiation. The weak SrK_α line is, in addition, overlapped on its left and right by the strong RbK_α and RbK_β lines, respectively. Contribution of the rubidium lines to the bulk intensity of the SrK_α line is determined experimentally with a set of samples of different rubidium contents, and the appropriate corrections are made. Four lepidolites ranging in age from 107 to 2635 million years and having rubidium–strontium ratios of 96 to 2350 are analyzed and the results compared to values given in the literature. The method is suitable for quick, inexpensive, and comparatively accurate determination of the Rb–Sr ratios and seems to allow age determinations of lepidolites of precambrian ages with accuracies similar to those obtained by isotope dilution methods.

Die Röntgen Fluoreszenz-Spektralanalyse von geochemischen Proben auf Elemente der Ordnungszahlen 25–40 (X-ray Fluorescence Spectral Analysis of Geochemical Samples for Elements of Atomic Numbers Z = 25 to 40).

K. H. Wedepohl (Mineralogisch-Petrographisches Institut, Universität Göttingen, Göttingen, Germany)

Z. Analyt. Chem. **180**, 246 (1961).

Usefulness of x-ray fluorescence analysis for trace element determination of the elements $Z = 25$ (manganese) to $Z = 40$ (zirconium) in rocks and minerals is discussed in detail. Proper choice of tubes, crystals, etc. for maximal detection sensitivity is discussed. For quantitative trace element analysis, the samples are finely ground and mixed with an internal standard. Arsenic trioxide is a particularly well-suited internal standard for the analysis of the elements Fe, Ni, Cu, while Zn and molybdenum oxide are used as standards for the determination of the elements Br, Rb, Sr, Y, and Zr. The average relative error of trace element determinations is approximately 3 to 10% and for the iron determination approximately 2.5%.

Theoretical Study of X-ray Fluorescent Determination of Traces of Heavy Elements in a Light Matrix. Application to Rocks and Soils

Z. H. Kalman and L. Heller (Department of Physics, Hebrew University, Jerusalem, and Geological Survey of Israel, Jerusalem, Israel)

Anal. Chem. **34**, 946 (1962).

An x-ray fluorescence method is described for quantitative determination of heavy elements in rocks and soil samples by directly analyzing finely ground powders. Relative fluorescent intensities of the elements in comparison to the intensity of the diffusely scattered tube radiation are measured in order to account for differences in bulk composition and absorption. These fluorescent intensities are largely independent of rock chemistry. Synthetically prepared samples of known trace element contents are used as standards and calibration curves for the determination of the elements Zn, Cu, Ni, Cr, and V are prepared. Sacrifices are made in the accuracy of the method in favor of speed and simplicity which makes possible a comparatively reliable analysis of a large number of samples in a comparatively short time.

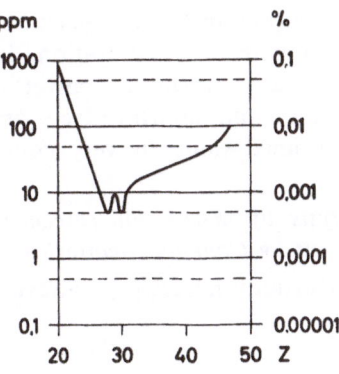

Fig. 23.1. Detection limits of elements $Z = 20$ (calcium) to $Z = 47$ (silver) in a SiO_2 matrix. Tungsten tube, 50 kV, 20 mA, LiF crystal, NaI scintillation counter (Wedepohl, 1961).

Quantitative röntgenspektralanalytische Bestimmung von Kalium, Rubidium, Strontium, Barium, Titan, Zirkonium und Phosphor (Quantitative X-ray Spectrochemical Determination of Potassium, Rubidium, Strontium, Barium, Titanium, Zirconium, and Phosphorous)

P. Hahn-Weinheimer and H. Ackermann (Mineralogisches Institut, Universität Frankfurt, Frankfurt am Main, Germany)

Z. Analyt. Chem. **194**, 81 (1963).

Local heterogeneities in K, Rb, Ti, Zr, P, Sr, and Ba contents of the Malsburg granite pluton are determined by x-ray fluorescence analysis. These measurements serve as a starting point for investigations of the regional heterogeneity of a large granite complex. Reproducible and accurate analytical results are obtained when the total rock sample is ground to below 37 μ. Systematic investigation of various grain-size fractions indicate that individual elements are selectively enriched in various fractions. Known amounts of the desired elements are added to a number of the samples, usually in the form of a rock-forming mineral, and the calibration curves are established. Standard rocks serve as a control. Relative errors are always smaller than the natural inhomogeneities of the rocks.

X-ray Spectrographic Analysis of Silica and Alumina Base Catalyst by a Fusion-Cast Disc Technique

J. E. Townsend (Cities Service Research and Development Co., Lake Charles, Louisiana)

Appl. Spectroscopy **17**, 37 (1963).

An x-ray method is developed to determine the elements Mo, Cr, Ni, V, Co, Cu, Cd, Zn, and Fe in silicon- and aluminum-containing catalysts. One gram of sample is decomposed with 30 g of borax at 1000°C and poured into a pellet. The surface of the pellet is then polished with carborundum paper. Synthetic samples of known metal contents and matrix compositions corresponding to those of the catalyst are prepared for calibration. In other cases, the particular elements are mixed to form a "universal standard." Calibration curves are straight lines and matrix effects are eliminated. Agreement with chemically determined values is very good.

Matrix Corrections in Trace Element Analysis by X-ray Fluorescence: Estimation of the Mass Absorption Coefficient by Compton Scattering

R. C. Reynolds Jr. (Department of Geology, Dartmouth College, Hanover, New Hampshire)

Am. Mineral. **48**, 1133 (1963).

An x-ray fluorescence method is described to determine quickly and quantitatively traces of the elements $Z = 28$ (nickel) to $Z = 47$ (silver) in rocks.

The samples are finely ground and measured, and one well-analyzed rock serves as standard. Relative intensity of the diffusely scattered incoherent x-ray radiation is measured for all samples to correct for differences in matrix absorption (Compton scattering of the MoK_α line). The intensity of the scattered radiation is approximately inversely proportional to the mass absorption coefficients of the matrix, and differences in the absorption may be compensated for mathematically.

X-ray Fluorescence Lead–Uranium Ratio Measurements in Allanite, Bankura District, India

S. B. Bhattacherjee and M. N. Kumar (University College of Science, Department of Pure Physics, Calcutta, India)

Anal. Chem. **36**, 1400 (1964).

X-ray fluorescence techniques are applied to determine the lead, thorium, and uranium contents in allanite (rare-earth-containing epidot). The samples are mixed with bismuth as an internal standard and the lines are registered on stripchart recordings. From the ratio $Pb/(U + 0.38Th) \times 7400$, the age of the mineral is calculated to be 1475 million years.

23.3. Determination of Coordination Number

The Kaolinite–Mullite Reaction Series: IV, The Coordination of Aluminum

G. W. Brindley and H. A. McKinstry (Department of Ceramic Technology, Pennsylvania State University, University Park, Pennsylvania)

J. Am. Ceramic Soc. **44**, 506 (1961).

The AlK_α emission wavelength depends upon the coordination of aluminum which allows distinction to be made between aluminum in four- and six-fold coordination with regard to oxygen. Furthermore, distinction may be made between four-fold coordinated aluminum *vs* oxygen and nitrogen, respectively. The wavelength emitted from aluminum metal serves as reference and wavelength shifts of the AlK_α line are measured with an x-ray fluorescence spectrometer (EDDT crystal). It is possible to distinguish between aluminum in four- and six-fold coordination in the case of thermally treated kaolinite minerals. Measurement of several aluminum containing substances is performed.

Properties of Soda Aluminosilicate Glasses: III, Coordination of Aluminium Ions

D. E. Day and G. E. Rindone (Department of Ceramic Technology, Pennsylvania State University, University Park, Pennsylvania)

J. Am. Ceramic Soc. **45**, 579 (1962).

X-ray fluorescence techniques are applied to determine the coordination

number of the aluminum ion in aluminosilicate glasses. Advantage is taken
of the fact that wavelength of emitted fluorescent radiation of aluminum
depends somewhat on the coordination number. On the basis of small
differences in wavelength, it is shown that aluminum occurs in four-fold
coordination in sodium-aluminosilicate glass as long as the aluminum
to sodium ratio is less than one. When the ratio of Al/Na is larger than
one, part of the aluminum occurs in six-fold coordination.

*Determining the Coordination Number of Aluminum Ions by X-ray Emission
Spectroscopy*

D. E. Day (Concrete Division, Waterways Experiment Station, Jackson,
Mississippi)

Nature **200**, 649 (1963).

An x-ray fluorescence technique for accurate measurement of the relative
wavelength of aluminum radiation is described. Aluminum metal serves as a
standard. The AlK_α wavelength varies slightly, depending upon the coordination. Data are reported for feldspars, clay minerals, aluminosilicates, alum, and other aluminum containing compounds. Chromium
tube, EDDT crystal, flow proportional counter, and helium atmosphere.

Petroleum and Coal Industry

X-ray fluorescence analysis is commonly applied in the petroleum and coal industry to the analysis of trace elements. Substances can often be analyzed directly provided the trace element contents are higher than 1 to 2 ppm. If this is not the case, the samples have to be ashed in the presence of sulfuric acid, benzosulfonic acids, and xylosulfonic acids, and the determination of the concentration is made in the ash. In order to compensate for variations in the composition of the petroleum, internal standards such as cobalt and chromium are often added. In the catalytical refining process of petroleum, it is of importance to know accurately the nickel, vanadium, and iron contents of the raw products, since these heavy metals poison the catalyst. When using platinum containing aluminum oxide catalysts, x-ray fluorescence analysis allows not only measurement of poisonous elements such as nickel and vanadium in the catalyst but also highly accurate determination of the platinum content in new and aged catalysts. The quality of refined products is often affected by the elements vanadium, sulfur, and chlorine, and their concentrations can easily be determined by x-ray fluorescence analysis. Fixed amounts of lead, bromine, and manganese-containing compounds are added to fuels to increase antiknocking properties, and their concentrations can be determined by measuring their fluorescent intensities. Lubricants contain small amounts of additives such as Zn, Ba, Ca, S, Cl, and P which serve to increase the lubricity: these elements can be determined quantitatively in the lubricant as well as in thin films which form on engine and other parts. In the following, reference is made to the literature.

24.1. Trace Element Determination in Petroleum Products
X-ray Spectrographic Method for the Determination of Vanadium and Nickel in Residual Fuels and Charging Stocks

E. N. Davis and B. C. Hoeck (Sinclair Research Laboratories, Inc., Harvey, Illinois)

Anal. Chem. **27**, 1880 (1955).

A method for trace element determination of nickel, vanadium, and iron is described. The procedure is quick and often requires not more than 20 g of substance; accuracy is comparable to other analytical methods. The sample is first ashed with sulfuric acid and the ash is then dissolved in dilute nitric acid. Part of the solution is dried on a glass plate which is rotated during measurement in order to compensate for inhomogeneities originating in the drying process. Fluorescent intensities of vanadium, nickel, and iron are measured. In order to eliminate the interference of iron on the determination of nickel in case of high iron contents, the iron is removed in a second part of the solution with thiocyanate and the nickel is measured in the iron free sample. Groups of calibration curves are plotted for varying amounts of iron and nickel and are used for quantitative determination of the concentrations.

X-ray Determination of Sulfur in Oils

W. R. Doughman, A. P. Sullivan, and R. C. Hirt (Research Division, American Cyanamid Co., Stamford, Connecticut)

Anal. Chem. **30**, 1924 (1958).

Sulfur in oils is determined by measuring the fluorescent intensity of the sulfur line and correcting it for the contribution of background radiation. Samples of well-known chemical compositions are used as standards. During measurement, the sample is housed in a specially constructed and cooled container. A second method of sulfur determination is based on absorption. Radioactive ^{55}Fe serves as a primary radiation-source which during decay emits characteristic radiation of 5.9 keV energy (MnK_α). In the concentration range of 0.1 to 1.2 wt.% sulfur, standard deviations in the order of 0.03% are obtained with both methods.

Quantitative Determination of Nickel in Oils by X-ray Spectrography

C. W. Dwiggins, Jr. and H. N. Dunning (Bureau of Mines, Petroleum Experiment Station, U.S. Department of the Interior, Bartlesville, Oklahoma)

Anal. Chem. **31**, 1040 (1959).

A quick method for the determination of nickel in oils using cobalt as an internal standard and without recourse to incineration of the samples is developed. Three gram samples, which contain a few ppm of nickel, are sufficient for analysis. An internal standard is used to correct for variations in matrix composition, efficiency of the x-ray tube, vaporization of the sample during measurement, differences in density, and interference with the iron line. Comparison with other methods indicates that the x-ray spectrographic determination is reliable, accurate, and applicable to a large variety of oils.

Determination of Sulfur and Chlorine in Petroleum Liquids by X-ray Fluorescence

T. C. Yao and F. W. Porsche [Research and Development Department, Standard Oil Co. (Indiana), Whiting, Indiana]

Anal. Chem. **31**, 2010 (1959).

X-ray fluorescence techniques are used for sulfur and chlorine determination in fuels, petroleum distillates and residues, additives, lubricants, insecticides, and weed killers. A commercial x-ray spectrograph with NaCl crystal, flow proportional counter, helium atmosphere, and pulse-height discriminator is applied. Lowest detectable concentration is 0.002 wt. %. When proper calibration procedures are worked out for each element, the x-ray method is as accurate as standard chemical techniques, but considerably faster, and the time required for analysis for each element is about 5 min.

Determination of Traces of Vanadium, Iron, and Nickel in Petroleum Oils by X-ray Emission Spectrography

C. C. Kang, E. W. Keel, and E. Soloman (Research and Development Laboratories, The M. W. Kellogg Co., Jersey City, New Jersey)

Anal. Chem. **32**, 221 (1960).

Vanadium, iron, and nickel contents of oils are determined without previous incineration of the sample using 4 ml of oil. Intensity of fluorescent lines is corrected mathematically for the absorption of sulfur, where the samples are assumed to have an average C—H ratio with about 12% H. It is further assumed that small compositional variations as well as oxygen and nitrogen contaminations are negligible. Estimated accuracy is ± 1 ppm, or $\pm 5\%$ relative.

Quantitative Determination of Traces of Vanadium, Iron, and Nickel in Oils by X-ray Spectroscopy

C. W. Dwiggins, Jr. and H. N. Dunning (Bureau of Mines, Petroleum Research Center, U.S. Department of the Interior, Bartlesville, Oklahoma)

Anal. Chem. **32**, 1137 (1960).

The samples are analyzed directly without previous incineration in order to avoid loss of volatile organic metal complexes. Internal standards of cobalt and chromium are added for quantitative determination of vanadium, iron, and nickel to compensate for varying sulfur contents. The method of the internal standard does not, however, work for oils containing appreciable amounts of brine; in such cases, a combined emission–absorption technique is applied. Two different possibilities for the correction of matrix effects in the trace element determination of oils are compared.

Determination of Sulfur in Gasoline by X-ray Emission Spectrography
R. A. Jones (Ethyl Corp. Research Laboratories, Detroit, Michigan)
Anal. Chem. **33**, 71 (1961).
The method is applied to the determination of sulfur contents of >0.002 wt. %. A sulfur-free diluent is added to part of the sample, and a diluent of known sulfur content is added to an equally large second portion of the sample. Parallel determination of the sulfur contents of the two solutions eliminates matrix effects, and corrections are made only for the contribution by the background and for differences in densities between the two solutions. Desired accuracy can be obtained in a minimum of time by applying optimal counting techniques. The technique has a broader range of application than the ASTM method and is also quicker and more accurate.

Direct Nickel Determinations in Petroleum Oils by X-ray at the 0.1 ppm Level
C. C. Hale and W. H. King, Jr. (Analytical Research Division, Esso Research & Engineering Co., Linden, New Jersey)
Anal. Chem. **33**, 74 (1961).
A multichannel unit with curved analyzing crystals is optimized for nickel determinations. The intensities of the NiK_α line and of the diffusely scattered background 0.01 Å away from the nickel line are measured simultaneously. Counting time is 30 min. The intensity ratio of the two lines is largely independent of the C—H-ratio and of the sulfur content of the sample. In order to avoid Ni losses during ashing, the oils are analyzed directly. The standard deviation, in the range of 0 to 0.6 ppm Ni, is 0.035 ppm Ni.

Determination of Traces of Nickel and Vanadium in Petroleum Distillates
J. E. Shott, Jr., Th. J. Garland, and R. O. Clark (Gulf Research & Development Co., Pittsburgh, Pennsylvania)
Anal. Chem. **33**, 506 (1961).

Ashing is necessary in order to determine Ni and V successfully in the concentration range of 2 ppm. Cobalt is added as internal standard and the samples are ashed in the presence of benzolsulfonic acid. The ash is then distributed equally on a sample holder by suspension and drying, and the relative fluorescent intensities are measured. With this method, contents of approximately 0.1 ppm can be determined when starting with an initial amount of 10 g of substance.

X-ray Fluorescence Method for Trace Metals in Refinery Fluid Catalytic Cracking Feedstocks

W. A. Rowe and K. P. Yates (Research Center, The Pure Oil Co., Crystal Lake, Illinois)

Anal. Chem. **35**, 368 (1963).

Procedures and results are described for determination of trace amounts of copper, iron, nickel, and vanadium in raw products for catalytic cracking. Cobalt is added as internal standard to 50 g of sample and ashed in the presence of xylosulfonic acid. The relative fluorescent intensities are then determined in relation to the cobalt content of the ash. A tube with a platinum anode is used in order to avoid undesirable interference with the characteristic lines of the x-ray tube. Standard deviations for copper, nickel, and vanadium are approximately 0.01 ppm, and for iron 0.05 ppm.

Problems of Direct Determination of Traces of Nickel in Oil by X-ray Emission Spectrography

E. L. Gunn (Research and Development, Humble Oil & Refining Co., Baytown, Texas)

Anal. Chem. **36**, 2086 (1964).

Traces of Ni (0.2 to 10 ppm) are determined directly in the raw products of catalytic cracking without previous ashing. Standard deviation is 0.1 ppm, and concentrations agree, within 0.1 ppm, with chemically determined values. In order to compensate for varying scattering efficiencies of the samples, the relative intensity of the NiK_α line to the diffusely scattered neighboring WL_λ line of the x-ray tube is determined. Differences in the composition due to varying sulfur contents or varying C—H ratios may affect the results but are neglected in the analysis of common raw products, and the method largely compensates for such differences. Sensitivity of the method can be improved by use of a suitable filter mounted on the x-ray tube.

24.2. Trace Element Determination in Catalysts

Determination of Trace Amounts of Iron, Nickel, and Vanadium in Catalysts by Fluorescent X-ray Spectrography

G. V. Dyroff and P. Skiba (Esso Laboratories, Research Division, Standard Oil Development Co., Linden, New Jersey)

Anal. Chem. **26**, 1774 (1954).

Poisonous effects of minor amounts of metals such as nickel and vanadium on cracking catalysts has been known in the petroleum industry for some time. Accordingly, efforts are made to develop suitable methods to determine these metals in the catalysts. Chemical, as well as spectrochemical, methods are applied with different success; however, both methods lack the accuracy required to distinguish between a good and poor catalyst. The x-ray fluorescence method described here is quick and accurate. For analysis 20 g of sample are ordinarily used. However, analyses may be carried out nondestructively with as little as 2 g of substance. Standards are prepared by mixing a known amount of metal to a synthetically prepared alumino-silicate catalyst. Interelement effects between iron, nickel, and vanadium are investigated and found to be negligible. The standard deviation, in the range of 0.1 to 1 % Fe, is 1.5 % relative, whereas for nickel and vanadium, in the concentration range 0.002 to 0.1 %, it is ± 0.001 %.

Determination of Platinum in Reforming Catalyst by X-ray Fluorescence

E. L. Gunn (Humble Oil & Refining Co., Baytown, Texas)

Anal. Chem. **28**, 1433 (1956).

An x-ray fluorescence technique is developed for determination of platinum in catalysts which are used in petroleum refining. An organic binder (5 wt. %) is added to the samples, which are then pressed into pellets. The standard deviation, in the concentration range of about 0.6 %, is 0.006 %. Pt. Comparison to chemically determined concentrations indicates excellent agreement. Time-dependent drift is small but measurable, and a procedure to correct for this effect is explained. Excitation potential of the x-ray tube is chosen so that the most advantageous ratio of line to background is obtained. Effects of contaminants (carbon, iron, water) in the catalyst are small and negligible. Thermal treatment of the aluminum containing catalyst does not appreciably effect the obtained values. Use of helium in place of air as a gas in the spectrometer chamber does not offer any appreciable advantages.

Quantitative Determination of Platinum in Alumina-Base Reforming Catalyst by X-ray Spectroscopy

A. J. Lincoln and E. N. Davis (Research and Development Division, Engelhard Industries, Inc., Newark, New Jersey, and Sinclair Research Laboratories, Inc., Harvey, Illinois)

Anal. Chem. **31**, 1317 (1959).

X-ray fluorescence techniques are sufficiently quick and accurate to determine platinum in platinum-containing catalysts which are applied in modern refining technology. Sample preparation is relatively simple: the samples are ground to a uniform powder and annealed. Two calibration procedures are discussed, which apply to excitation by molybdenum or tungsten tubes, respectively (a molybdenum tube is preferred). Relative fluorescent intensities in relation to an external standard are measured to compensate for drift in the apparatus. The standard deviation for platinum contents of 0.6 % is 0.0025 %. Time span from receipt of sample to analytical result is approximately 3 h; however, the physical presence of the operator is required for only 30 min.

24.3. Determination of Admixtures in Fuels

X-ray Fluorescence Analysis of Ethyl Fluid in Aviation Gasoline

L. S. Birks, E. J. Brooks, H. Friedman, and R. M. Roe (U.S. Naval Research Laboratory, Washington, D.C.)

Anal. Chem. **22**, 1258 (1950).

X-ray fluorescence analysis is applied to determination of lead and bromine in aviation gasoline, and the characteristic PbL_α and BrK_α lines are measured. When counting the lead line for one minute, a probable error of ± 0.06 ml/gallon is obtained for a content of 4 ml tetraethyllead per gallon of aviation fuel. For the bromine line, the probable error is found to be ± 0.16 ml per gallon or 1.8 ml ethylene bromide per gallon. The relative intensities of the lead and bromine lines are a measure of the relative contents and are independent of any changes in the composition. Presence of elements such as chlorine have practically no effect on the lead and bromine determinations. Since there is a certain degree of interaction between lead and bromine in the sample, groups of calibration curves for fixed mixing ratios of lead to bromine are constructed.

Determination of Bromine in Liquids by X-ray Fluorescent Spectroscopy

G. T. Kokotailo and G. F. Damon (Research and Development Department, Socony Vacuum Laboratories, Paulsboro, New Jersey)

Anal. Chem. **25**, 1185 (1953).

Bromine content of liquid hydrocarbons is determined. Selenium is added to the sample as internal standard in order to compensate for the effects of contaminants in the starting material. A group of calibration curves is plotted for various densities of hydrocarbons whereby the weight % of bromine is plotted vs the intensity ratio of the BrK_α/SeK_α lines. The technique allows determination of bromine in the concentration range of 0 to 0.4 wt. % with an accuracy of ± 0.01 %.

Determination of Tetraethyllead in Gasoline by X-ray Fluorescence
F. W. Lamb, L. M. Niebylski, and E. W. Kiefer (Research Laboratories, Ethyl Corp., Detroit, Michigan)
Anal. Chem. **27**, 129 (1955).
Aim of this study is to take advantage of the rapidity and accuracy of x-ray fluorescence techniques to determine tetraethyllead in fuels. Study of the matrix effects indicates that one and the same calibration curve may be used, independent of whether tetraethyllead is present in the fuel by itself or together with bromine containing compounds. Errors due to the presence of sulfur are small and the effect of additional admixtures such as phosphorus may be neglected. Effects of variations in C—H ratios may be corrected for by measuring the density. The substance is measured in a water-cooled container and the time required for analysis is approximately 5 to 10 min per sample; the average error is ± 0.026 ml tetraethyllead per gallon.

Determination of Manganese in Gasoline by X-ray Emission Spectrography
R. A. Jones (Research Laboratories, Ethyl Corp., Detroit, Michigan)
Anal. Chem. **31**, 1341 (1959).
A procedure for the determination of manganese in fuels is developed. It is found that corrections have to be made for variations in the composition of the fuels and of their additives. The internal standard principle is modified by dipping an iron rod of known and fixed geometry into the sample. The intensity ratio of manganese to iron is practically independent of the composition of the fuel and, hence, only one calibration curve is required. Samples of known manganese contents serve as standards, and one analysis requires approximately 15 min. The average standard deviation, in the range of 0.1 to 1 g Mn/gal, is between 0.003 to 0.007 g/gal. Comparison to manganese contents determined by flame photometry indicates agreement.

Über eine Methode zur röntgenfluoreszenzspektrographischen Bleibestimmung in Benzin (On a Procedure for Determination of Lead in Fuels by X-ray Fluorescence)
H. Preis and A. Esenwein (EMPA, Zurich, Switzerland)

Schweizer Archiv **26**, 317 (1960).

In order to eliminate undesirable effects of bromine on the lead determination, 1 vol. % ethylene bromide is added to the sample (i.e., approximately 30 to 100 times the normal bromine concentration). When considering the relative fluorescent intensities of lead in relation to the bromine radiation it is possible to reduce effects resulting from variations in densities in the bromine and sulfur contents of the fuels to a point that use of a single calibration curve for all types of fuels results in accuracies of analysis which approach presently accepted ASTM norms.

Determination of Lead in Gasoline by X-ray Fluorescence Using an Internal Intensity Reference

E. L. Gunn (Humble Oil and Refining Company, Research and Development, Baytown, Texas)

Appl. Spectroscopy **19**, 99 (1965).

A procedure for the determination of lead in fuels is developed. A platinum rod is dipped into the liquid that is to be analyzed in order to compensate for variations in the compositions of the fuels, and relative fluorescent intensities of lead to platinum are measured. With this procedure, the lead content may be determined with only one calibration curve, and agreement with chemically determined values is good.

24.4. Determination of Additives in Lubricating Oils

Determination of Barium, Calcium, and Zinc in Lubricating Oils

E. N. Davis and R. A. von Nordstrand (Sinclair Research Laboratories, Inc., Harvey, Illinois)

Anal. Chem. **26**, 973 (1954).

This procedure allows determination of barium, calcium, and zinc contents of lubricating oils, and the analysis of one element requires 3 to 12 min. Accuracy is found to be equivalent to chemical methods of analysis. Application of helium as a gas to the spectrometer chamber allows analysis of barium and calcium, elements which ordinarily cannot be measured. Interelement effects between the three elements are considered mathematically and experimentally. Depending upon the concentration of the admixed barium, groups of calibration curves are constructed for the determination of zinc. Relative errors for barium and calcium contents larger than 0.05 % are 2 to 3 %, while for zinc contents larger than 0.005 % they are only 1 to 2 %.

The X-ray Emission Analysis of Thin Films Produced by Lubricating Oil Additives

T. P. Schreiber, A. C. Ottolini, and J. L. Johnson (General Motors Research Laboratories, Warren, Michigan)

Appl. Spectroscopy **17**, 17 (1963).

X-ray emission techniques are found to be useful for study of thin films that originate when additives in lubricating oils react with metals. These additives increase the lubricating properties of the oil via adsorption to, or reaction with, metal surfaces and form thin films of up to 1500 Å thickness. The x-ray technique makes possible determination of zinc, barium, phosphorus, sulfur, and chlorine in concentrations from 0 to 75 $\mu g/cm^2$. Synthetic films serve as standards.

Röntgen-Emissionsspektralanalyse von Schmierölen (*X-ray Emission Spectral Analysis of Lubricating Oils*)

R. Louis (Esso A. G. Forschungslaboratorium, Hamburg–Harburg, Germany)

Z. Analyt. Chemie **201**, 336 (1964).

The nature and extent of matrix effects in x-ray emission spectral analysis of lubricating oils are studied. Selection of internal standards is made on the basis of theoretical considerations and is aimed towards elimination of these effects. The usefulness of the standard is tested experimentally. Analytical procedures are worked out on the basis of these results for the determination of zinc, barium, calcium, chlorine, sulfur, and phosphorus in fresh and aged lubricating oils. The oils may be analyzed directly with high accuracy and without time consuming sample preparation. Quantitative detection limit depends upon the element and ranges from 1 to 100 ppm; approximately 20 to 30 min are required for one determination.

Apparative Probleme der Röntgenanalyse von Mineralölprodukten (*Instrumental Problems in X-ray Analysis of Mineral Oil Products*)

R. Louis (Forschungslaboratorium der Esso A. G., Hamburg–Harburg, Germany)

Erdöl Kohle **17**, 360 (1964).

Instrumental problems in x-ray fluorescence analysis, with special emphasis on application to the study of mineral oil products, are treated. X-ray tubes, analyzing crystals, and collimators are discussed and the characteristics of the crystals (dispersion, resolution, reflection) are described in detail. Furthermore, a detailed discussion is given of discriminators, counters, and the conditions which affect radiation on its path from the sample chamber to the counter.

Nachweisgrenzen bei der Röntgen-Emissionsspektralanalyse von Mineralölen
(*Detection Limits in X-ray Emission Spectral Analysis of Mineral Oils*)
R. Louis (Esso A. G., Forschungslaboratorium, Hamburg–Harburg, Germany)

Z. Analyt. Chemie **208**, 34 (1965).

Detection limits for elements of atomic numbers $Z = 82$ to $Z = 15$ are determined experimentally with the calibration curve method near the concentration 0. When making full use of all instrumental possibilities, detection limits for all elements to and inclusive chlorine of less than 1 ppm, and for sulfur and phosphorus between 1 and 3 ppm are obtained when counting 30 min per element. Dependence of detection limit on wavelength and counting time is determined by assuming that counting errors for longer measurement times are only determined by statistical deviations. Even in case of high concentrations, matrices and elements affect detection limits only little; however, they do affect the detection sensitivity (counts/sec ppm). Use of internal standards in trace element analysis is therefore recommended.

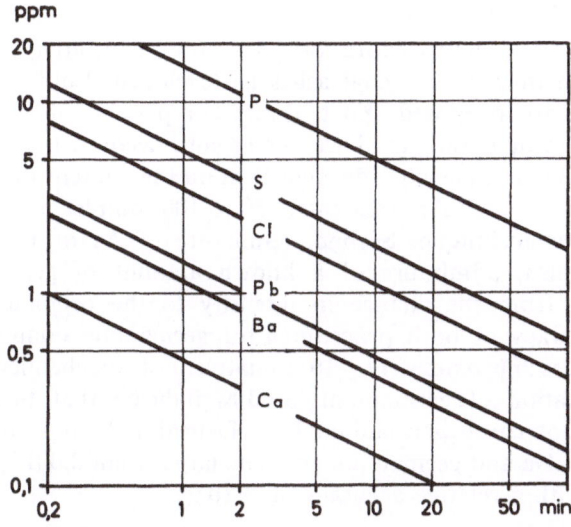

Fig. 24.1. Detection limits of some elements in mineral oils as a function of counting times. Extrapolation to higher counting times is made under the assumption that measurement errors are only dependent upon statistical variations in the pulse rate (Louis, 1965).

Rontgenfluoreszenzanalyse und ihre Anwendung in der Petroleumindustrie (*X-ray Fluorescence Analysis and Its Application in the Petroleum Industry*)

R. Jenkins (Esso Research Ltd., Abingdon, England)

Erdöl Z. **2**, 59 (1963).

A historical review of the development of the x-ray method is given in the introduction. Typical examples of applications such as the Pb determination in fuels, trace metal determination in alumina catalysts and raw products, study of deposits and residues in oil-fired boilers, determination of sulfur in petroleum products, study and identification of deposits in combustion engines, and of corrosion products in zinc die-cast bearings are mentioned.

24.5. Trace Element Determination in Coals

Quantitative Analyses by Fluorescent X-ray Spectrography. Determination of Germanium in Coal and Coal Ash

W. J. Campbell, H. F. Carl, and Ch. E. White (Eastern Experiment Station, Bureau of Mines, U.S. Department of the Interior and University of Maryland, College Park, Maryland)

Anal. Chem. **29**, 1009 (1957).

An x-ray fluorescence procedure for quick and quantitative determination of germanium in coal and coal ashes is developed. Four procedures of measurement are proposed. First, direct comparison of the fluorescent intensities in sample and standard, taking into account that the intensity of the GeK_α line is somewhat dependent upon the content of SiO_2, Al_2O_3 and Fe_2O_3 in the coal. Approximately 50 to 100 samples may be analyzed daily with this technique. Second, admixture of As or Ga as internal standards. Third, admixture of a known amount of germanium, and extrapolation from the change in intensity to the original germanium content. Accuracy of both procedures for germanium contents of more than 0.1% Ge is approximately $\pm 10\%$ relative. Fourth, chemical separation and concentration of germanium. Studies indicate that burning below 600°C does not cause germanium loss. Detection limit is on the order of 2 to 4 ppm Ge and germanium contents larger than 0.001% Ge may be determined with a relative accuracy of $\pm 10\%$.

The Determination of Sulfur and Vanadium in Carbon Materials by X-ray Fluorescence Analysis

R. H. Black and W. J. Forsyth (Aluminium Laboratories Limited, Arvida, Quebec, Canada)

Norelco Rep. **6**, 53 (1959).

An x-ray fluorescence procedure is developed to determine quantitatively the sulfur and vanadium contents in coal, petroleum and coal coke, anthracite, and tar. Solid samples are finely ground, and the tar is slightly heated to a point where it can be poured. Synthetically mixed and well-analyzed samples serve as standards. The fluorescent intensity of the vanadium is corrected for the effects of the sulfur by making use of a group of calibration curves for various sulfur contents. Comparison to chemically determined values indicates agreement within less than 5% relative for S and V.

Chapter 25

Chemical Industry

A major area of application of x-ray fluorescence analysis in the chemical industry is the determination of heavy trace elements in organic substances, such as the toxic elements Pb, As, Cu, Ni, Hg, and Se in pharmaceutics and dyes, and contaminants such as Ti, Mn, Co, Zn, Sb, Cl, and Al which occur in the catalytical production of plastics. These contaminations can largely be determined without complicated sample preparation, and detection limits on the order of 10^{-6} (1 ppm) may be achieved. Matrix effects on the fluorescent intensities in the simultaneous analysis of various substances can usually be calculated numerically on the basis of the bulk chemical composition without taking recourse to specific standards, thus making possible a broad application to trace element analysis. X-ray fluorescence analysis may also be used to determine higher contents of heavy elements such as S, P, Cl, Br, and I, or of metal in organic matrices. Experiments to determine the C—H ratio of hydrocarbons from scattering intensities have been reported. A unique procedure for quantitative determination of organic molecules, which cannot be determined by x-ray fluorescence analysis, involves determination of the contents of heavy elements such as halogens, sulfur, or metals which are sometimes stoichiometrically associated with the organic molecules. With this procedure, it is possible, for example, to monitor organic molecules in production control via their heavy associated elements. Molecules which do not contain heavy elements can be determined quantitatively after specific reactions have taken place, such as reactions with bromine, formation of metal complexes, or formation of salts. Reference to the literature is made in the following sections.

25.1. Determination of Trace Elements

The Simultaneous Determination of Traces of Selenium and Mercury in Organic Compounds by X-ray Fluorescence

E. C. Olson and J. W. Shell (The Upjohn Company, Kalamazoo, Michigan)

Anal. Chem. Acta **23**, 219 (1960).
An x-ray fluorescence method is developed to determine lowest contents of selenium and mercury in organic substances. Samples are either finely ground and 600 mg of it are pressed into a sample holder, or the substance is dissolved and the measurement is made in the solution. Known amounts of selenium and mercury are added to selenium- and mercury-free samples which serve as standards. A tungsten tube is used for excitation; the SeK_α line is overlapped by two characteristic tube lines $WL_{\gamma1,2}$ (a molybdenum tube would therefore be better suited for excitation). The relative line intensities of $(SeK_\alpha + WL_\gamma)/WL_{\beta1}$ or $HgL_\alpha/WL_{\beta1}$, respectively, are measured and correlated to the selenium and mercury contents in order to compensate for variations in the tube efficiency. Accuracy of measurement is ± 1 ppm Se or Hg in the concentration range of 2 to 40 ppm.

Determination of Traces of Metals in Organic Products by Means of X-ray Fluorescence Spectroscopy
R. Westrik (Centraal Laboratorium, Staatsmijnn in Limburg, Geleen, Netherlands)
Rev. univers. Mines [9] **18**, 279 (1961).
A very accurate and quick method for the determination of traces of titanium and iron in polyethylene is described. Polyethylene contains traces of contaminants which are derived from the catalysts used in the catalytical production process. These contaminants affect color and quality of the product. Since polyethylene constitutes a matrix consisting of light elements, high detection limits can be expected; these limits are 1 ppm for titanium and 0.1 ppm for iron. Liquid paraffins of identical bulk composition, to which known amounts of titanium or iron have been added, serve as standards. The fluorescent line of the iron is overlapped by an interfering iron line originating from the apparatus, and it is necessary to determine the intensity of the interfering line using an iron-free sample. Experiments to determine the aluminum contents by similar procedures are mentioned.

Determination of Catalyst Residues in Polyolefins by X-ray Emission Spectrometry
G. D. Smith and R. L. Maute (Monsanto Chemical Co., Texas City, Texas)
Anal. Chem. **34**, 1733 (1962).
An x-ray fluorescence method for direct determination of aluminum, chlorine, titanium, and iron in polyolefins is developed. The procedure does not require prior concentration of the elements and is well suited for any plastic. The fluorescent intensity of the particular element is measured

and corrected for the contribution by the background radiation. Synthetic samples, which contain known amounts of the elements in question, serve as standards. Analyses are made in the concentration ranges of 1 to 100 ppm Ti, 0.5 to 100 ppm Fe, 10 to 300 ppm Cl, and 5 to 500 ppm Al. Precision and accuracy is comparable to or better than chemical procedures; in addition, x-ray fluorescence analysis is 6 to 8 times as fast.

Die Abhängigkeit der Fluoreszenzintensität vom Massenabsorptionskoeffizienten der Matrix bei der Spurenbestimmung durch Röntgenfluoreszenz (*Dependence of Fluorescent Intensity on Mass Absorption Coefficients of Matrices in the Trace Element Determination by X-ray Fluorescence*)

R. Müller (Physik. Laboratorien, CIBA A.G., Basel, Switzerland)

Spectrochim. Acta **20**, 143 (1964).

In this paper, the basic question of dependence of fluorescent intensity of trace elements on the composition of the matrix is discussed. It is shown that the fluorescent intensity of an element is inversely proportional to the absorption coefficient of the matrix. It is also shown that this relationship may be used to carry out trace element analysis in completely different substances without specific standards. Solutions which are easy to prepare

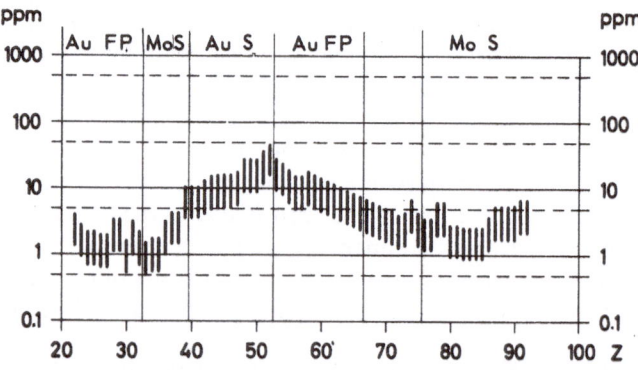

Fig. 25.1. Detection limits of the elements $Z = 22$ (titanium) to $Z = 92$ (uranium) in organic substances. The detection limits are high or low depending upon the mass absorption coefficients; plotted are the detection limits for substances with $\mu(1\ \text{Å}) = 1.6$ (hydrocarbons) to $\mu(1\ \text{Å}) = 14.0$ (sulfur- and chlorine-containing pharmaceutics). Note is made of the most favorable x-ray tube (Au or Mo) and counter (flow proportional or scintillation counter), LiF crystal, 50 kV, 20 mA. The K_α lines are measured for the elements $Z = 22$ (titanium) to $Z = 52$ (tellurium), while the L_α lines are measured for the heavy elements (author's measurements).

may be used for calibration, and matrix effects are considered mathematically. Examples are given for trace element determinations of iron, copper, and vanadium in pharmaceuticals, dyes, and other organic substances.

X-ray Fluorescence Determination of Cobalt, Zinc, and Iron in Organic Matrices

S. A. Bartkiewicz and E. A. Hammatt (Esso Research Laboratories and Humble Oil & Refining Co., Baton Rouge, Louisiana)

Anal. Chem. **36**, 833 (1964).

A method for the determination of cobalt, zinc, and iron in organic liquids is developed. Copper is added as internal standard in order to compensate for variations in sample compositions. Organic liquids of known cobalt, zinc, and iron contents serve as standards. Detection limits are approximately 10 ppm Co, 5 ppm Zn, and 5 ppm Fe; in the concentration range of 0.1 wt. %, accuracy for Co and Zn is approximately 0.003 wt. %.

25.2. Determination of Major Elements

Über die Brombestimmung durch Röntgenfluoreszenz (On the Determination of Bromine by X-ray Fluorescence)

H. Herrmann (Farbwerke Hoechst AG., Frankfurt am Main, Germany)

Z. analyt. Chem. **181**, 122 (1961).

A procedure for the determination of bromine in organic substances is described. The substance is decomposed, dissolved, and a known amount of selenium is added to the solution as internal standard. Relative fluorescent intensity of bromine is then independent of variations in sample compositions and drift. Agreement with chemically determined bromine contents is very good.

Determination of Aluminum in Organoaluminum Compounds by X-ray Fluorescence

H. F. Smith and R. A. Royer (Research and Development Department, Continental Oil Co., Ponca City, Oklahoma)

Anal. Chem. **35**, 1098 (1963).

An x-ray fluorescence method for determination of aluminum in highly reactive substances is described. High reactivity of the substance requires special sample preparation. The fluorescent intensity of the aluminum is directly proportional to the aluminum content. Highest sensitivity is obtained when using a chromium x-ray tube. Analyses cover concentration ranges from 0.05 to 10% Al. Agreement with chemically determined aluminum contents is excellent.

X-ray Fluorescence Spectrometric Analysis of the Copper(II) and Mercury(II) Complexes of 6-Chloro-2 methoxy-9-thiolacridine

K. E. Daugherty, R. J. Robinson, and J. I. Mueller (Department of Chemistry and Ceramic Engineering Division, University of Washington, Seattle, Washington)

Anal. Chem. **36**, 1098 (1964).

An x-ray fluorescence method for quantitative analysis of two coexisting copper and mercury complexes of 6-chloro-2 methoxy-9-thiolacridine is described. The ratio of the CuK_α to HgL_α line is measured for the copper determination and plotted *vs* the copper content, while for the mercury determination the reverse ratio is measured and evaluated. Nearly linear calibration curves are obtained when plotting on semilog paper. Relative error, in the concentration ranges of 0 to 6 mg Cu and 0 to 20 mg Hg is $\pm 3\%$. It appears that this method can be applied to analysis of a large variety of complex-salt mixtures.

X-ray Fluorescence Spectrometric Analysis of Iron(III), Cobalt(II), Nickel(II), and Copper(II) Chelates of 8-Quinolinol

K. E. Daugherty, R. J. Robinson, and J. I. Mueller (Department of Chemistry and Ceramic Engineering Division, University of Washington, Seattle, Washington)

Anal. Chem. **36**, 1869 (1964).

A method is developed to analyze various metal complexes of 8-quinolinol simultaneously. The relative fluorescent intensities of one metal to that of the associated metal are measured in two-component mixtures. Nearly linear calibration curves are obtained when plotted on semilog paper. Analysis of four-component mixtures containing iron, cobalt, nickel, and copper complexes of 8-quinolinol is difficult; however, results are satisfactory when the ratio of FeK_α/NiK_α is measured for the iron determination; the ratio of CuK_α/CoK_α for the copper determination; the ratio of NiK_α/CoK_α for the nickel determination; and the ratio of CoK_α/CuK_α for the cobalt determination.

X-ray Fluorescence Spectrometric Analysis of Rubidium(I) and Cesium(I) Salts of 5-Nitrobarbituric Acid

K. E. Daugherty, M. W. Goheen, R. J. Robinson, and J. I. Mueller (Department of Chemistry and Ceramic Engineering Division, University of Washington, Seattle, Washington)

Anal. Chem. **36**, 2372 (1964).

X-ray fluorescence methods for the determination of various metal complexes occurring simultaneously is developed and applied to the measurement of organic rubidium and cesium complexes. The relative fluorescent intensity of rubidium to cesium is measured for the analysis of rubidium, and the reverse ratio is determined for the analysis of cesium. Bent calibration curves originate when plotting data on semilog paper.

Quantitative Determination of Low Atomic Number Elements Using Intensity Ratio of Coherent to Incoherent Scattering of X-rays

C. W. Dwiggins, Jr. (Petroleum Research Center, Bureau of Mines, U.S. Department of the Interior, Bartlesville, Oklahoma)

Anal. Chem. **33**, 67 (1961).

A new procedure for determination of elements of low atomic numbers is developed. The method is applied to determination of hydrogen and carbon in hydrocarbons and particularly in petroleum, when the matrix contains, in addition, other elements. In this procedure the scattering efficiency of the matrix or a characteristic tube line is measured; the ratio of the coherently scattered portion to the incoherently scattered portion of the radiation is a measure for the C—H ratios in the scattering substance. A procedure is developed to correct for the effects of sulfur and nitrogen on scattering efficiencies. A commercial x-ray unit is used for analysis, and accuracy and reliability of the method appear to be equal or better than common microanalytical methods.

Quantitative Determination of Carbon in Solid Hydrocarbons Using the Intensity Ratio of Incoherent to Coherent Scattering of X-rays

C. J. Toussaint and G. Vos (Chemistry Department, Analytical Section, Euratom, Ispra, Italy)

Appl. Spectroscopy **18**, 171 (1964).

The carbon content of solid hydrocarbons is determined by measuring the intensity ratio of the coherently scattered to the incoherently scattered background radiation. The scattered portion depends to a different degree on the scattering matrix so that its composition can be analyzed. The intensity of the coherently and incoherently scattered MoK_α line is measured and samples of known C—H ratios serve as standards. Analyses in the range of 90 to 97% C agree well with microanalytically determined numbers.

25.3. Determination of Organic Substances via X-Ray Analysis of an Associated Element

X-ray Fluorescent Determination of Organic Substances through Inorganic Association

G. I. Papariello, H. Letterman, and W. J. Mader (Research Department, CIBA Pharmaceutical Co., Summit, New Jersey)

Anal. Chem. **34**, 1251 (1962).

An x-ray fluorescence method for determination of organic substances through association with an element that can be measured by x-ray fluorescence techniques is described. Requirements for the association reaction are discussed which are necessary to make possible the quantitative determination of the associated organic substances. Three procedures for association are developed: reaction with bromine, metal complex formation, and formation of salts. Examples for the determination of organic substances via these procedures are discussed. This indirect method is selective, specific, and sensitive.

A Simple, Indirect Sensitive Procedure for the Determination of Nitrogen (Ammonia) at the Microgram and Submicrogram Level

J. C. Mathies, P. K. Lund, and W. Eide (Pacific Northwest Research Foundation and Swedish Hospital, Seattle, Washington)

Anal. Biochemistry **3**, 408 (1962).

An x-ray fluorescence method for the determination of microgram and submicrogram amounts of nitrogen present in ammonia is described. The diffusion procedure after Conway is applied and ammonia is bound with a filter paper that has been dipped in Nessler's reagent. Excessive Nessler reagent is removed with water and a brown insoluble residue remains on the filter paper: this is the mercury containing reaction product of ammonia with Nessler's reagent. The ammonia content can then be determined by measuring the mercury fluorescence.

Chapter 26

Medicine and Biology

X-ray fluorescence techniques combined with very simple procedures of sample preparation such as drying of a few droplets of liquid on filter paper allow measurement of calcium, potassium, chlorine, sulfur, and phosphorus in blood serum and in cell fluids. X-ray fluorescence is often more accurate and considerably faster than other analytical techniques. The hemoglobin contents of the blood may be determined by measuring the fluorescent intensity of the associated iron in the hemoglobin. In the case of toxicologic tests, the bromine content of blood serum, urine, and of cells or spine fluids may be determined reliably. Ashes of tissue, serum, and bones may be tested for heavy elements such as iron, zinc, copper, and strontium, and the calcium–strontium ratio may be determined. X-ray fluorescence further allows one to determine the sulfur content in electrophoretically separated protein fractions of the serum and to detect pathologic changes. X-ray fluorescence analysis further allows the determination of contents of specific and biologically effective substances as long as they are associated with a heavy metal. Examples for these determinations are cobalt-containing vitamin B_{12}, iodine-containing hormones, or zinc-containing substances. Instrumental modifications made possible quantitative determination of the substitution of thymine by a bromine containing pyrimidine in 35 mg DNA (deoxyribonucleic acid); detection limits achieved are on the order of 10^{-3} μg (10^{-9} g) of bromine. In general, detection limits are in the neighborhood of 0.1 to 1 μg.

It is important in modern industrial hygiene to analyze the dust and heavy-metal content of air. X-ray fluorescence makes possible the qualitative and quantitative determination of the content of heavy metals such as cobalt, chromium, iron, lead, mercury, nickel, platinum, vanadium, and zinc in dust. Plants and animal foods may be investigated for their heavy-metal contents: sample preparation is simple because it usually involves only drying and grinding. With this procedure, calcium, potassium, selenium, zinc, molybdenum, and strontium are measured in various plants. Reference to the literature is made in the following sections.

26.1. Medicine

X-ray Fluorescence as a Tool for the Analysis of Submicrogram Quantities of the Elements in Biological Systems

S. Natelson and S. L. Bender (Biochemistry Laboratories of the Roosevelt Hospital, New York and Philips Electronics Inc., Mt. Vernon, New York)

Microchem. J. **3**, 19 (1959).

Principles and applications of x-ray fluorescence are discussed. A procedure for the determination of iodine-containing hormones in blood serum is presented which requires determination of 0.2 μg of iodine. It is shown that the x-ray fluorescence method is also suitable for the determination of phosphorus, sulfur, chlorine, calcium, and potassium in blood serum. Other possible areas of application are the determination of cobalt in vitamin B_{12}, iron in hemoglobin, and halogens, sulfur, and phosphorus in organic compounds.

X-ray Spectroscopy in the Clinical Laboratory. I. Calcium and Potassium

S. Natelson, M. R. Richelson, B. Sheid, and S. L. Bender (Roosevelt Hospital, New York, Philips Electronics Inc., Mt. Vernon, and Brooklyn College, Brooklyn, New York)

Clinical Chem. **5**, 521 (1959).

A procedure for the determination of calcium and potassium in smallest amounts of blood serum is described. Five to 25 μl of serum are dried from drops on a filter paper. Measurements are carried out with a flow proportional counter in a helium atmosphere. Reproducibility is approximately $\pm 5\%$ (2σ) when counting for one minute. Agreement with data obtained by flame photometry (potassium) and after Clark–Collip (calcium) is excellent. The x-ray fluorescence method for the determination of calcium is quicker and simpler than the commonly used methods in a clinical laboratory; although the photometric determination of potassium is still quicker than the x-ray fluorescence method, sample preparation for the latter technique can easily be done automatically.

X-ray Spectroscopy in the Clinical Laboratory. II. Chlorine and Sulfur: Automatic Analysis of Ultramicro Samples

S. Natelson and B. Sheid (Roosevelt Hospital, New York and Brooklyn College, Brooklyn, New York)

Clinical Chem. **6**, 299 (1960).

Chlorine and sulfur are determined in serum. Serum (20 μl) is dropped in well-defined droplets on a strip of filter paper using a ring of wax to prevent the liquid from spreading out, the filter paper is automatically brought into position for measurement and a recorder registers the intensity. This apparatus is described extensively. The chlorine determination by this technique compares favorably with common titration methods. Reproducibility is $\pm 3.4\%$ (2σ). The sulfur values do not differ from the ones obtained by ashing, and reproducibility is $\pm 2.2\%$ (2σ). Sulfur–protein ratios in normal and pathologically modified serum are investigated and reasons for differences are discussed.

X-ray Spectroscopy in the Clinical Laboratory. III. Sulfur Distribution in the Electrophoretic Protein Fractions of Human Serum

S. Natelson and B. Sheid (Roosevelt Hospital, Department of Biochemistry, New York, New York)

Clinical Chem. **6**, 314 (1960).

A procedure is developed to determine the sulfur content in individual protein fractions of serum. First, 0.02 ml of the particular serum is separated into the individual protein fractions using paper electrophoretic methods; the relative sulfur contents are then measured directly in the individual fractions. Finally, the total sulfur content of the serum as well as the protein content of the individual fraction is determined and the sulfur content of the various proteins is calculated. For certain diseases, the sulfur content of the protein deviates strongly from a normal value.

X-ray Spectroscopy in Biology and Medicine. I. Total Iron (Hemoglobin) Content in Human Whole Blood

P. K. Lund and J. C. Mathies (Pacific Northwest Research Foundation, Inc., Laboratory of Pathology, Swedish Hospital, Seattle, Washington)

Norelco Rep. **7**, 127 (1960).

A quick, accurate, and nondestructive method for clinical testing of the hemoglobin content of the blood is developed. In this procedure $25\,\mu l$ of blood is dropped onto a round filter paper and dried. Samples of known hemoglobin content serve as standards. This method of analysis is well suited for automatic measurement.

X-ray Spectroscopy in Biology and Medicine. II. Calcium Content of Human Blood Serum

J. C. Mathies and P. K. Lund (Pacific Northwest Research Foundation, Inc., and Laboratory of Pathology, Swedish Hospital, Seattle, Washington)

Norelco Rep. **7**, 130 (1960).

A quick microprocedure for calcium determination in serum is developed. Serum ($50\,\mu l$) is deposited in small cavities of an aluminum carrier and dried quickly. Since sulfur and chlorine affect the fluorescent intensity of calcium, the accuracy of the calcium determination may be improved when the approximate protein (sulfur) and chlorine contents of the sample are known. Corrections in the measurements are made when strong deviations from the normal compositions are encountered. Calcium which is added to the serum can be detected quantitatively. Calcium contents determined with this method agree with chemically determined numbers.

X-ray Spectroscopy in Biology and Medicine. III. Bromide (Total Bromine) in Human Blood, Serum, Urine, and Tissues)

J. C. Mathies and P. K. Lund (Pacific Northwest Research Foundation, Inc., and Laboratory of Pathology, Swedish Hospital, Seattle, Washington)

Norelco Rep. **7**, 134 (1960).

A quick and nondestructive microprocedure for the determination of bromine in serum, urine, and spinal fluid is developed. Twenty-five microliters of liquid are dried on a round filter paper and the fluorescent intensity of the bromine is measured. Bromine-free liquids to which a known amount of sodium bromide is added serve as standards. The method makes possible detection of as little as 4–5 μg of bromine in tissues and liquids.

X-ray Spectroscopy in the Clinical Laboratory. IV. Phosphorus; Total Blood Iron as a Measure of Hemoglobin Content

S. Natelson and B. Sheid (Department of Biochemistry, Roosevelt Hospital, New York, New York)

Clinical Chem. **7**, 115 (1961).

X-ray spectroscopy is applied to the determination of total phosphorus content of serum and to the iron determination (hemoglobin) of the total blood. The samples are deposited as drops on filter paper, dried, and then analyzed. An apparatus for automatic measurement of a series of samples is described. Comparison to values determined by ashing indicate no appreciable deviations other than the fact that the iron contents can be determined more accurately by the conventional technique, even though the x-ray technique is certainly adequate for requirements in a clinical laboratory. Hemoglobin determination by x-ray fluorescence is more significant than colorimetrically determined values since measurements with the former technique are not affected by other blood cells.

X-ray Spectrometric Determination of Strontium in Human Serum and Bone

S. Natelson and B. Sheid (Department of Biochemistry, Roosevelt Hospital, New York, New York)

Anal. Chem. **33**, 396 (1961).

A useful and accurate technique for the determination of strontium in serum and bones is developed. The samples are ashed and the ash is dissolved in a hydrochloric solution. Part of the solution is dried, drop by drop, on a filter paper in a specially constructed apparatus, and the fluorescent intensity of strontium is measured. The calcium to strontium ratios in bone are 1204 to 1231 and in serum 390. Normal human serum contains approximately 26 μg Sr per 100 ml serum.

X-ray Fluorescence Analysis Using Ion-Exchange Resin for Sample Support

R. L. Collin (Cancer Research Institute, New England, Deaconess Hospital, Boston, Massachusetts)

Anal. Chem. **33**, 605 (1961).

In connection with a study of bones, an x-ray fluorescence method is developed to detect trace elements in solutions. The solution is placed in an ion-exchange column. The ion-exchange material is pelleted and then analyzed. Rubidium is added to the original solution in order to increase the accuracy. Feasibility of the method is discussed using the example of the strontium determination in calcium containing solutions. The relative standard deviation in the concentration range of 2 to 100 ppm Sr is 5.2%. Solutions of strongly different compositions may be studied with this method provided a background correction is made.

Die röntgenfluorimetrische und aktivierungsanalytische Brombestimmung in biologischem Material (Bromine Determination in Biological Materials Using X-ray Fluorescence and Activation Techniques)

L. Beyermann (Institut für anorganische Chemie und Kernchemie, Universitat Mainz, Mainz, Germany)

Z. analyt. Chem. **183**, 199 (1961).

A highly sensitive method for the bromine determination in approximately 1 ml of serum or 0.1 g of organ substance is developed. The samples are dried, ground, and 100 mg of the dried powder is pelleted after 50 mg of wax are added. Serum samples of known bromine contents serve as standards. The procedure is well suited for the analysis of serum and muscle samples.

X-ray Fluorescence Analysis in Biology

Th. Hall (Associate of the Sloan–Kettering Institute for Cancer Research, New York, New York)

Science **134**, 499 (1961).

A review is given of the problems in medicine and biology which may be studied using x-ray fluorescence analysis. Procedures which do not require involved sample preparation, such as serum and blood study and investigation of tissues and bones, are mentioned. Difficulties are then discussed which arise when accurate localization of heavy-element compounds in the organism is required. The feasibility of a microprobe which simultaneously allows the viewing of the analyzed position in the electron microscope is discussed.

Assay for the Elements Chromium, Manganese, Iron, Cobalt, Copper, and Zinc Simultaneously in Human Serum and Sea Water by X-ray Spectrometry

S. Natelson, D. R. Leighton, and C. Calas (Department of Biochemistry, Roosevelt Hospital, New York, New York)

Microchem. J. **6**, 539 (1962).

X-ray spectrometry may be applied to the analysis of the elements chromium to zinc in biologically important systems. Human serum is ashed and the ash is dissolved in diluted hydrochloric acid, while ocean water is dried and dissolved in diluted hydrochloric acid. Sodium chloride is precipitated while the other chlorides remain in solution. Part of the solution is again dried and dissolved in methanol. The methanol is dropped onto filter paper and measured. A thin titanium foil is placed on the tungsten tube in order to suppress the chromium radiation of the tube. Typical serum analyses yield the following values: Cr, Mn, and Co are not detectable; Fe, 72 μg/100 ml; Ni, 5.5 μg/100 ml; Cu, 148 μg/100 ml; and Zn, 126 μg/100 ml. An ocean water analysis gives the following values: Cr, Mn, and Co could not be detected; Fe 40 μg/100 ml; Ni, 3.2 μg/100 ml; Cu, 122 μg/100 ml; and Zn, 100 μg/100 ml. The detection limit is 2 μg/100 ml or approximately 0.02 ppm, when using 5 ml of serum or ocean water.

Determination of Zinc, Copper, and Iron in Biological Tissues. An X-ray Fluorescence Method

G. V. Alexander (Department of Biophysics and Nuclear Medicine, School of Medicine, University of California, Los Angeles, California)

Anal. Chem. **34**, 951 (1962).

An x-ray fluorescence method for the determination of zinc, copper, and iron in ashes of biological substances is developed. Approximately 100 mg of ash is mixed with nickel carbonate and sodium carbonate (internal standards), ground, and pressed into a sample holder. Tests indicate that the intensity ratio of the zinc, copper, and iron lines to those of the internal standard are independent of the matrix composition. It is shown by experiment that biological substances may be ashed without loss of zinc, copper, and iron. Details of the method and examples of applications are presented.

Bromine Analysis in 5-Bromouracil-labeled DNA by X-ray Fluorescence
L. Zeitz and R. Lee (Division of Biophysics, Sloan-Kettering Institute for Cancer Research, New York, New York)
Science **142**, 1670 (1963).
A procedure and an instrument are developed to determine quantitatively the replacement of thymine by bromine-containing pyrimidine in DNA. This involves a bromine determination in purified DNA. A spectrometer is constructed in which approximately 35 μg of DNA are irradiated by the primary tube radiation. The intensity of the bromine fluorescent line is measured by a proportional counter and associated pulse-height analyzer. The intensity of the diffusely scattered radiation of the molybdenum line is measured as reference; this intensity is approximately proportional to the total amount of sample. Organic substances of identical bulk chemical composition and known bromine content serve as standards. Detection limit is approximately 10^{-9} g of bromine in a sample of 35×10^{-6} g DNA. Feasibility of analysis of iodine containing substances together with analysis of bromine containing materials is discussed.

X-ray Fluorescence Analysis of Biological Tissues
G. V. Alexander (Laboratory of Nuclear Medicine and Radiation Biology, Department of Biophysics, and Nuclear Medicine, University of California, Los Angeles, California)
Appl. Spectroscopy **18**, 1 (1964).
Feasibility of application of x-ray fluorescence analysis in the study of biological substances is discussed and a general review of the principles of the method are given. The method is explained with the aid of several examples. Finally, the relation between x-ray fluorimetric and optical methods are pointed out which together are very suitable for the study of biological substances.

26.2. Analysis of Plants, Animal Feed, Water, and Air

Analysis of Dried Plant Material by X-ray Emission Spectrography
C. S. Brandt and V. A. Lazar (U.S. Plant, Soil and Nutritions Laboratory, Agricultural Research Service, U.S. Department of Agriculture, Ithaca, New York)

Agricultural and Food Chem. **6**, 306 (1958).

A method for the analysis of a number of elements in dried and ground plant material is described. Concentration of an element is determined by making use of the diffusely scattered tube radiation as a comparative value, where the intensity ratio is approximately proportional to the concentration. Chemically analyzed samples serve as standards. Standards have to be measured daily in order to obtain satisfactory results. Molybdenum and zinc contents of various plants are presented.

Fluorescent X-ray Determination of Selenium in Plant Material

R. Handley (Department of Soils and Plant Nutritions, University of California, Berkeley, California)

Anal. Chem. **32**, 1719 (1960).

A method for the determination of selenium in plants is described. The material is dried, ground, and pressed into a sample holder. The measured intensity of the SeK_α line is corrected for the contribution by background radiation. In order to compensate for different absorption of the various materials, the intensity ratio of $(SeK_\alpha-U)/U$ is formed. This ratio is proportional to the selenium content. Standards are prepared by adding known amounts of selenium to selenium-free samples. Selenium contents of plants are presented and compared to chemically determined values.

X-ray Spectrography of Plant Material

P. B. Vose (Welsh Plant Breeding Station, University College of Wales, Aberystwyth, Wales)

Laboratory Practice **10**, 30 (1961).

The x-ray fluorescence technique is applied to the analysis of plants. The material is dried, ground, and measured either in the form of a powder or as a pellet. Natural samples of known contents or synthetically prepared materials serve as standards. Natural materials may only be applied as standards when their compositions can be determined accurately by another technique; in the case of synthetic materials, it is often difficult to prepare homogeneous representative standards. A third possibility for calibration is that of mixing a known amount of the desired element to the sample that is to be analyzed. In this case, the sample is suspended in acetone and the required amount of strontium is added. After drying, the relative fluorescent intensity of the strontium is measured, corrected for background radiation, and related to the intensity of the background radiation in the neighborhood of the strontium line. A second sample is treated analogously with the exception that no strontium is added. The original strontium content may then be determined by extrapolation from the relative intensity increase after adding strontium.

Plant Analysis by X-ray Fluorescence Spectrography: Determination of Calcium and Potassium

D. F. Ball and D. F. Perkins (Nature Conservancy, Bangor Research Station, Bangor, Caernarvonshire, Wales)

Nature **194**, 1163 (1962).

After drying and grinding of the plant material, cellulose is added and the powder is pressed into a pellet. The effects of variations in sample compositions on fluorescent intensities are investigated. Calibration curves are presented for the range of 5 to 25 mg potassium and 4 to 10 mg of calcium per gram of dried substance.

The Determination of Strontium in Tap Water by X-ray Fluorescence Spectrometry

R. G. Stone (Department of Scientific and Industrial Research, Laboratory of the Government Chemist, London, England)

The Analyst **88**, 56 (1963).

A method for the determination of the natural strontium content in tap water is described as a first step towards the determination of the strontium-90 content. A larger amount of tap water is concentrated to 100 ml and calcium and strontium are precipitated simultaneously as oxalates and transformed into nitrates. A known amount of rubidium is added as internal standard and 2 drops of the sample are dried on a mylar film of the sample container. The intensity ratio of SrK_α/RbK_α is measured, and the internal standard compensates for variations in the calcium content of the tap water. Samples of known strontium content are prepared as standards and natural strontium contents of tap water are found to range between 0.005 and 0.4 ppm.

X-ray Analysis of Foundry Dusts for Quartz and Iron in Relation to Silicosis and Siderosis. Fluorescent Spectral Analysis for Iron

G. L. Clark and H. C. Terford (Department of Chemistry and Chemical Engineering, University of Illinois, Urbana, Illinois)

Anal. Chem. **26**, 1413, 1416 (1954).

Because of the importance of iron determinations in dust and air of steel mills for the diagnosis of siderosis and because of inefficiency of most chemical methods, an x-ray fluorescence procedure is developed. Determination of the iron content is only one part of the study which also involves x-ray diffraction determination of α-quartz and medical tests of workers for silicosis and siderosis. As an internal standard, 20 wt. % of nickel are mixed with the dust samples. The relative fluorescent intensity of the iron is proportional to the iron content.

Application of X-ray Emission Spectrography to Airborne Dusts in Industrial Hygiene Studies

R. C. Hirt, W. R. Doughman, and J. B. Gisclard (Research Division, American Cyanamid Co. Stamford, Connecticut, and Central Medical Department American Cyanamid Co., New York, New York)

Anal. Chem. **28**, 1649 (1956).

The x-ray emission method is applied to quantitative determination of the concentration of heavy metals in the concentration range of 1 to 100 μg for the purpose of studies in industrial hygiene. When working with catalysts, ores, and dyes, heavy-metal-containing dust transported in air may be a health hazard. Dust contained in air is concentrated on a fiberglass plate which is investigated directly in the x-ray spectrograph for the desired elements. The fluorescent intensity is corrected for the contribution by background radiation, and synthetic samples of known heavy metal contents serve as standards. Calibration curves are constructed for concentrations as low as 1 to 5 μg and for the elements cobalt, chromium, iron, lead, mercury, nickel, platinum, vanadium, and zinc. Typical examples of applications are dust from catalysts (platinum, vanadium), ores (cobalt), and dye (chromium). Extension of the method to other elements is feasible.

The Determination of Microgram Quantities of Uranium in Airborne Dust Samples by Means of X-ray Spectrography

J. A. Murray and Th. H. Bartlett (Metallurgical and Chemical Departments, Western Nuclear Inc., Denver, Colorado)

Norelco Rep. **11**, 132 (1964).

Uranium content of air and of dust is determined by x-ray fluorescence in the control of uranium ore mills. Dust-bearing air is blown through a filter paper and the uranium content of the material deposited on the film is measured directly. Solutions of known uranium content are deposited on the filter paper and serve as standards; linear calibration curves result. Comparison to fluorimetrically determined contents indicates that sensitivity of x-ray fluorescence analysis is equally high but that the x-ray fluorescence technique is faster and simpler. Detection limit is 0.1 μg U/200 liters of air.

Chapter 27

Analysis of Small Amounts of Substance and of Small Areas

X-ray fluorescence analysis has been successfully applied to the determination of qualitative and quantitative compositions of very minute amounts of sample. The ratios of major components may be determined unequivocally in microgram amounts. The substance is concentrated in the middle of a filter paper, mylar foil, or a similar carrier. In order to lower the effects of background radiation, small filter paper disks may be used which are just large enough to carry the amount of liquid that is to be analyzed. The filter paper disks are held in the proper position by thin mylar or tape strips which are clamped into the sample holder. The liquid may also be dried directly on a mylar strip. Usually, linear calibration curves result. Qualitative and quantitative analysis of smallest amounts of sample is applied successfully in mineralogy and metallurgy; furthermore, x-ray fluorescence provides a simple and reliable identification of heavy metal-containing compounds in the analysis of paper chromatograms. Properly designed devices make possible aperturing of the primary x-ray radiation to a beam of 100 μ in diameter which allows one to scan individual areas in metal and ore sections, as well as in welding and soldering products, point by point. Reference to the literature is made in the following sections. Natelson *et al.* and Lund and Mathiés have applied x-ray fluorescence analysis to study of smallest amounts of substance in clinical tests; (Chapter 26).

Trace Analysis by X-ray Emission Spectrography
H. G. Pfeiffer and P. D. Zemany (General Electric Research Laboratory, Schenectady, New York)
Nature **174**, 397 (1954).
Feasibility of the x-ray fluorescence technique to analysis of smallest amounts of substance is illustrated with the example of the determination

of zinc. The zinc containing solution is deposited in drops onto filter paper and application of a Whatman's filter paper yields the lowest scattered background. A linear calibration curve is constructed for the concentration range 0 to 16 μg Zn. Detection limit is 0.1 μg Zn for 120 sec of counting time.

Note on the Analysis of Small Quantities of Material by X-ray Fluorescence
N. W. H. Addink (Philips Research Laboratories, Eindhoven, Netherlands)
Rev. univers. Mines [9] **15**, 530 (1959).
Interaction between various components may best be reduced by either analyzing in highly diluted solutions or in very thin sample layers. Under these conditions, the fluorescent intensity is proportional to the concentration.

A Possible Solution to the Matrix Problem in X-ray Fluorescence Spectroscopy
E. J. Felten, I. Fankuchen, and J. Steigmann (Department of Physics, Polytechnic Institute of Brooklyn, New York)
Anal. Chem. **31**, 1771 (1959).
Suitability of the filter paper technique for quantitative analysis is studied. Solutions of two neighboring transition elements are prepared, the filter paper is soaked in the solutions, and the intensity ratio of the two components is then measured. This ratio changes slightly depending upon the number of stacked filter papers (interactions increase with increasing sample thickness). When extrapolating the intensity ratio to zero sample thickness, however, this limiting value is directly proportional to the molar mixing ratio of the two components. (Comment: In the more general case, where the two elements are not directly next to each other in the periodic chart, the intensity ratio is not directly proportional to the molar mixing ratio but a constant factor has to be added.)

Reliability of Trace Determinations by X-ray Emission Spectrography, X-ray Microscopy and Microanalysis (1960) p. 321.
H. A. Liebhafsky, H. G. Pfeiffer, and P. D. Zemany (General Electric Research Laboratory, Schenectady, New York)
Theoretical principles of errors in x-ray fluorescence analysis are discussed using the example of the determination of smallest amounts of substance. A 3-mm wide strip of mylar foil is used as a carrier and clamped into the sample container; precisely 0.002 ml of solution are then positioned with a micropipette into the center of the strip and the solution is dried.

Use of a small mylar strip lowers the background radiation so that the detection limit for manganese is 0.003 μg Mn (tungsten tube, 50 kV, 20 mA, LiF crystal, proportional counter). Results of 49 measurements in the concentration range 0.001 to 0.003 μg Mn are presented and the detection limit is verified.

Untersuchungen zur quantitativen Auswertung von Papierchromatogrammen in der Spurenanalyse durch Röntgenfluoreszenzspektroskopie (Quantitative Evaluation of Paper Chromatograms in Trace Element Analysis by X-ray Fluorescence Spectroscopy)

E. Jackwerth and H. J. Kloppenburg (Institut für Spektrochemie und angewandte Spektroskopie, Dortmund, Germany)

Z. analyt. Chem. **179**, 186 (1961).

Paper chromatograms are evaluated quantitatively by x-ray fluorescence analysis. Interfering effects by third components are largely eliminated by chromatographic separation. Two types of measurement are investigated, namely analysis via evaluation of impulse diagrams for continuously moving paper strips, and determination of pulse rates for various amounts of substance with stationary chromatograms. Standard deviation for measurements of various chromatograms for the same amount of substance is approximately 3%. Preliminary experiments indicate detection limits of less than 1 μg for zinc and gallium. Applicability of the technique to the determination of organic substances via association with a heavy element is discussed.

Röntgenfluoreszenzanalyse kleinster Mengen und Bereiche (X-ray Fluorescence Analysis of the Smallest Amounts of Substance and Areas)

K. Tögel (Siemens und Halske AG, Wernerwerk für Messtechnik, Karlsruhe)

Siemens Z. **36**, 497 (1962).

Instrumental arrangements for the analysis of smallest amounts of substance and areas are discussed in detail. A device is constructed which allows the reduction of the primary x-ray beam to a diameter of 100 μ. With this technique, individual phases and inclusions in an As–Co–Ni ore are analyzed nondestructively and identified unequivocally. Analysis of a polished section of brass allows quantitative determination of differences in the chemical composition of α and β-brass when very weak fluorescent lines have to be measured. Analysis of the spectra with an analyzing crystal is avoided and the individual lines are identified via their quantum energies. Furthermore, the filter method may be applied to determine the spectral intensity in a narrow wavelength range.

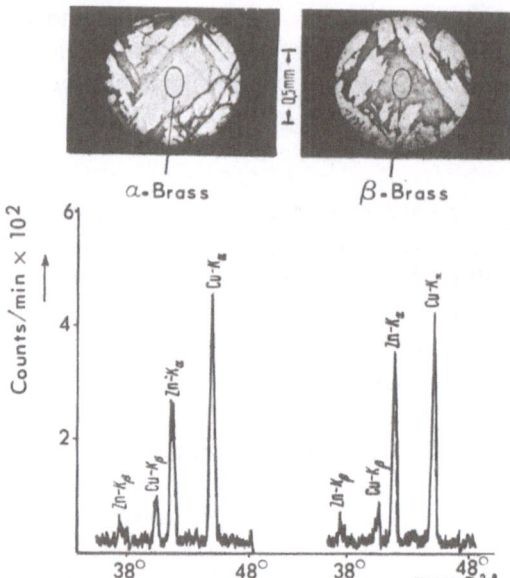

Fig. 27.1. Analysis of small areas. Apertures are used to produce an x-ray beam of 0.14 × 0.2 mm in size which allows determination of the chemical composition of α- and β-brass in a polished section. Photomicrograph and strip chart recording. Areas analyzed are marked (courtesy of Tögel, 1962a).

Quantitative Röntgenfluoreszenzanalyse geringster Substanzmengen mit der Filterdifferenzmethode (Quantitative X-ray Fluorescence Analysis of the Smallest Amounts of Substance with the Filter Difference Method)

P. Mecke (Forschungslaboratorium der Siemens & Halske AG, Munich, Germany)

Z. analyt. Chem. **193**, 241 (1963).

Thin iron–nickel films are studied for their quantitative composition. In order to detect local changes in chemistry and film thickness; a special apparatus is constructed which allows spatial resolution of 100 μ. The spectral intensity in the range of the NiK_α line is determined by the filter difference method using a nickel or an iron filter, respectively; the spectral intensity in the neighborhood of the FeK_α line is determined by subtraction of the intensity of the NiK_α line, and absorption of the NiK_α line by the nickel filter is considered. With this procedure, iron films of 200 to 300 Å thickness and nickel films of approximately 1300 Å thickness are analyzed which corresponds to an iron concentration of 0.001 μg.

Microanalysis on Controlled Spot Test Paper by X-ray Fluorescence

J. L. Johnson and B. E. Nagel (General Motors Research Laboratories, Warren, Michigan)

Michrochim. Acta **3**, 525 (1963).

A method for quantitative determination of major and minor components of the smallest amounts of substance is developed. An area of approximately 100 mm^2 on a filter paper is marked by a ring of wax. Twenty milliliters of liquid are deposited on the filter paper and the fluorescent intensities are measured. Solutions of known contents serve as standards. Analysis of a gold–tin hard solder involves dissolving of approximately 250 μg of substance in aqua regia, separation of the precipitated silver, and measurement of gold and tin by x-ray fluorescence in the solution. Detection limits are 3.6 μg Au and 1.9 μg Sn. One to 3 mg of substance are dissolved for the analysis of minor components. In order to guarantee even distribution of the liquid on the filter paper, only very little of the wetting liquid is at first deposited on the filter paper. Copper content of cadmium sulfide crystals, and copper and manganese contents of iron "whiskers" of approximately 1 mg are determined. Concentration of minor elements is between 0.1 and 2 %, and the detection limits are 0.2 to 0.4 μg Cu and 0.2 μg Mn.

Determination of Microgram Quantities of Chloride in High-Purity Titanium by X-ray Spectrochemical Analysis

J. S. Rudolph and R. J. Nadalin (Westinghouse Research Laboratories, Pittsburgh, Pennsylvania)

Anal. Chem. **36**, 1815 (1964).

The chlorine content of ultrapure titanium is determined by dissolving 3 g of metal in hydrofluoric acid; in this procedure nitric acid is added to oxidize the titanium. Chlorine is then precipitated with silver nitrate and the precipitate is filtered and washed. The fluorescent intensity of the silver which is associated with chlorine is measured directly in the precipitate. Samples of known chlorine content are prepared for standardization and treated with silver nitrate just as is the sample. A linear calibration curve is obtained for the concentration range of 25 to 2000 μg (15 to 2000 ppm Cl). Analyses indicate that chlorine content of titanium varies with the grain size. Comparison to results obtained by other analytical techniques indicates good agreement.

Text References and Articles Abstracted in Part III

Achey, F. A. and Serfass, E. J. (1958); "A new approach to the measurement of coating thickness by fluorescent x-ray absorption," *J. Electrochem. Soc.* **105**, 204. (Abstract in Chap. 22.)

Addink, N. W. H. (1959); "Note on the analysis of small quantities of material by x-ray fluorescence," *Rev. univers. Mines* [9] **15**, 530. (Abstract in Chap. 27.)

Addink, N. W. H., Kraay, H., and Witmer, A. W. (1962); "The putting to advantage of the absorbing qualities of diluting agents for obtaining 45 degree calibration curves in x-ray fluorescence analysis," *Colloqu. Spectros. internat.* **1961/3**, 368.

Adler, I. and Axelrod, J. M. (1955a); "Internal standards in fluorescent x-ray spectroscopy," *Spectrochim. Acta* **7**, 91.

Adler, I. and Axelrod, J. M. (1955b); "Determination of thorium by fluorescent x-ray spectrometry," *Anal. Chem.* **27**, 1002. (Abstract in Sec. 18.5.)

Advances in X-Ray Analysis. Vol. 1–14; edited by W. M. Mueller, G. R. Mallett and M. J. Fay, Plenum Press, New York.

Alexander, G. V. (1962); "Determination of zinc, copper, and iron in biological tissues. An x-ray fluorescence method," *Anal. Chem.* **34**, 951. (Abstract in Sec. 26.1.)

Alexander, G. V. (1964); "X-ray fluorescence analysis of biological tissues," *Appl. Spectroscopy* **18**, 1. (Abstract in Sec. 26.1.)

Alley, B. J. and Higgins, J. H. (1963); "Empirical corrections for variable absorption of soft x-rays by Mylar," *Norelco Rep.* **10**, 77.

Andermann, G. and Kemp, J. W. (1958); "Scattered x-ray as internal standards in x-ray emission spectroscopy," *Anal. Chem.* **30**, 1306.

Andermann, G. (1961); "Improvements in the x-ray emission analysis of cement row mix," *Anal. Chem.* **33**, 1689. (Abstract in Sec. 23.1.)

Andermann, G. and Allen, J. D. (1961); "X-ray emission analysis of finished cements," *Anal. Chem.* **33**, 1695. (Abstract in Sec. 23.1.)

Arsuffi, G. (1960); "Die Röntgenspektralanalyse. Neue Zürcher Zeitung, Beilage Technik, Nr. 1165, vom 6. 4. 1960.

Ashley, R. W. and Jones, R. W. (1959); "Application of x-ray fluorescence methods to the analysis of zircaloy," *Anal. Chem.* **31**, 1632. (Abstract in Sec. 20.1.)

Ball, D. F. and Perkins, D. F. (1962); "Plant analysis by x-ray fluorescence spectrography: determination of calcium and potassium," *Nature* **194**, 1163. (Abstract in Sec. 26.2.)

Bareham, F. R. and Fox, J. G. M. (1959/60); "X-ray fluorescence analysis and its application to copper alloys," *J. Inst. Metals* **88**, 344. (Abstract in Sec. 20.1.)

Barkla, C. G. and Sadler, C. A. (1907); "Secondary x-rays and the atomic weight of nickel," *Phil. Mag.* [6] **14**, 408.

Barkla, C. G. and Sadler, C. A. (1909); "The absorption of Röntgen-rays," *Phil. Mag.* [6] **17**, 739.

Barkla, C. G. (1911); "The spectra of the fluorescent Röntgen radiations," *Phil. Mag.* [6] **22**, 396.

Bartkiewicz, S. A. and Hammatt, E. A. (1964); "X-ray fluorescence determination of cobalt, zinc, and iron in organic matrices," *Anal. Chem.* **36**, 833. (Abstract in Sec. 25.1.)

Beattie, H. J. and Brissey, R. M. (1954); "Calibration method for x-ray fluorescence spectrometry," *Anal. Chem.* **26**, 980. (Abstract in Sec. 19.1.)

Berger, R. and Deceuleneer, P. (1961); "Application de la fluorescence X au controle analytique dans une aciérie de moulage," *Rev. univers. Mines* [9] **15**, 207. (Abstract in Sec. 19.1.)

Beyermann, L. (1961); "Die röntgenfluorimetrische und aktivierungsanalytische Brombestimmung in biologischem Material," *Z. analyt. Chem.* **183**, 199. (Abstract in Sec. 26.1.)

Bhattacherjee, S. B. and Kumar, M. N. (1964); "X-ray fluorescence lead–uranium ratio measurements in Allanite. Bankura District, India," *Anal. Chem.* **36**, 1400. (Abstract in Sec. 23.2.)

Birks, L. S. and Brooks, E. J. (1950); "Hafnium–zirconium and tantalum–columbium systems. Quantitative analysis by x-ray fluorescence," *Anal. Chem.* **22**, 1017. (Abstract in Sec. 18.1.)

Birks, L. S., Brooks, E. J., Friedman, H., and Roe, R. M. (1950); "X-ray fluorescence analysis of ethyl fluid in aviation gasoline," *Anal. Chem.* **22**, 1258. (Abstract in Sec. 24.3.)

Birks, L. S. and Brooks, E. J. (1951); "Analysis of uranium solutions by x-ray fluorescence," *Anal. Chem.* **23**, 707. (Abstract in Sec. 18.5.)

Birks, L. S. (1959); *X-ray Spectrochemical Analysis*, Interscience Publ., New York and London.

Birks, L. S. and Batt, A. P. (1963); "Use of a multichannel analyzer for electron probe microanalysis," *Anal. Chem.* **35**, 778.

Birks, L. S., Labrie, R. J., and Criss, J. W. (1966); "Energy dispersion for quantitative x-ray spectrochemical analysis," *Anal. Chem.* **38**, 701.

Black, R. H. and Forsyth, W. J. (1959); "The determination of sulfur and vanadium in carbon materials by x-ray fluorescence analysis," *Norelco Rep.* **6**, 53. (Abstract in Sec. 24.5.)

Blavier, P., Hans, H., Tyou, P., and Houbart, I. (1960); "Elemental analysis of alnico type alloys by x-ray fluorescence," *Cobalt* **7**, 33. (Abstract in Sec. 20.1.)

Blochin, M. A. (1957); *Physik der Röntgenstrahlen* (translated from the Russian), VEB Verlag Technik, Berlin.

Blochin, M. A. (1963); *Methoden der Röntgenspektralanalyse* (translated from the Russian), B. G. Teubner Verlagsges, Leipzig.

Bohr, N. (1913); "On the constitution of atoms and molecules I and II," *Phil. Mag.* **26**, 1, 476.

Bragg, W. H. and Bragg, W. L. (1913); "The reflection of x-rays by crystals," *Proceed. Royal Soc. London* [A] **88**, 428.

Brandt, C. S. and Lazer, V. A. (1958); "Analysis of dried plant material by x-ray emission spectrograph," *Agricultural and Food Chem.* **6**, 306. (Abstract in Sec. 26.2.)

Brindley, G. W. and McKinstry, H. A. (1961); "The kaolinite–mullite reaction series: IV, the coordination of aluminium," *J. Am. Ceramic Soc.* **44**, 506. (Abstract in Sec. 23.3.)

Brissey, R. M. (1953); "Analysis of high temperature alloys by x-ray fluorescence," *Anal. Chem.* **25**, 19. (Abstract in Sec. 19.1.)

Brodkorb, E. (1961); "Bestimmung von Kupfer in Schwefelkies durch Röntgenfluoreszenzanalyse," *Siemens Z.* **35**, 451. (Abstract in Sec. 23.2.)

Brooks, E. J. and Birks, L. S. (1957); "Compton scattering interference in fluorescent x-ray spectroscopy," *Anal. Chem.* **29**, 1556.

Bruch, J. (1962); "Die Anwendung der Röntgenfluoreszenzanalyse bei der Untersuchung verschiedener Ferrolegierungen," *Archiv Eisenhüttenwesen*, **33**, 5. (Abstract in Sec. 19.1.)

Burke, W. E., Hinds, L. S., Deodato, G. E., Sager, E. D., and Borup, R. E. (1964); "Internal standard x-ray spectrographic procedure for the determination of calcium, barium, zinc, and lead in hydrocarbons," *Anal. Chem.* **36**, 2404.

Burnham, H. D., Hower, J., and Jones, L. C. (1957); "Generalized x-ray emission spectrographic calibration applicable to varying compositions and sample forms," *Anal. Chem.* **29**, 1827. (Abstract in Sec. 19.1.)

Cameron, J. F. and Rhodes, J. R. (1960); "Measurement of tinplate thickness using fluorescent x-rays excited by a radioactive source," *Brit. J. Appl. Phys.* **11**, 49. (Abstract in Sec. 19.1.)

Cameron, J. F. and Rhodes, J. R. (1961); "X-ray spectrometry with radioactive sources," *Nucleonics* **19**, 53.

Campbell, W. J. and Carl, H. F. (1954); "Quantitative analysis of niobium and tantalum in ores by fluorescent x-ray spectroscopy," *Anal. Chem.* **26**, 800. (Abstract in Sec. 18.2.)

Campbell, W. J. and Carl, H. F. (1955); "Combined radiometric and fluorescent x-ray spectrographic method of analyzing for uranium and thorium," *Anal. Chem.* **27**, 1884. (Abstract in Sec. 18.5.)

Campbell, W. J. and Carl, H. F. (1956); "Fluorescent x-ray spectrographic determination of tantalum in commercial niobium oxides," *Anal. Chem.* **28**, 960. (Abstract in Sec. 18.2.)

Campbell, W. J., Carl, H. F., and White, C. E. (1957); "Quantitative analysis by fluorescent x-ray spectrography. Determination of germanium in coal and coal ash," *Anal. Chem.* **29**, 1009. (Abstract in Sec. 24.5.)

Campbell, W. J. and Brown, J. D. (1964); "X-ray absorption and emission," *Anal. Chem.* **36**, 312 R.

Carter, G. F. (1962); "X-ray fluorescence analysis of light elements, aluminium–chromium," *Appl. SpectroscoP*, **16**, 159.

Cass, D. E. and Kelly, J. H. (1963); "Some advances in the application of x-ray fluorescence to the continuous measurement of tin coating weight," *Norelco Rep*, **10**, 49. (Abstract in Chap. 22.)

Chodos, A. A. and Engel, C. G. (1961); "Fluorescent x-ray spectrographic analysis of amphibolite rocks," *Am. Mineral.* **46**, 120. (Abstract in Sec. 23.2.)

Claisse, F. (1957); "Accurate x-ray fluorescence analysis without internal standard," *Norelco Rep.* **4**, 3, 95.

Claisse, F. and Samson, C. (1962); "Effets des hétérogénités en fluorescence des rayons X. Lab.," Ministere des richesses naturelles, Report S-67, Quebec.

Clark, G. L. and Terford, H. C. (1954); "X-ray analysis of foundry dusts for quartz and iron in relation to silicosis and siderosis. Fluorescent spectral analysis of iron," *Anal. Chem.* **26**, 1414, 1416. (Abstract in Sec. 26.2.)

Collin, R. L. (1961); "X-ray fluorescence analysis using ion exchange resin for sample support," *Anal. Chem.* **33**, 605. (Abstract in Sec. 26.1.)

Compton, A. H. and Allison, S. K. (1935); *X-rays in Theory and Experiment*, Van Nostrand Co. Inc., Toronto, New York, London.

Cook, G. B., Mellish, C. E., and Payne, J. A. (1960); "Measurement of thin metal layers. Fluorescent x-ray production by radioisotope sources," *Anal. Chem.* **32**, 590. (Abstract in Chap. 22.)

Coster, D. and Hevesy, G. (1923); "On the missing element of atomic number 72," *Nature* **111**, 79. (Abstract in Sec. 18.1.)

Criss, J. W. and Birks, L. S. (1968); "Calculation methods for fluorescent x-ray spectrometry," *Anal. Chem.* **40**, 1080.

Cullen, T. J. (1962); "Coherent scattered radiation internal standardization in x-ray spectrometric analysis of solutions," *Anal. Chem.* **34**, 812.

Daugherty, K. E., Robinson, R. J., and Mueller, J. I. (1964a); "X-ray fluorescence spectrometric analysis of the copper (II) and mercury (II) complexes of 6-chloro-2-methoxy-9-thiolacridine," *Anal. Chem.* **36**, 1098. (Abstract in Sec. 25.2.)

Daugherty, K. E., Robinson, R. J., and Mueller, J. I. (1964b); "X-ray fluorescence spectrometric analysis of iron(III), cobalt(II), nickel(II), and copper(II) chelates of 8-quinolinol," *Anal. Chem.* **36**, 1869. (Abstract in Sec. 25.2.)

Daugherty, K. E., Goheen, M. W., Robinson, R. J., and Mueller, J. I. (1964); "X-ray fluorescence spectrometric analysis of rubidium(I) and cesium(I) salts of 5-nitrobarbituric acid," *Anal. Chem.* **36**, 2372. (Abstract in Sec. 25.2.)

Davis, C. M. and Clark, G. R. (1958); "X-ray spectrographic analysis of nickel-containing alloys with varied sample forms," *Appl. Spectroscopy* **12**, 123. (Abstract in Sec. 19.1.)

Davies, O. L. (1957); *Statistical Methods in Research and Production*, Oliver and Boyd, London and Edinburgh.

Davies, T. A. (1961); "X-ray fluorescence spectroscopy in metallurgical research," *Rev. univers. Mines* [9] **15**, 228. (Abstract in Sec. 19.1.)

Davis, E. N. and Nordstrand, R. A. (1954); "Determination of barium, calcium and zinc in lubricating oils," *Anal. Chem.* **26**, 937. (Abstract in Sec. 24.4.)

Davis, E. N. and Hoeck, B. C. (1955); "X-ray spectrographic method for the determination of vanadium and nickel in residual fuels and charging stocks," *Anal. Chem.* **27**, 1880. (Abstract in Sec. 24.1.)

Day, D. E. and Rindone, G. E. (1962); "Properties of soda aluminosilicate glasses: III, coordination of aluminum ions," *J. Am. Ceramic Soc.* **45**, 579. (Abstract in Sec. 23.3.)

Day, D. E. (1963); "Determining the coordinating number of aluminum ions by x-ray emission spectroscopy," *Nature* **200**, 649. (Abstract in Sec. 23.3.)

Dickens, P., König, P., and Dippel, Th. (1963); "Bestimmung der Zusammensetzung von Stahlwerkschlacke während des Schmelzens mit einem selbsttätigen Röntgenfluoreszenzspektrometer," *Archiv. Eisenhüttenwesen*, **34**, 519. (Abstract in Sec. 19.2.)

Doe, B. R., Chodos, A. A., Rose, A. W., and Godijn, E. (1961); "The determination of iron in sphalerite by x-ray fluorescence spectrometry," *Am. Mineral.* **46**, 1056. (Abstract in Sec. 20.2.)

Dolby, R. M. (1959); "Some methods for analysing unresolved proportional counter curves of x-ray line spectra," *Proc. Phys. Soc. London* **73**, 81.

Dothie, H. J. (1962); " 'Filterscan,' a novel technique in x-ray fluorescence analysis," *Nature* **196**, 984.

Doughman, W. R., Sullivan, A. P., and Hirt, R. C. (1958); "X-ray determination of sulfur in oils," *Anal. Chem.* **30**, 1924. (Abstract in Sec. 24.1.)

Dwiggins, C. W. and Dunning, H. N. (1959); "Quantitative determination of nickel in oils by x-ray spectrography," *Anal. Chem.* **31**, 1040. (Abstract in Sec. 24.1.)

Dwiggins, C. W. and Dunning, H. N. (1960); "Quantitative determination of traces of vanadium, iron, and nickel in oils by x-ray spectrography," *Anal. Chem.* **32**, 1137. (Abstract in Sec. 24.1.)

Dwiggins, C. W. (1961); "Quantitative determination of low atomic number elements using intensity ratio of coherent to incoherent scattering x-rays," *Anal. Chem.* **33**, 67. (Abstract in Sec. 25.3.)

Dyroff, G. V. and Skiba, P. (1954); "Determination of trace amounts of iron, nickel, and vanadium on catalysts by fluorescent x-ray spectrography," *Anal. Chem.* **26**, 1774. (Abstract in Sec. 24.2.)

Eckhardt, F. J. and Fesser, H. (1961); "Anwendung der Röntgenspektralanalyse bei Untersuchungen auf dem Gebiet der Bodenforschung. 2. Informationstagung über 'Anwendung der Röntgenspektralanalyse'," *Darmstadt*, Febr. 1961, p. 78.

Eichhoff, H. J., Beck, K., and Kiefer, S. K. (1965); "Halbquantitative röntgenfluores-zenzspektroskopische Analyse von gelösten und pulverförmigen Proben," *G-I-T, Fachzeitschr. f. d. Laboratorium* **9**, 687.

Fagel, J. E., Liebhafsky, H. A., and Zemany, P. D. (1958); "Determination of tungsten and molybdenum by x-ray emission spectrography," *Anal. Chem.* **30**, 1918. (Abstract in Sec. 18.3.)

Felten, E. J., Fankuchen, I., and Steigman, J. (1959); "A possible solution to the matrix problem in x-ray fluorescence spectroscopy," *Anal. Chem.* **31**, 1771. (Abstract in Chap. 27.)

Fischer, D. W. and Baun, W. L. (1965); "Diagram and nondiagram lines in K-spectra of Al and O from metallic and anodized aluminium," *J. Appl. Phys.* **36**, 534.

Friedlander, S. and Goldblatt, A. (1959); "A comparison of precision for solid, liquid, and powder sampling techniques in the x-ray fluorescence analysis of high temperature alloys," *Appl. Spectroscopy* **13**, 91.

Friedrich, W., Knipping, P., and Laue, M. (1912); "Interferenz-Erscheinungen bei Röntgenstrahlen," *Sitzungsber. math. phys. Kl. Bayerischen Akad. Wiss. Jg.* 1912, 303.

Fünfer, E. and Neuert, H. (1954); *Zählrohre und Szintillationszähler*, Verlag G. Braun, Karlsruhe.

Gilfrich, J. V. and Sullivan, D. C. (1963); "Determination of aluminium in aluminium–iron alloys," *Norelco Rep.* **10**, 127. (Abstract in Chap. 21.)

Gilfrich, J. V. and Birks, L. S. (1968); "Spectral distribution of x-ray tubes for quantitative x-ray fluorescence analysis," *Anal. Chem.* **40**, 1077.

Gillam, E. and Heal, H. T. (1952); "Some problems in the analysis of steels by x-ray fluorescence," *Brit. J. Appl. Phys.* **3**, 353. (Abstract in Chap. 19.1.)

Ginsberg, H. and Pfundt, H. (1962); "Bestimmungsgrenzen für kleine und kleinste Beimengungen in Reinaluminium und Raffinal bei Anwendung der Röntgenspektro-skopie," *Z. Metallkunde* **53**, 695. (Abstract in Chap. 21.)

Glocker, R. (1949); *Materialprüfung mit Röntgenstrahlen*. 3. Aufl. Springer Verlag, Berlin, Göttingen and Heidelberg.

Glocker, R. and Schreiber, H. (1928); "Quantitative Röntgenspektralanalyse mit Kalterregung des Spektrums," *Ann. Phys. Chem.* **85**, 1089.

Guinier, A. (1961); "Possibilités et limitations de l'analyse par spectrographie X," *Rev. univers. Mines* [9], **17**, 143.

Gunn, E. L. (1956); "Determination of platinum in reforming catalyst by x-ray fluorescence," *Anal. Chem.* **28**, 1433. (Abstract in Sec. 24.2.)

Gunn, E. L. (1957); "Fluorescent x-ray spectral analysis of powdered solids by matrix dilution," *Anal. Chem.* **29**, 184.

Gunn, E. L. (1964); "Problems of direct determination of traces of nickel in oil by x-ray emission spectrography," *Anal. Chem.* **36**, 2086. (Abstract in Sec. 24.1.)

Gunn, E. L. (1965); "Determination of lead in gasoline by x-ray fluorescence using an internal intensity reference," *Appl. Spectroscopy* **19**, 99. (Abstract in Sec. 24.3.)

Hahn-Weinheimer, P. and Ackermann, H. (1963); "Quantitative röntgenspektral-analytische Bestimmung von Kalium, Rubidium, Strontium, Barium, Titan, Zirkonium und Phosphor," *Z. analyt. Chem.* **194**, 81. (Abstract in Sec. 23.2.)

Haftka, F. J. (1959); "Röntgenfluoreszenzanalyse von Pulvern," *Rev. univers. Mines* [9], **15**, 549.

Hakkila, E. A. and Waterbury, G. R. (1960); "X-ray fluorescence spectrographic determination of impurities and alloying elements in tantalum container materials," *Talanta*, **6**, 46. (Abstract in Sec. 18.2.)

Hakkila, E. A., Hurley, R. G., and Waterbury, G. R. (1964); "X-ray fluorescence spectrometric determination of zirconium and molybdenum in the presence of uranium," *Anal. Chem.* **36**, 2094. (Abstract in Sec. 18.1.)

Hale, C. C. and King, W. H. (1961); "Direct nickel determination in petroleum oils by x-ray at the 0,1 ppm level," *Anal. Chem.* **33**, 74. (Abstract in Sec. 24.1.)

Hall, Th. (1961); "X-ray fluorescence analysis in biology," *Science* **134**, 499. (Abstract in Sec. 26.1.)

Hamos, V. L. (1945); "On the determination of very small quantities of substances by the x-ray microanalyses," *Arkiv. Math. Astrom. Fysik* **31A**, 1.

Handbook of Chemistry and Physics, edited by Ch. D. Hodgman, Chemical Rubber Publ. Co., Cleveland (Ohio).

Handbuch der Physik, Vol. 30, Röntgenstrahlen; edited by S. Flügge, Springer Verlag, Berlin, Göttingen and Heidelberg, 1957.

Handley, R. (1960); "Fluorescent x-ray determination of selenium in plant material," *Anal. Chem.* **32**, 1719. (Abstract in Sec. 26.2.)

Hans, A. (1959); "Analyse rapide du cuivre, du zinc et du cobalt par fluorescence X dans les poudres de minerais et produits de concentration," *Rev. univers. Mines* [9], **15**, 545. (Abstract in Sec. 20.2.)

Heidel, R. H. and Fassel, V. A. (1958); "X-ray fluorescent spectrometric determination of yttrium in rare earth mixtures," *Anal. Chem.* **30**, 176. (Abstract in Sec. 18.4.)

Heidel, R. H. and Fassel, V. A. (1961); "Fluorescent x-ray spectrometric determination of scandium in ores and related materials," *Anal. Chem.* **33**, 913. (Abstract in Sec. 18.4.)

Heinrich, K. F. J. and McKinley, T. D.; "The determination of impurities in elemental niobium and its compounds by x-ray spectroscopy," report of Du Pont de Nemours Co. Pigments Department.

Heritage, R. J., Young, A. S., and Bott, I. B. (1963); "Composition studies of evaporated nickel–iron layers," *Brit. J. Appl. Phys.* **14**, 439. (Abstract in Chap. 22.)

Herrmann, M. (1961); "Über die Brombestimmung durch Röntgenfluoreszenz," *Z. analyt. Chem.* **181**, 122. (Abstract in Sec. 25.2.)

Herzog, L. F. (1960); "Age determination by x-ray fluorescence rubidium-strontium ratio measurement in lepidolite," *Science* **132**, 293. (Abstract in Sec. 23.2.)

Hevesy, G. and Alexander, E. (1933); *Praktikum der chemischen Analyse mit Röntgenstrahlen*, Akad. Verlagsges, Leipzig.

Hill, R. D., Church, E. L., and Mihelich, J. W. (1952); "The determination of gamma-ray energies from beta-ray spectroscopy and a table of critical x-ray absorption energies," *Rev. Sci. Instr.* **23**, 523.

Hirt, R. C., Doughman, W. R., and Gisclard, J. B. (1956); "Application of x-ray-emission spectrography to airborne dusts in industrial hygiene studies," *Anal. Chem.* **28**, 1649. (Abstract in Sec. 26.2.)

Hooper, P. R. (1964); "Rapid analysis of rocks by x-ray fluorescence," *Anal. Chem.* **36**, 1271. (Abstract in Sec. 23.2.)

Houk, W. W. and Silverman, L. (1959); "Determination of iron, chromium and nickel by fluorescent x-ray analysis," *Anal. Chem.* **31**, 1069. (Abstract in Sec. 19.1.)

Hower, J. (1959); "Matrix corrections in the x-ray spectrographic trace element analysis of rocks and minerals," *Am. Mineral.* **44**, 19. (Abstract in Sec. 23.2.)

Jackwerth, E. and Kloppenburg, H. J. (1961); "Untersuchungen zur quantitativen Auswertung von Papierchromatogrammen in der Spurenanalyse durch Röntgenfluoreszenzspektroskopie," *Z. analyt. Chem.* **179**, 186. (Abstract in Chap. 27.)

Jahnke, E., Emde, F., and Lösch, F. (1960); *Tafeln höherer Funktionen*, Teubner, Stuttgart.

Jenkins, R. (1963); "Die Röntgenfluoreszenzanalyse und ihre Anwendung in der Petroleumindustrie," *Erdöl-Zeitschrift H.* **2**, 59. (Abstract in Sec. 24.4.)

Jenkins, R. and Hurley, P. W. (1965); "Effects of surface finish in the x-ray fluorescence analysis of bulk materials," 12. *Colloquium spectroscopicum internationale, Exeter,* Hilger & Watts Ltd., London, p. 444.

Johnson, C. M. and Stout, P. R. (1958); "Interferences from Compton scattering from matrices of low atomic number," *Anal. Chem.* **30**, 1921.

Johnson, J. L. and Nagel, B. E. (1963); Microanalysis of controlled spot test paper by x-ray fluorescence," *Microchim. Acta* **3**, 525. (Abstract in Chap. 27.)

Jones, R. A. (1959); "Determination of manganese in gasoline by x-ray emission spectrography," *Anal. Chem.* **31**, 1341. (Abstract in Sec. 24.3.)

Jones, R. A. (1961); "Determination of sulfur in gasoline by x-ray emission spectrography," *Anal. Chem.* **33**, 71. (Abstract in Sec. 24.1.)

Jones, R. W. and Ashley, R. W. (1959); "X-ray fluorescence analysis of stainless steel in aqueous solutions," *Anal. Chem.* **31**, 1629. (Abstract in Sec. 19.1.)

Kaiser, H. and Specker, H. (1956); "Bewertung und Vergleich von Analysenverfahren," *Z. analyt. Chemie* **149**, 46.

Kalman, Z. H. and Heller, L. (1962); "Theoretical study of x-ray fluorescent determination of traces of heavy elements in a light matrix," *Anal. Chem.* **34**, 946. (Abstract in Sec. 23.2.)

Kang, C. C., Keel, E. W., and Soloman, E. (1960); "Determination of traces of vanadium, iron, and nickel in petroleum oils by x-ray emission spectrography," *Anal. Chem.* **32**, 221. (Abstract in Sec. 24.1.)

Karttunen, J. O. (1963); "The separation and fluorescent x-ray spectrometric determination of zirconium, molybdenum, ruthenium, rhodium, and palladium in solution in uranium-base fission alloys," *Anal. Chem.* **35**, 1045. (Abstract in Sec. 18.1.)

Kehl, W. L. and Russell, R. G. (1956); "Fluorescent x-ray spectrographic determination of uranium in waters and brines," *Anal. Chem.* **28**, 1350. (Abstract in Sec. 18.5.)

Kemp, J. W., Hasler, M. F., and Jones, J. L. (1954); "Outline of fluorescent x-ray spectroscopy," *ARL.-Spectrographers' News Letters* **7**/3.

Kessler, H. and Rammensee, H. (1960); "Probleme und Möglichkeiten der Röntgenfluoreszenzspektroskopie bei der Untersuchung von Aluminium, insbesondere Aluminiumgußlegierungen," *Z. Metallkunde* **51**, 548. (Abstract in Chap. 21.)

Kilday, B. A. and Michaelis, R. E. (1962); "Determination of lead in leaded steels by x-ray spectroscopy," *Appl. Spectroscopy* **16**, 137. (Abstract in Sec. 19.1.)

King, A. G. and Dunton, P. (1955); "Quantitative analysis for thorium by x-ray fluorescence," *Science* **122**, 72. (Abstract in Sec. 18.5.)

Kirchmayr, H. R. and Mach, D. (1964); Die Röntgenfluoreszenzanalyse seltener Erdmetall-Mangan-Legierungen," *Z. Metallkunde* **55**, 247. (Abstract in Sec. 18.4.)

Koh, P. K. and Caugherty, B. (1952); "Metallurgical applications of x-ray fluorescent analysis," *J. Appl. Phys.* **23**, 427.

Kokotailo, G. T. and Damon, G. F. (1953); "Determination of bromine in liquids by x-ray fluorescent spectroscopy," *Anal. Chem.* **25**, 1185. (Abstract in Sec. 24.3.)

Kopineck, H. J. (1956); "Zur Schnellbestimmung des Eisens in der Schlacke," *Archiv Eisenhüttenwesen* **27**, 753. (Abstract in Sec. 19.2.)

Kopineck, H. J. (1962); "Zur Röntgenfluoreszenz-Spektralanalyse," *Archiv Eisenhüttenwesen*, **33**, 327.

Kopineck, H. J. and Schmitt, P. (1961); "Zur Anwendung der Röntgenfluoreszenz-Spektralanalyse in der Eisenhüttenindustrie," *Archiv Eisenhüttenwesen* **32**, 19. (Abstract in Sec. 19.2.)

Kopineck, H. J. and Schmitt, P. (1965); "Röntgenspektrometrische Analyse nicht-metallischer Stoffe im Eisenhüttenlaboratorium," *Archiv Eisenhüttenwesen* **36**, 87.

Krächter, H. and Jäger, W. (1957); "Quantitative Ermittlung von Legierungselementen im Stahl mit der Röntgen-Fluoreszenz-Spektralanalyse," *Archiv Eisenhüttenwesen* **28**, 633. (Abstract in Sec. 19.1.)

Kriege, O. H. and Rudolph, J. S. (1963); "The determination of microgram quantities of zirconium in iron, cobalt, and nickel alloys by x-ray fluorescence," *Talanta* **10**, 215. (Abstract in Sec. 19.1.)

Kulenkampff, H. (1922); "Uber das kontinuierliche Röntgenspektrum," *Ann. Phys.* **69**, 548.

Kulenkampff, H. and Schmidt, L. (1943); "Die Energieverteilung im Spektrum der Röntgen-Bremsstrahlung," *Ann. Phys.* **43**, 494.

Lachance, G. R. and Traill, R. J. (1966); "A practical solution to the matrix problem in x-ray analysis," *Canadian Spectroscopy* **11**, 43.

Laffolie, H. de (1961); "Röntgenspektroskopie im Eisenhüttenlaboratorium," *DEW-Techn. Ber.* **1**, 161. (Abstract in Sec. 19.1.)

Laffolie, H. de (1962a); "Beitrag zur röntgenspektrometrischen Analyse legierter Stähle," *Archiv Eisenhüttenwesen* **33**, 101. (Abstract in Sec. 19.1.)

Laffolie, H. de (1962b); "Beitrag zur röntgenspektrochemischen Analyse niedrig legierter Stähle," *DEW-Techn. Ber.* **2**, 119. (Abstract in Sec. 19.1.)

Lamb, F. W., Niebylski, L. M., and Kiefer, E. W. (1955); "Determination of tetra-ethyllead in gasoline by x-ray fluorescence," *Anal. Chem.* **27**, 129. (Abstract in Sec. 24.3.)

Landolt-Börnstein; *Zahlenwerte und Funktionen*, 6. Aufl., edited by A. Eucken, Springer Verlag, Berlin, Göttingen and Heidelberg (1950).

Laue, M. (1912); "Eine quantitative Prüfung der Theorie für die Interferenz-Erscheinungen bei Röntgenstrahlen," *Sitzungsber. math. phys. Kl. Bayerischen Akad. Wiss. Jg.* **1912**, 363.

Laue, M. (1913); "Kritische Bemerkungen zu den Deutungen der Photogramme von Friedrich und Knipping," *Phys. Z.* **14**, 421.

Leroux, J. and Mahmud, M. (1967a); "Improvement of x-ray spectrographic analysis by filtration of the *L* lines from the primary beam," *Canadian Spectroscopy* **12**, No. 4.

Leroux, J., Mahmud, M., and Davey, A. B. C. (1967b); "Fluorescence yield of a given element as a function of the spectral distribution of the exciting primary beam," *Canadian Spectroscopy* **12**, No. 5.

Lewis, G. I. and Goldberg, E. D. (1956); "X-ray fluorescence determination of barium, titanium, and zinc in sediments," *Anal. Chem.* **28**, 1282. (Abstract in Sec. 23.2.)

Liebhafsky, H. A. and Winslow, E. H. (1958); "X-ray absorption and emission," *Anal. Chem.* **30**, 580.

Liebhafsky, H. A., Pfeiffer, H. G., Winslow, E. H., and Zemany, P. D. (1960); *X-ray Absorption and Emission in Analytical Chemistry*, John Wiley & Sons Inc., New York and London.

Liebhafsky, H. A., Pfeiffer, H. G., and Zemany, P. D. (1960); "Reliability of trace determinations by x-ray emission spectrography," in *X-ray Microscopy and Microanalysis*, edited by E. Engström, V. E. Cosslett, and H. H. Pattee, Elsevier Publ. Co. Amsterdam, p. 321. (Abstract in Chap. 27.)

Liebhafsky, H. A., Winslow, E. H., and Pfeiffer, H. (1960); "X-ray absorption and emission," *Anal. Chem.* **32**, 240 R.

Liebhafsky, H. A., Winslow, E. H., and Pfeiffer, H. (1962); "X-ray absorption and emission," *Anal. Chem.* **34**, 282 R.

Lihl, F. and Fischhuber, J. (1962); "Zur Frage der röntgenfluoreszenzanalytischen Bestimmung des Siliziums in Aluminium-Silizium-Legierungen," *Z. Metallkunde* **53**, 186. (Abstract in Chap. 21.)

Lincoln, A. J. and Davis, E. N. (1959); "Quantitative determination of platinum in alumina base reforming catalyst by x-ray spectroscopy," *Anal. Chem.* **31**, 1317. (Abstract in Sec. 24.2.)

Linder, A. (1957); *Statistische Methoden.* 2. Aufl. Birkhäuser, Basel and Stuttgart.

Loevinger, R. and Berman, M. (1951); "Efficiency criteria in radioactivity counting," *Nucleonics* **9/1**, 26.

Longobucco, R. (1962); "Determination of major and minor constituents in ceramic materials by x-ray spectrometry," *Anal. Chem.* **34**, 1263. (Abstract in Sec. 23.2.)

Lonsdale, K. (1948); "Geiger counter measurements of Bragg and diffuse scattering of x-rays by single crystals," *Acta Cryst.* **1**, 12.

Louis, R. (1964a); "Apparative Probleme der Röntgenanalyse von Mineralölprodukten," *Erdöl und Kohle* **17**, 360. (Abstract in Sec. 24.4.)

Louis, R. (1964b); "Röntgen-Emissionsspektralanalyse von Schmierölen," *Z. analyt. Chem.* **201**, 336. (Abstract in Sec. 24.4.)

Louis, R. (1965); "Nachweisgrenzen bei der Röntgen-Emissionsspektralanalyse von Mineralölen," *Z. analyt. Chem.* **208**, 34. (Abstract in Sec. 24.4.)

Lublin, P. (1959); "Determination of cladding thicknesses of nuclear fuel elements by x-rays," *Norelco Rep.* **6**, 57. (Abstract in Chap. 22.)

Lucas-Tooth, H. J. and Price, B. J. (1961); "A mathematical method for the investigation of interelement effects in x-ray fluorescent analyses," *Metallurgia* **64**, 149. (Abstract in Sec. 20.1.)

Luke, C. L. (1963a); "Determination of refractory metals in ferrous alloys and high-alloy steel by the borax disk x-ray spectrochemical method," *Anal. Chem.* **35**, 56. (Abstract in Sec. 19.1.)

Luke, C. L. (1963b); "Trace analysis of metals by borax disk x-ray spectrometry," *Anal. Chem.* **35**, 1551.

Lund, P. K. and Mathies, J. C. (1960); "X-ray spectroscopy in biology and medicine I. Total iron (hemoglobin) content in human whole blood," *Norelco Rep.* **7**, 127. (Abstract in Sec. 26.1.)

Lytle, F. W. and Heady, H. H. (1959); "X-ray emission spectrographic analysis of high-purity rare earth oxides," *Anal. Chem.* **31**, 809. (Abstract in Sec. 18.4.)

Mack, M. and Spielberg, N. (1958); "Statistical factors in x-ray intensity measurements," *Spectrochim. Acta* **12**, 169.

Maneval, D. R. and Lovell, H. L. (1960); "Determination of lanthanum, cerium, praseodymium, and neodymium as major components by x-ray emission spectroscopy," *Anal. Chem.* **32**, 1289. (Abstract in Sec. 18.9.)

Marti, W. (1961); "Determination of the interelement effect in the x-ray fluorescence analysis of Cr in steels," *Spectrochim. Acta* **17**, 379. (Abstract in Sec. 19.1.)

Marti, W. (1962); "On the determination of the interelement effect in the x-ray fluorescence analysis of steels," *Spectrochim. Acta* **18**, 1499. (Abstract in Sec. 19.1.)

Mathies, J. C. and Lund, P. K. (1960a); "X-ray spectroscopy in biology and medicine: II. Calcium content of human blood serum," *Norelco Rep.* **7**, 130. (Abstract in Sec. 26.1.)

Mathies, J. C. and Lund, P. K. (1960b); "X-ray spectroscopy in biology and medicine. III. Bromide (total bromine) in human blood, serum, urine, and tissues," *Norelco Rep.* 7, 134. (Abstract in Sec. 26.1.)

Mathies, J. C., Lund, P. K., and Eide, W. (1962); "A simple, indirect sensitive procedure for the determination of nitrogen (ammonia) at the microgram and sub-microgram level," *Anal. Biochemistry* 3, 408. (Abstract in Sec. 26.1.)

Maurette, M. and Despujols, J. (1960); "Sur le dosage du plomb et du zink dans les minerais pauvres par spectrométrie de fluorescence X," *J. Chim. physique* 57, 1099. (Abstract in Sec. 20.2.)

Mecke, P. (1963); "Quantitative Röntgenfluoreszenzanalyse geringster Substanzmengen mit der Filterdifferenzmethode," *Z. analyt. Chem.* 193, 301.

Menis, O., Halteman, E. K., and Garcia, E. E. (1963); "X-ray emission analysis of pluto-nium and uranium compound mixtures,"*Anal. Chem.* 35, 1049. (Abstract in Sec. 18.5.)

Meyer, J. W. (1961); "Determination of iron, calcium, and silicon in calcium silicates by x-ray fluorescence," *Anal. Chem.* 33, 692. (Abstract in Sec. 23.1.)

Milledge, H. J. (1962); "Mass absorption coefficients μ/ρ of the elements for the range of wavelengths 0.30 (0.05) 2.75 Å, with values of Z/A and A/N, in *International Tables for X-ray Crystallography*, Vol. 3, Sec. 3.2; edited by K. Lonsdale, The Kynoch Press, Birmingham.

Mitchell, Betty J. (1958); "X-ray spectrographic determination of tantalum, niobium, iron, and titanium oxide mixtures. Using simple arithmetic corrections for inter-element effects," *Anal. Chem.* 30, 1894. (Abstract in Sec. 18.2.)

Mitchell, Betty J. (1960); "X-ray spectrographic determination of zirconium, tungsten, vanadium, iron, titanium, tantalum, and niobium oxides. Application of the correction factor method," *Anal. Chem.* 32, 1652. (Abstract in Sec. 18.2.)

Mitchell, Betty J. (1961); "Prediction of x-ray fluorescent intensities and interelement effects," *Anal. Chem.* 33, 917.

Mitchell, Betty J. and O'Hear, H. J. (1962); "General x-ray spectrographic solution method for analysis of iron-, chromium-, and/or manganese-bearing materials," *Anal. Chem.* 34, 1620. (Abstract in Sec. 19.1.)

Mitra, G. B. and Wilson, A. J. C. (1960); "Variation with particle size of the effective x-ray absorption coefficient of heterogeneous slabs," *Brit. J. Appl. Phys.* 11, 43.

Molloy, M. W. and Kerr, P. F. (1960); "X-ray spectrochemical analysis: an application to certain light elements in clay minerals and volcanic glass," *Am. Mineral.* 45, 911. (Abstract in Sec. 23.2.)

Mortimore, D. M. and Romans, P. A. (1952); "X-ray spectroscopy as a control method in the production of zirconium and hafnium," *J. Opt. Soc. Am.* 42, 673. (Abstract in Sec. 18.1.)

Mortimore, D. M., Romans, P. A.,and Tews, J. L. (1954); "X-ray spectroscopic deter-mination of columbium and tantalum in rare-earth ores," *Appl. Spectroscopy* 8, 24. (Abstract in Sec. 18.2.)

Moseley, H. G. J. (1913); "The high-frequency spectra of the elements," *Phil. Mag.* 26, 1024.

Moseley, H. G. J. (1914); "The high-frequency spectra of the elements. II," *Phil. Mag.* 27, 703.

Müller, R. (1962a); "Die Berechnung der Eichkurven Röntgenfluoreszenzintensität/ Konzentration in Gemischen," *Spectrochim. Acta* 18, 123.

Müller, R. (1962b); "Die Röntgenfluoreszenzintensität der Elemente bei Verwendung einer Molybdän- und einer Wolframröhre," *Spectrochim. Acta* 18, 1515.

Müller, R. (1963); "Der Einfluß der Korngröße und der Kornbeschaffenheit auf die Fluoreszenzintensität von pulverförmigen Proben," *Vortrag gehalten im Kolloqu. angewandte Röntgenographie*, Wien, in press.

Müller, R. (1964); "Die Abhängigkeit der Fluoreszenzintensität vom Massenabsorptionskoeffizient der Matrix bei der Spurenbestimmung durch Röntgenfluoreszenz," *Spectrochim. Acta* **20**, 143. (Abstract in Sec. 25.2.)

Murray, J. A. and Bartlett, Th. H. (1964); "The determination of microgram quantities of uranium in airborne dust samples by means of x-ray spectrography," *Norelco Rep.* **11**, 132. (Abstract in Sec. 26.2.)

Natelson, S. and Bender, S. L. (1959); "X-ray fluorescence as a tool for the analysis of submicrogram quantities of the elements in biological systems," *Microchem. J.* **3**, 19. (Abstract in Sec. 26.1.)

Natelson, S., Richelson, M. R., Sheid, B., and Bender, S. L. (1959); "X-ray spectroscopy in the clinical laboratory. I. Calcium and potassium," *Clinical Chem.* **5**, 521. (Abstract in Sec. 26.1.)

Natelson, S. and Sheid, B. (1960a); "X-ray spectroscopy in the clinical laboratory. II. Chlorine and sulfur; automatic analysis of ultramicro samples," *Clinical Chem.* **6**, 299. (Abstract in Sec. 26.1.)

Natelson, S. and Sheid, B. (1960b); "X-ray spectroscopy in the clinical laboratory. III. Sulfur distribution in the electrophoretic protein fractions of human serum," *Clinical Chem.* **6**, 314. (Abstract in Sec. 26.1.)

Natelson, S. and Sheid, B. (1961a); "X-ray spectroscopy in the clinical laboratory. IV. Phosphorus; total blood iron as a measure of hemoglobin content," *Clinical Chem.* **7**, 115. (Abstract in Sec. 26.1.)

Natelson, S. and Sheid, B. (1961b); "X-ray spectrometric determination of strontium in human serum and bone," *Anal. Chem.* **33**, 396. (Abstract in Sec. 26.1.)

Natelson, S., Leighton, D. R., and Calas, C. (1962); "Assay for the elements chromium, manganese, iron, cobalt, copper, and zinc simultaneously in human serum and sea water by x-ray spectrometry," *Microchem. J.* **6**, 539. (Abstract in Sec. 26.1.)

Neff, H. (1959a); *Grundlagen und Anwendung der Röntgenfeinstruktur-Analyse*, R. Oldenbourg, München.

Neff, H. (1959b); "Über die gegenseitige Beeinflussung von Legierungskomponenten bei der Röntgen-Fluoreszenz-Analyse," *Materialprüf.* **1**, 108.

Neff, H. (1959c); "Messen von Röntgenstrahlen mit verbesserten Detektoren und Zählmethoden," *Siemens Z.* **33**, 655.

Neff, H. (1961); "Uber grundlegende, präparative und substanzmengenmäßige Grenzen bei der Röntgenfluoreszenzanalyse," *Rev. univers. Mines* [9] **15**, 164.

Neff, H. (1963); "Optimierung von Aufnahmebedingungen für Röntgenanalysen," *Archiv Eisenhüttenwesen* **34**, 903.

Niekerk, J. N. van, Sterlow, F. W. E., and Wybenga, F. T. (1961); "X-ray spectrographic determination of thorium in low concentration," *Appl. Spectroscopy* **15**, 121. (Abstract in Sec. 18.5.)

Noakes, G. E. (1954); "An absolute method of x-ray fluorescence analysis applied to stainless steels," *ASTM. Spec. Tech. Publ.* **157**, 57.

Noddack, W., Tacke, J., and Berg, O. (1925); "Die Ekamangane," *Naturwiss.* **13**, 567.

Norton, D. A. (1957); "X-ray fluorescence as applied to cyrtolite," *Am. Mineral.* **42**, 492. (Abstract in Sec. 23.2.)

Olson, E. C. and Shell, J. W. (1960); "The simultaneous determination of traces of selenium and mercury in organic compounds by x-ray fluorescence," *Anal. Chim. Acta* **23**, 219. (Abstract in Sec. 25.1.)

Orrell, E. W. and Gidley, P. J. (1964); "The analysis of aluminosilicates by x-ray fluorescence," *Trans. Brit. Ceramic Soc.* **63**, 19. (Abstract in Sec. 23.2.)

Papariello, G. I., Letterman, H., and Mader, W. J. (1962); "X-ray fluorescent determination of organic substances through inorganic association," *Anal. Chem.* **34**, 1251. (Abstract in Sec. 25.3.)

Parrish, W. (1956a); "X-ray intensity measurements with counter tubes," *Philips Tech. Rev.* **17**, 206.

Parrish, W. (1956b); "X-ray spectrochemical analysis," *Philips Tech. Rev.* **17**, 269.

Parrish, W. and Kohler, T. R. (1956); "Use of counter tubes in x-ray analysis," *Rev. Sci. Instr.* **27**, 795.

Patrick, R. F. (1952); "Some applications of the fluorescent x-ray spectrometer in ceramics," *J. Am. Ceramic Soc.* **35**, 189. (Abstract in Sec. 20.1.)

Pfeiffer, H. G. and Zemany, P. D. (1954); "Trace analysis by x-ray emission spectrography," *Nature* **174**, 397. (Abstract in Chap. 27.)

Pfundt, H. (1960); "Automatische Analysenmethoden, insbesondere in der Leichtmetallindustrie," *Erzbergbau u. Metallhüttenwesen* **13**, 110. (Abstract in Chap. 21.)

Pfundt, H. (1961); "Die Anwendung der Röntgenfluoreszenzanalyse auf dem Gebiete der Leichtmetalle," *Rev. univers. Mines* [9], **17**, 203. (Abstract in Chap. 21.)

Pfundt, H. (1964); "Die quantitative Aluminiumoxidanalyse mit Hilfe der Röntgenspektralanalyse," *Metall* **18**, 1067. (Abstract in Chap. 21.)

Pish, G. and Huffman, A. A. (1955); "Quantitative determination of thorium and uranium in solutions by fluorescent x-ray spectrometry," *Anal. Chem.* **27**, 1875. (Abstract in Sec. 18.5.)

Pluchery, M. M. (1963); "Recherche d'une methode absolue de dosage par fluorescence X applicable aux lames minces de fer-nickel obtenues par evaporation sous vide," *Spectrochim. Acta* **19**, 533. (Abstract in Chap. 22.)

Preis, H. and Esenwein, A. (1959); "Über ein Auswertungsverfahren bei der quantitativen Röntgenfluoreszenz-Spektralanalyse von Mehrstoffsystemen," *Schweizer Archiv* **25**, 415. (Abstract in Sec. 20.1.)

Preis, H. and Esenwein, A. (1960); "Über eine Methode zur röntgenfluoreszenzspektrographischen Bleibestimmung in Benzin," *Schweizer Archiv* **26**, 317. (Abstract in Sec. 24.3.)

Rabillon, R. (1961); "Quelques applications de la spectrométrie X á l'analyse de produits divers," *Rev. univers. Mines* [9], **17**, 291.

Reith, M. and Weisert, E. D. (1956); "Rapid chemical analysis with the x-ray spectrograph," *Metall. Progress* **70**, July, p. 83. (Abstract in Sec. 20.1.)

Renaud, M. (1963); "Le calcul du transfert de rayonnement en fluorescence X," *Comptes rendus* **256**, 3086; 3837; **257**, 3379.

Reynolds, R. C. (1963); "Matrix corrections in trace element analysis by x-ray fluorescence: Estimation of the mass absorption coefficient by Compton scattering," *Am. Mineral.* **48**, 1133. (Abstract in Sec. 23.2.)

Richtmyer, F. K. and Kennard, E. H. (1942); *Introduction to Modern Physics*, McGraw-Hill Book Co., New York.

Robinson, J. M. M. and Gertiser, E. P. (1964); "X-ray spectrographic analysis of raw materials for the cement industry," *ASTM Materials Research and Standards* **4**, 228. (Abstract in Sec. 23.1.)

Rhodin, T. N. (1955); "Chemical analysis of thin films by x-ray emission spectrography," *Anal. Chem.* **27**, 1857. (Abstract in Chap. 22.)

Röntgen, W. C. (1898a); "Über eine neue Art von Strahlen I und II," *Ann. Phys. Chem.* **64**, 1; 12.

Röntgen, W. C. (1898b); "Weitere Beobachtungen über die Eigenschaften der X-Strahlen," *Ann. Phys. Chem.* **64**, 18.

Roos, C. E. (1957); "K-Fluorescence yield of several metals," *Phys. Rev.* **105**, 931.

Rose, H. J., Adler, I., and Flanagan, F. J. (1963); "X-ray fluorescence analysis of the light elements in rocks and minerals," *Appl. Spectroscopy* **17**, 81. (Abstract in Sec. 23.2.)

Ross, P. A. (1928); "A new method of spectroscopy for faint X-radiation," *J. Opt. Soc. Am.* **16**, 433.

Rothmann, H., Schneider, H., Niebuhr, J., and Pothmann, C. (1962); "Röntgenspektrometrische Bestimmung von Tantal und Niob und einiger Begleitelemente in tantalniobhaltigen Stoffen," *Archiv. Eisenhüttenwesen* **33**, 17. (Abstract in Sec. 18.2.)

Rowe, W. A. and Yates, K. P. (1963); "X-ray fluorescence method for trace metals in refinery fluid catalytic cracking feedstocks," *Anal. Chem.* **35**, 369. (Abstract in Sec. 24.2.)

Rudolph, J. S. and Nadalin, R. J. (1964); "Determination of microgram quantities of chloride in high purity titanium by x-ray spectrochemical analysis," *Anal. Chem.* **36**, 1815. (Abstract in Chap. 27.)

Sagel, K. (1959); *Tabellen zur Röntgen-Emissions- und Absorptions-Analyse*, Springer Verlag, Berlin, Göttingen, and Heidelberg.

Sandström, A. E. (1957); "Experimental methods of x-ray spectroscopy," in *Handbuch der Physik*, Vol. 30, edited by S. Flügge, Springer Verlag, Berlin, Göttingen, and Heidelberg.

Sarian, S. and Weart, H. W. (1963); "X-ray fluorescence analysis of polyphase metals. Use of borax glass matrix," *Anal. Chem.* **35**, 115. (Abstract in Sec. 20.4.)

Sawatzky, A. and Jones, S. (1967); "Correction for nonlinearity in x-ray counting systems of electron-probe microanalyzers," *J. Appl. Physics* **38**, 4758.

Schaafs, W. (1957); "Erzeugung von Röntgenstrahlen," in *Handbuch der Physik*, Vol. 30, edited by S. Flügge, Springer Verlag, Berlin, Göttingen and Heidelberg.

Schloemer, H. (1960); "Über den Einsatz der Röntgenfluoreszenz-Analyse in der Zementchemie," *Zement-Kalk-Gips* **13**, 522. (Abstract in Sec. 23.1.)

Schreiber, H. (1929); "Quantitative chemische Analyse mittels des Röntgenemissionsspektrums," *Z. Physik* **58**, 619.

Schreiber, T. P., Ottolini, A. C., and Johnson, J. L. (1963); "The x-ray emission analysis of thin films produced by lubricating oil additives," *Appl. Spectroscopy* **17**, 17. (Abstract in Sec. 24.4.)

Seemann, H. J., Schmidt, G., and Stavenow, F. (1961); "Untersuchungen zur quantitativen Röntgenfluoreszenzanalyse," *Z. Naturforschg.* **16a**, 25.

Sherman, J. (1954); "The correlation between fluorescent x-ray intensity and chemical composition," *ASTM Spec. Techn. Publ.* **157**, 27.

Sherman, J. (1955); "The theoretical derivation of fluorescent x-ray intensities from mixtures," *Spectrochim. Acta* **7**, 283.

Short, M. A. (1960); "Detection and correction of nonlinearity in x-ray proportional counters," *Rev. Sci. Instr.* **31**, 618.

Shott, J. E., Garland, Th. J., and Clark, R. O. (1961); "Determination of traces of nickel and vanadium in petroleum distillates," *Anal. Chem.* **33**, 507. (Abstract in Sec. 24.1.)

Siegbahn, M. (1931); *Spektroskopie der Röntgenstrahlen*. 2. Aufl. Springer, Berlin.

Siemens, H. (1962); "Bestimmungen von Elementen geringer Konzentration in Bleiglanz mit Hilfe der Röntgenfluoreszenzanalyse," *Z. Erzbergbau u. Metallhüttenwesen* **15**, p. 163.

Silverman, L., Houk, W. W., and Moudy, L. (1957); "Determination of uranium dioxide in stainless steels. X-ray fluorescent spectrographic solution technique," *Anal. Chem.* **29**, 1762. (Abstract in Sec. 18.5.)

Silverman, L., Houk, W. W., and Moudy, L. A. (1960); "Determination of uranium dioxide in stainless steels by the x-ray fluorescence method," *Advances in X-Ray Analysis*, Vol. 1, Plenum Press, New York, p. 271.

Smith, G. D. and Maute, R. L. (1962); "Determination of catalyst residues in polyolefins by x-ray emission spectrometry," *Anal. Chem.* **34**, 1733. (Abstract in Sec. 25.1.)

Smith, H. F. and Royer, R. A. (1963); "Determination of aluminium in organoaluminium compounds by x-ray fluorescence," *Anal. Chem.* **35**, 1098. (Abstract in Sec. 25.2.)

Van Someren, E. H. S., Lachman, F., and Birks, F. T.; *Spectrochemical Abstracts* Vol. 1–10, covering 1932–1964, Hilger & Watts Ltd., London.

Spielberg, N. and Ladell, J. (1960); "Crystallographic aspects of extra reflections in x-ray spectrochemical analysis," *J. Appl. Phys.* **31,** 1659.

Stephenson, S. T. (1957); "The continuous x-ray spectrum," in *Handbuch der Physik* Vol. 30, edited by S. Flügge, Springer Verlag, Berlin, Göttingen, and Heidelberg.

Stetter, A. and Kern, H. (1964); "Spektrometrische Untersuchung nichtmetallischer Stoffe in Eisenhüttenlaboratorien III. Die Untersuchung von Eisensinter und Hochofen-schlacke mit einem programmgesteuerten Röntgenfluoreszenz-Spektrometer," *Archiv Eisenhüttenwesen* **35,** 867. (Abstract in Sec. 19.2.)

Stoecker, W. C. and McBride, C. H. (1961); "X-ray spectrographic determination of thorium in uranium ore concentrates," *Anal. Chem.* **33,** 1709. (Abstract in Sec. 18.5.)

Stone, R. G. (1963); "The determination of strontium in tapwater by x-ray fluorescence spectrometry," *The Analyst* **88,** 56. (Abstract in Sec. 26.2.)

Stone, R. R. and Potts, K. T. (1963); "Thin films analysis by x-ray fluorescence," *Norelco Rep.* **10,** 94. (Abstract in Chap. 22.)

Stoner, G. A. (1962); "Rapid, automatic x-ray analysis of magnesium alloys," *Anal. Chem.* **34,** 123. (Abstract in Chap. 21.)

Stoops, R. F. and McKee, K. H. (1961); "Sampling errors in the x-ray fluorescent determination of titanium in a high temperature alloy," *Anal. Chem.* **33,** 589. (Abstract in Sec. 20.2.)

Strasheim, A. and Wybenga, F. T. (1964); "The determination of certain noble metals in solution by means of x-ray fluorescence spectroscopy," *Appl. Spectroscopy* **18,** 16.

Sun, S-Ch. (1959); "Fluorescent x-ray spectrometric estimation of aluminium, silicon, and iron in the flotation products of clays and bauxites," *Anal. Chem.* **31,** 1322. (Abstract in Sec. 23.1.)

Sundkvist, G. J., Lundgren, F. O., and Lidström, L. J. (1964); "Automatic x-ray determination of lead and zinc in the tailings of an ore dressing plant," *Anal. Chem.* **36,** 2091 (Abstract in Sec. 20.2.)

Tanemura, T. (1961); "Nondispersive x-ray spectroanalysis with filter and proportional counter," *Rev. Sci. Instr.* **32,** 364.

Taylor, J. and Parrish, W. (1955); "Absorption and counting-efficiency data for x-ray detectors," *Rev. Sci. Instr.* **26,** 367.

Tertian, R., Galling, F., and Géninasca, R. (1961); "Dosage précis et rapide de l'uranium dans ses composés par fluorescence X—Application aux carbures d'uranium," *Rev. univers. Mines* [9], **17,** 298. (Abstract in Sec. 18.5.)   ·

Tertian, R. (1968a); "Vers une méthode absolue et générale d'analyse chimique quanti-tative élémentaire," *Comptes rendus* **266,** 617.

Tertian, R. (1968b); "Contrôle de l'effet de matrice en fluorescence X et principe d'une méthode quasi-absolue d'analyse quantitative en solution solide ou liquide," *Spectro-chim. Acta* **23B,** 305.

Tertian, R. (1968c); "A rapid and accurate x-ray determination on the rare earths elements in solid or liquid materials using the double dilution method," 17*th* Annual Denver X-Ray Conference, Aug. 1968.

Tögel, K. (1958); "Spektrochemische Analyse mit Röntgenstrahlen," *Siemens Z.* **32,** 371.

Tögel, K. (1960); "Über die Röntgenfluoreszenzanalyse im Vakuum auf Elemente der Ordnungszahlen 12 bis 22," *Siemens Z.* **34,** 726.

Tögel, K. (1962a); "Röntgenfluoreszenzanalyse kleinster Mengen und Bereiche," *Siemens Z.* **36,** 497. (Abstract in Chap. 27.)

Tögel, K. (1962b); "Grundlagen der Präparationstechnik bei der Röntgenfluoreszenz-analyse," *Siemens Z.* **36,** 597.

Tomboulian, D. H. (1957); "The experimental methods of soft x-ray spectroscopy and the valence band spectra of the light elements," in *Handbuch der Physik*, Vol. 30, edited by S. Flügge, Springer Verlag, Berlin, Göttingen, and Heidelberg.

Tomkins, M. L., Borun, G. A., and Fahlbusch, W. A. (1962); "Quantitative determination of tantalum, tungsten, niobium, and zirconium in high-temperature alloys by x-ray fluorescent solution method," *Anal. Chem.* **34**, 1260. (Abstract in Sec. 18.2.)

Toussaint, C. J. and Vos, G. (1964); "Quantitative determination of carbon in solid hydrocarbons using the intensity ratio of incoherent to coherent scattering of x-rays," *Appl. Spectroscopy* **18**, 171. (Abstract in Sec. 25.3.)

Townsend, J. E. (1963); "X-ray spectrographic analysis of silica and alumina base catalyst by a fusion-cast disc technique," *Appl. Spectroscopy* **17**, 37. (Abstract in Sec. 23.2.)

Traill, R. J. and Lachance, G. R. (1965); "A new approach to x-ray spectrochemical analysis," *Geol. Surv. Canada*, Paper 64–57.

Turnley, W. S. (1960); "X-ray fluorescence analysis of plutonium," *Talanta* **6**, 189. (Abstract in Sec. 18.5.)

Ulrey, C. T. (1918); "An experimental investigation of the energy in the continuous x-ray spectra of certain elements," *Phys. Rev.* **11**, 401.

Victoreen, J. A. (1949); "The calculation of x-ray mass absorption coefficients," *J. Appl. Physics* **20**, 1141.

Volborth, A. (1965); "Dual grinding and x-ray analysis of all major oxides in rocks to obtain true composition," *Appl. Spectroscopy* **19**, 1. (Abstract in Sec. 23.2.)

Vose, P. B. (1961); "X-ray spectrography of plant material," *Laboratory Practice* **10**, 30. (Abstract in Sec. 26.2.)

Wang, M. S. (1962); "Rapid sample fusion with lithium tetra borate for emission spectroscopy," *Appl. Spectroscopy* **16**, 141.

Wassmann, K. (1964); "Einsatz des Röntgenfluoreszenzverfahrens zur Schnellanalyse von Messingschmelzen," *Metall* **18**, 1178. (Abstract in Sec. 20.2.)

Wedepohl, K. H. (1961); "Die Röntgen-Fluoreszenz-Spektralanalyse von geochemischen Proben auf Elemente der Ordnungszahlen 25–40," *Z. analyt. Chem.* **180**, 246. (Abstract in Sec. 23.2.)

Welday, E. E., Baird, A. K., McIntyre, D. B., and Madlem, K. W. (1964); "Silicate sample preparation for light-element analysis by x-ray spectrography," *Am. Mineral.* **49**, 889. (Abstract in Sec. 23.2.)

Westrik, R. (1961); "Determination of traces of metals in organic products by means of x-ray fluorescence spectroscopy," *Rev. univers. Mines* [9], **18**, 279. (Abstract in Sec. 25.1.)

Weyl, R. (1961); "Zur zerstörungsfreien Messung von Zusammensetzung und Schichtdicke kleiner Bereiche in dünnen Schichten mit Hilfe der Röntgenfluoreszenz," *Z. angew. Physik* **13**, 283. (Abstract in Chap. 22.)

Wilson, H. M. and Wheeler, G. V. (1957); "The determination of uranium in solution by x-ray spectrometry," *Appl. Spectroscopy* **11**, 128. (Abstract in Sec. 18.5.)

Witmer, A. W. and Addink, N. H. W. (1965); "Quantitative x-ray fluorescence analysis without standard samples (thin-layer method)," *Science u. Industry* **12/2**, 1.

Wood, R. E. and Bingham, E. R. (1961); "X-ray spectrographic determination of copper in low grade porphyry ores," *Anal. Chem.* **33**, 1344. (Abstract in Sec. 20.2.)

Yao, T. C. and Porsche, F. W. (1959); "Determination of sulfur and chlorine in petroleum liquids by x-ray fluorescence," *Anal. Chem.* **31**, 2010. (Abstract in Sec. 24.1.)

Zeitz, L. and Lee, R. (1963); "Bromine analysis in 5-bromouracil-labeled DNA by x-ray fluorescence," *Science* **142**, 1670. (Abstract in Sec. 26.2.)

Zemany, P. D. (1959); "Use of 55 Fe for measuring titanium coating thickness," *Rev. Sci. Instr.* **30**, 292. (Abstract in Chap. 22.)

Zemany, P. D. (1960); "Line interference corrections for x-ray spectrographic determination of vanadium, chromium, and manganese in low-alloy steels," *Spectrochim. Acta* **16,** 736. (Abstract in Sec. 19.1.)

Index